Physical Chemistry of Soils
The Fundamentals of Soil Science

Pellegrino Conte
Department of Agricultural, Food and Forest Sciences
University of Palermo
Palermo (Italy)

CRC Press
Taylor & Francis Group
Boca Raton London New York

CRC Press is an imprint of the
Taylor & Francis Group, an **informa** business
A SCIENCE PUBLISHERS BOOK

Cover credit: **Cover illustration by Alessia Ferraresi.**

First edition published 2026
by CRC Press
2385 NW Executive Center Drive, Suite 320, Boca Raton FL 33431

and by CRC Press
4 Park Square, Milton Park, Abingdon, Oxon, OX14 4RN

CRC Press is an imprint of Taylor & Francis Group, LLC

Library of Congress Cataloging-in-Publication Data (applied for)

ISBN: 978-1-041-31678-7 (hbk)
ISBN: 978-1-041-31680-0 (pbk)
ISBN: 978-1-003-78021-2 (ebk)

DOI: 10.1201/9781003780212

Typeset in Times New Roman
by Prime Publishing Services

Dedication

To my wife, Alessia.
Without her unwavering patience, support, and encouragement, this book would not have been possible.

Preface

This book is the culmination of at least eight years of dedicated work. The idea to write a book on the physical chemistry of soils emerged unexpectedly in 2016, long before the COVID-19 pandemic reshaped our world. It all began when the editorial office of CRC Press reached out to me, asking if I had an idea for a book on the physics of soils. At that time, I hesitated—not because I doubted the connection between soil chemistry and physics, which is well established, but because the request specifically emphasized soil physics.

As a chemist, I was accustomed to approaching soils primarily from a molecular and reactive standpoint, focusing on interactions at the atomic and chemical levels. However, as I reflected on the project, I recognized that soil behavior arises from a seamless interplay between physical and chemical principles. Structure, reactivity, and dynamics cannot be fully understood in isolation; a comprehensive perspective encompassing both disciplines is essential. This realization reinforced my conviction that a book bridging these domains was not only feasible but necessary.

This led to the conception of the *Physical Chemistry of Soils*. My goal was not merely to compile existing knowledge but to unify these two scientific perspectives, providing a comprehensive framework for soil science. Traditionally, soil science is studied through different lenses—agronomy, environmental science, geology, and engineering, among others. However, in my three decades of teaching and research, I have noticed a prevailing tendency, particularly in Italy, where most of my academic work takes place: soil physical chemistry is often addressed primarily through agronomic applications. While this focus is undeniably important, it does not fully capture the richness and complexity of soil behavior at a fundamental level.

The need for another book on soil science, broadly speaking, arises from this very gap. Many existing texts focus on the agricultural significance of soils or their geotechnical aspects, but few comprehensively integrate the fundamental physical and chemical processes governing their behavior. This book aims to provide a nuanced perspective that unites molecular interactions, thermodynamics, and physicochemical principles to offer a deeper understanding of soil systems.

Beyond being a reference for researchers and students, I hope this book fosters interdisciplinary thinking. Understanding soils demands a multidisciplinary perspective that transcends traditional boundaries and integrates knowledge from chemistry, physics, geology, and environmental science. I intend to provide a resource that facilitates this synthesis, allowing readers to appreciate the complexity of soils not just as static systems but as dynamic, interactive environments governed by fundamental physicochemical principles.

Science does not advance in silos but through the convergence of knowledge, where different disciplines inform and enrich one another. The study of soil physical chemistry is no exception; it benefits from drawing upon diverse methodologies, theoretical approaches, and experimental insights. To support this broader intellectual exploration, I have included a "Further Reading" section at the end of each chapter.

These sections are not merely bibliographic references but curated pathways for readers who wish to delve deeper into specific topics introduced in the chapters. They are designed to bridge gaps in understanding, provide historical context, and highlight key advancements that have shaped our knowledge of soil systems. Whether you are a student seeking a more structured foundation, a researcher looking for specialized studies, or simply a curious reader eager to expand your perspective, these readings offer an opportunity to engage with the wider scientific discourse surrounding soil physical chemistry.

I sincerely hope this book contributes to that dialogue, helping readers navigate the fascinating world of soil physical chemistry with clarity, curiosity, and a spirit of continuous learning. By fostering a deeper engagement with the subject, I trust that this book—and the additional readings it encourages—will inspire new questions, insights, and connections across disciplines. I am deeply grateful to all those who have supported me throughout this journey—colleagues, students, and the editorial team who believed in this project. Writing this book has been a journey of discovery for me, and I hope it serves as an insightful guide for you, the reader, as well.

Pellegrino Conte

Contents

Chapter 1

The Scientific Method and the Soil Science

1.1 WHAT IS SCIENCE?

According to Evandro Agazzi[1], science is a form of knowledge "distinct from [...] emotional, artistic, religious and other such attitudes towards reality". Nowadays, every field of research, from mathematics to physics, from chemistry to biology, from economics to politics, from philosophy to history, is admitted as part of the scientific knowledge provided that rigor and objectivity are applied. In other words, "science" refers not to what is studied but to how something is investigated.

For the aims of the present book, which is dedicated to the physics and chemistry of soils, it can be simple to understand the meaning of the term "rigor"—in that, it refers to the mathematical language used to model the phenomena occurring in soils—whereas the meaning of objectivity is more difficult to achieve.

The general meaning of the word "objectivity" in any dictionary states: "judgment based on observable phenomena and uninfluenced by emotions or personal prejudices."[2] Due to this, objectivity means that a phenomenon appears the same to all observers. Clearly, this cannot be entirely accurate[3]. As a simple example, let us consider the color change of phenolphthalein from uncolored to red at the turning point during an acid-base titration. Different observers may see different shades of red, thereby leading to different results depending on what they interpret as "red." In order to reduce the discrepancies among what different observers perceive as "red," the color can be defined as light with a wavelength ranging between 650 and 700 nm. Without a clear definition of the nature of

[1] See the Further Reading section

[2] This definition has been reported from https://www.vocabulary.com/dictionary/objectivity

[3] This definition has been reported from https://www.vocabulary.com/dic

DOI: 10.1201/9781003780212-1

the red color, the only thing an observer can state about the phenomenon at the titration's turning point is that a color change has occurred.

The simple example reported above illustrates that observing a phenomenon cannot be independent of the observer's perspective[4]. Objectivity about the observation of a phenomenon not only requires the use of instruments that are unaffected by human senses, but it also necessitates the definition of constraints (referred to as boundary conditions) within which the phenomenon can be considered repeatable and reproducible[5].

Once the phenomenon is made repeatable and reproducible, a first hypothesis about its nature can be formulated. This hypothesis represents the model (also referred to as "theory"), supported by mathematical rigor, used to explain how the phenomenon occurs. The model must explain not only the observations already made about the phenomenon but also those yet to be realized. New experiments must then be designed to validate the model and check its predictive ability. If the new experiments support the predictions made by the model, the latter becomes stronger. The more predictions the model supports, the stronger it is, thereby becoming a real theory[6] (Figure 1.1). If even a single experiment fails to support the predictions made by a model, the model must either be revised to account for the unexpected results or rejected altogether. It is important to keep in mind that any model discussed here is merely a representation of reality. Any other model rigorously supported by mathematics and possessing the same predictive ability can also be used to describe the nature of the phenomenon. Among the different models having similar predictive capabilities, the simplest one is usually applied. In other words, the model using the lowest number of explanations and the fewest parameters to verify its reliability is certainly preferred[7].

According to what has been discussed above, the evolution of science is based on the continuous change and/or refinement of existing theories explaining repeatable and reproducible phenomena. This means that what we call "scientific truth" is neither absolute nor dogmatic. On the contrary, its

---

[4] Consider, for example, the images that appear when a drug such as LSD—lysergic acid diethylamide—is taken. These images seem real to those under the influence of the drug, yet they are far from reality. In Chapter 2, we will see that even the instruments used in a typical analytical chemistry lab are subject to errors. However, a scientist's skill lies in minimizing these errors and carefully interpreting the data from analyses to construct a model that explains the observed phenomenon.

[5] The reader is addressed to Chapter 2 to understand the differences between "repeatability" and "reproducibility."

[6] Examples of scientific theories include the theory of evolution, quantum electrodynamics (QED), the Big Bang theory, and theories about dinosaur extinctions. In scientific discourse, the term "theory" does not imply uncertainty, as it often does in everyday language. Instead, it refers to a well-substantiated explanation of natural phenomena, supported by evidence and rigorous testing. However, like all scientific knowledge, theories remain open to refinement or revision as new evidence emerges.

[7] This is also known as "Ockham's razor" named after a 14th-century friar who devised methodological principles for simplifying traditional ontological and epistemological theories..

power lies in the strength of the speculative schemes designed to organize the phenomena representing the observed reality. However, a detailed excursion into the philosophy of science is beyond the aims of the book. Detailed information on the scientific way of thinking and on how science is constructed can be found in the Further Reading section of the present chapter. Here, the intention is to show, from a practical point of view, that any form of scientific knowledge is based on the "scientific method" represented in Figure 1.1 and briefly summarized above.

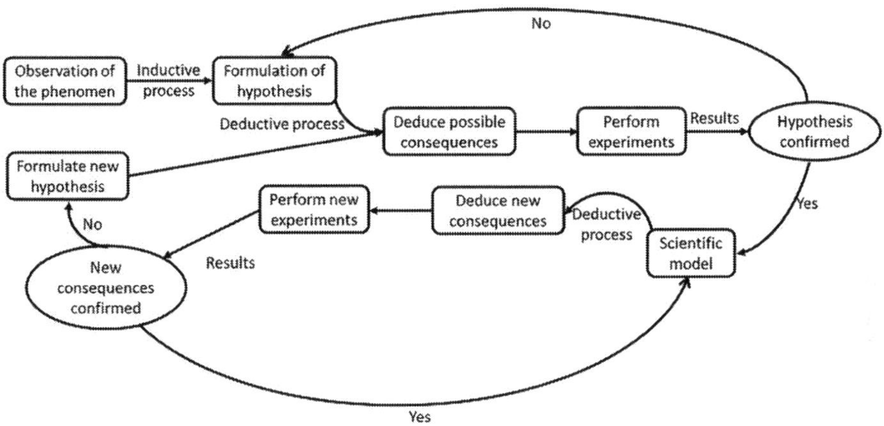

Figure 1.1 The scientific method. A phenomenon is observed. Then, a hypothesis about its nature is formulated via an inductive process. This hypothesis is used to deduce (deductive process) possible consequences that are either confirmed or not through experiments. If the hypothesis is confirmed, it becomes a scientific model from which new consequences can be deduced. These latter are either confirmed or not through experiments. If the experiments do not confirm the hypothesis, a new one must be formulated, and the cycle continues.

1.2 CAN SOIL BE A SCIENCE, AND WHAT IS NEEDED TO STUDY SOIL SCIENCE?

The first part of the question in the title of this section is intended as a provocation, as the discussion so far makes it clear that the answer can only be affirmative. Science is not defined by the object of study but rather by the rigorous methodologies and principles applied to its investigation. Therefore, soil unquestionably falls within the realm of science, provided that its study adheres to objectivity and systematic inquiry.

A more meaningful question is: what does soil science encompass?

As will be reiterated throughout this book, soil is an extraordinarily complex system that sustains life in all its forms. It plays a crucial role in ecosystems by regulating water cycles, serving as a habitat for countless microorganisms, and acting as a reservoir of essential nutrients. The soil is particularly significant for humans due to its role in food and textile production. A deep understanding of its physical, chemical, and biological properties is fundamental to improving agricultural practices, ensuring sustainability, and addressing environmental challenges such as soil erosion, contamination, and climate change.

Thus, soil science is a broad and interdisciplinary field integrating principles from physics, chemistry, and biology. In terms of physics, students will explore topics such as soil texture, water potential, bulk and particle density, colloid behavior, and thermal properties—all of which influence soil fertility and its ability to support plant life[8]. From a chemical perspective, the study extends to both inorganic and organic components, focusing on the reactions governing soil formation, evolution, and degradation, as well as nutrient cycling and soil solution equilibria.

Given its interdisciplinary nature, soil science requires a strong foundation in mathematics, physics, general and organic chemistry, analytical chemistry, biochemistry, and physical chemistry (e.g., thermodynamics and reaction kinetics). Mastery of these subjects is essential for understanding and modeling soil behavior, interpreting experimental data, and applying scientific principles to real-world soil management and conservation efforts.

EXERCISES

1. Describe the key steps of the scientific method and how they apply to soil science.
2. Explain the difference between a hypothesis, a model, and a theory in the context of scientific research.
3. Discuss the importance of rigor and objectivity in scientific investigations, providing specific examples from soil science.
4. Why are repeatability and reproducibility crucial for validating scientific hypotheses and building robust theories?
5. Define "soil fertility" and list the factors that contribute to it.
6. How does physics contribute to our understanding of soil processes? Provide specific examples.
7. What are the key differences between inorganic and organic chemistry in the context of soil science?
8. How is knowledge of thermodynamics relevant to the study of soils?
9. Given their complexity and variability, what are the main challenges in applying the scientific method to the study of soils?
10. Research and discuss a scientific article that illustrates the application of the scientific method to a specific problem in soil science.

[8] The locution "soil fertility" is fully explained in Chapter 13.

FURTHER READING

Agazzi, E. 2014. Scientific Objectivity and Its Contexts. Springer International Publishing Switzerland. ISBN: 9783319046594

Chalmers, A.F. 1976. What is this Thing Called Science? 3rd Ed. University of Queensland Press. ISBN: 0872204529

Kuhn, T.S. 1957. The Copernican Revolution. Planetary Astronomy in the Development of Western Thought. Harvard University Press, Cambridge. ISBN: 0674171039

Kuhn, T.S. 1970. The Structure of Scientific Revolutions. The University of Chicago Press, 2nd ed. ISBN: 0226458040

Popper, K.R. 1959. The Logic of Scientific Discovery. Routledge, 1st Ed. in English. ISBN: 0415278449

Chapter **2**

How to Handle Numbers in Soil Science?

2.1 INTRODUCTION

In the popular language, an error is something that has been done wrong. Conversely, in the scientific language, an error is something related to the variability of a quantitative measurement. In other words, the term "error" refers to how well something is known. As an example, let us consider the tachymeter in Figure 2.1.

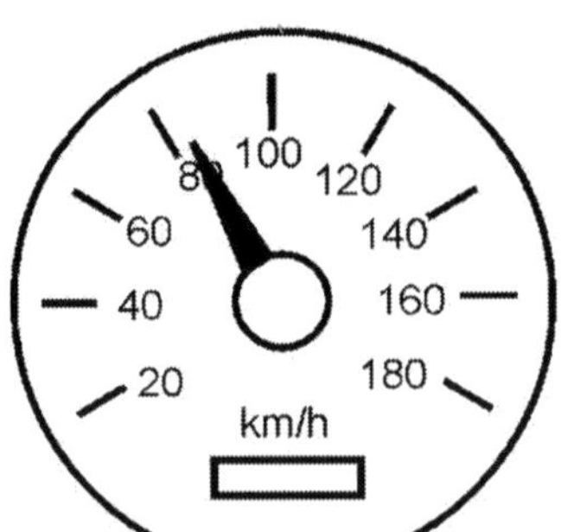

Figure 2.1 A tachymeter measures the speed of a vehicle. In the present case, it reports a velocity of around 80 km per hour (km/h or km h^{-1}).

The tachymeter's pointer shows that the vehicle speed is not exactly 80 km h^{-1} as it points between 80 and 100 km h^{-1}. Can we say which is the true speed value? No. We can only provide an approximate value. Let us say 82 or 83 km h^{-1}. However, it can also be 84 or 85 km h^{-1}. There is no possibility of precisely reporting the speed value, which can range between 82 and 85 km h^{-1}. In science, when a measurement (such as that from the tachymeter in Figure 2.1)

DOI: 10.1201/9781003780212-2

is said to be affected by an error, it means that an exact absolute value cannot be determined. Instead, the measurement falls within a range of values known as the "uncertainty" or "confidence interval." The smaller the uncertainty, the higher the precision of the measurement. Therefore, a fundamental objective in scientific research is to minimize measurement uncertainties to improve accuracy and reliability.

The present chapter aims at teaching how to handle and interpret experimental data and results. It is not intended to be exhaustive but only a small compendium of rules to apply when a thesis or a paper is written based on experimental measurements. For deep details on the mathematics of the errors, the reader is referred to the Further Reading section.

2.2 OBSERVABILITY, MEASURABILITY, REPEATABILITY, AND REPRODUCIBILITY

In chemistry, as well as in all scientific fields, observation and analysis of phenomena are key to identifying predictive mathematical models that describe all naturally occurring events. In particular, the fundamental basis for the scientific method lies in the observability, measurability, repeatability, and reproducibility of phenomena.

Observability refers to the monitoring (i.e., detection) of an event (or an object) by using any type of instrument or method. Undetectability of a phenomenon (either an event or an object) only means that experimental sensitivity (i.e., the smallest amount that an instrument, a method, or a system can reveal under specified conditions) is too low for any possible detection. For example, human eyes can be considered our personal instruments to observe natural phenomena or objects. We can see a fish swimming in a lake. However, our visual limit prevents us from "seeing" forms of lives smaller than a few millimeters. To "see" more, we need to sample water, go to the lab, and use special instruments, such as optical microscopes, which enhance our visual limits from a few millimeters to a few nanometers. To observe objects smaller than a few nanometers, we must switch to other microscopes, such as the scanning electron ones. In other words, the enhancement of instrumental sensitivity allows the observability of phenomena on progressively smaller scales.

The simple observation of a phenomenon is not enough to elaborate suitable models to predict its behavior. For example, we can walk in a forest and observe leaves' decomposition. If we wish to elaborate a model to understand what happens to the leaves on the forest floor, we need to ask questions such as: How quickly does decomposition occur? Are the decomposition mechanisms biotic, abiotic, or both? To answer the questions, key parameters important for identifying the decomposition process must be identified and quantified. One of these is the elemental content (i.e., carbon (C), hydrogen (H), and nitrogen (N) amounts) monitored versus the time, thus providing a kinetic of leaves transformation on the forest floor (this answers the first question about the rate of leaves decomposition). To reveal the decomposition mechanisms, isolation and quantification of the degradation products must be achieved. These can be related to biotic and/or abiotic degradation pathways.

According to the aforementioned considerations, elaborating a model to rationalize a phenomenon requires not only observability but also measurability (i.e., quantitative evaluation) of the key parameters describing the observed naturally occurring event. Once numerical data (i.e., measurements) are obtained, hypotheses about the nature and mechanisms of the phenomenon can be formulated. Such hypotheses must be confirmed or refuted by reproducing the phenomenon (when possible[1]) in the lab and collecting measurements many times. Hence, repeatability and reproducibility are additional but important issues for the enhancement of the model robustness (i.e., the ability of a model to work efficaciously while its parameters and/or assumptions are altered, thereby proving that it can operate without failure under a variety of conditions). Repeatability refers to the ability of a single researcher to repeat several times the same measurements under exactly the same conditions by always obtaining the same results. Reproducibility refers to the ability of different researchers in different worldwide spread labs to reproduce the same experiments and retrieve the same measurements with results that are always comparable among them.

Unrepeatability and unreproducibility of measurements can be due to failures in the hypotheses formulated to rationalize the phenomenon or to errors in measurements. If errors in the measurements can be excluded, hypotheses must be reformulated, and measurements must be repeated in order to ensure repeatability and reproducibility.

2.3 CAUSES OF ERRORS IN MEASUREMENTS

The example discussed in Figure 2.1 shows that one possible source of uncertainty is related to the calibration of the instrument. The term "calibration" refers to the act of determining, rectifying, or checking the gradations of an analytical instrument. In this concern, the tachymeter shown in Figure 2.1 can only measure speed rates ranging in intervals of 20 km h^{-1}. In order to ameliorate the quality of the measurement (i.e., reduce the uncertainty source), a different instrumental calibration is needed. For this particular example, the new calibration should reduce the speed rate interval to a smaller value (e.g., from 20 to 5 km h^{-1}).

[1] While laboratory experiments are a powerful tool for testing scientific hypotheses, not all systems can be reproduced under controlled conditions. Large-scale or highly complex phenomena, such as astrophysical events (e.g., supernovae and black hole formation), geological processes (e.g., plate tectonics and volcanic eruptions), and aspects of evolutionary biology, occur on timescales, spatial dimensions, or under conditions that make direct laboratory reproduction impractical or impossible. However, these systems still belong to science because they can be studied indirectly, such as observational data, computational models, and natural experiments. By applying rigorous methodologies, scientists can analyze patterns, compare predictions with real-world data, and validate or refine their theories, even without direct experimental replication.

Calibration problems can also occur when an instrument has not been used for a long time or has become worn, resulting in non-coherent instrumental feedback. As an example, the electron capture detector (ECD) used in gas chromatography analyses contains a source of radioactivity, that is ^{63}Ni. Radioactivity is continuously emitted even if the detector is not in use. For this reason, the ordinary maintenance of the ECD system must include regular calibration to maintain its functionality.

Instrumental calibration can be influenced by environmental conditions such as temperature, pressure, and humidity variations. For example, Figure 2.2A illustrates the effect of temperature on the capacity of a volumetric flask, while Figure 2.2B presents an FTIR background spectrum showing signals from atmospheric carbon dioxide and water.

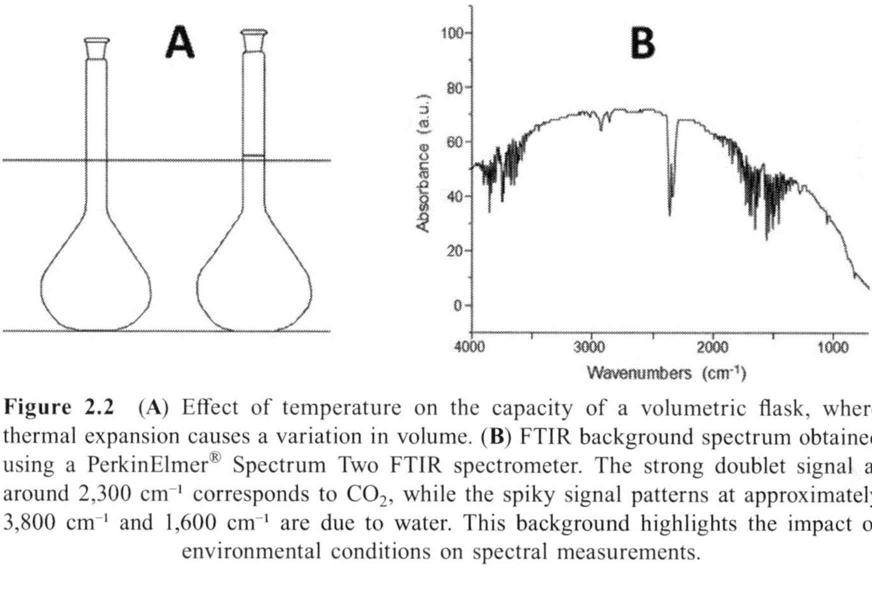

Figure 2.2 (**A**) Effect of temperature on the capacity of a volumetric flask, where thermal expansion causes a variation in volume. (**B**) FTIR background spectrum obtained using a PerkinElmer® Spectrum Two FTIR spectrometer. The strong doublet signal at around 2,300 cm^{-1} corresponds to CO_2, while the spiky signal patterns at approximately 3,800 cm^{-1} and 1,600 cm^{-1} are due to water. This background highlights the impact of environmental conditions on spectral measurements.

After heating to a temperature of, for example, 105°C, thermal expansion causes volume distortion, rendering the flask unusable for precise volume measurements (Figure 2.2A). This highlights the importance of using glass with a low thermal expansion coefficient during manufacturing and ensuring proper handling in the lab. Specifically, lab glassware should be air-dried rather than oven-dried to prevent such distortions. Similarly, the presence of atmospheric carbon dioxide and water (Figure 2.2B) can interfere with FTIR analysis by masking important signals. To obtain accurate spectral data, it is essential to subtract the background spectrum before acquiring measurements.

Figure 2.3 illustrates an example of parallax error, which occurs due to the apparent shift in the position of the burette pointer when the observer's viewpoint changes. To minimize parallax errors while reading any scale, the observer must ensure they are looking directly at the pointer from a perpendicular angle.

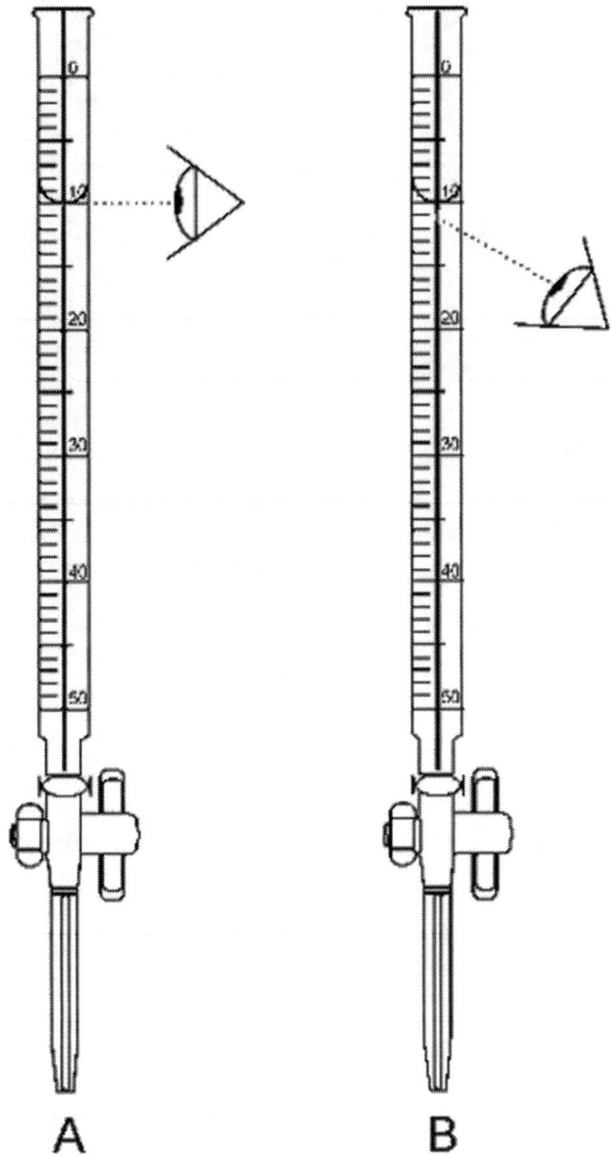

Figure 2.3 Example of parallax error. (**A**) The correct volume reading is obtained when the observer's eye is positioned perpendicular to the burette scale. (**B**) An incorrect volume reading occurs when the eye is not aligned perpendicularly with the burette scale. (Image created using ACDLabs© freeware software.)

Table 2.1 summarizes the most common sources of uncertainty in experimental observations. However, it is important to note that no list of errors can ever be exhaustive, as each experiment is unique. Therefore, sources of error must be carefully evaluated, identified, and accounted for on a case-by-case basis as each experimental design is developed and implemented. In fact, the most effective way to minimize errors in experiments depends on the investigator's approach, which should be as meticulous and careful as possible.

Table 2.1 Non-exhaustive list of possible sources of errors in chemistry experiments

Possible Sources of Errors
— Inadvertent contamination of a solution by dropping another substance into it or by using dirty instruments.
— Inconsistent sampling, thus, makes the sample not representative of the whole system.
— Investigator's careless errors.
— Wrong evaluation of any physical change occurring during an experiment (example: color change in titrations).
— Wrong instrumental calibration due to instrumental wear and/or rupture.
— High instrumental noise.
— Low instrumental promptness or low investigator's promptness in stopping the experiment.
— Wrong environmental conditions include hot rooms, high—humidity places, insufficient illumination, or planes with the wrong inclination.
— Presence of vibrations affects the efficiency of an instrument (such as during weight measurements on lab balances).
— Parallax error due to the wrong position of the observer when reading scale meters.
— Choice of the wrong instrumentation for a specific experiment.
— Choice the wrong units to report the results of an experiment.
— Instrumental drifts.

2.4 RANDOM AND SYSTEMATIC ERRORS

Let us consider a sample weighed seven times by using two different balances, as indicated in Table 2.2.

Table 2.2 Example of random and systematic errors. This table compares weight measurements (in grams) obtained using a calibrated and a non-calibrated balance across multiple trials

Trial	Weight (g) Measured Using a Calibrated Balance (CB)	Weights (g) Measured Using a Non-Calibrated Balance (NCB)
Trial 1	2.000	2.015
Trial 2	2.002	2.017
Trial 3	1.950	1.965
Trial 4	1.998	2.013
Trial 5	2.010	2.025
Trial 6	2.000	2.015
Trial 7	1.987	2.002

One of the two balances was correctly calibrated before the measurements (CB), whereas the second one was not (NCB). Both the central and the right-handmost columns in Table 2.2 show that statistical fluctuations (in either direction)

occur in the acquired data due to precision limitations in both CB and NCB devices. The aforementioned uncertainties result from the inability of the investigator to take the same measures in the same way to achieve exactly the same number. For example, unpredictable small vibrations and minor changes in room humidity or temperature can affect all the measurements, thereby producing numbers slightly different from each other. The errors due to statistical fluctuations in either direction are referred to as “random errors.” Random errors affect both experimental repeatability and reproducibility. In fact, once an instrument has been correctly calibrated (either in the investigator’s lab or in labs spread worldwide), its feedback must always return the same results modulated by statistical fluctuations, independently of its location. The smaller the statistical fluctuations, the better the quality of the experimental data. The quantitative evaluation of the random errors will be accounted for in the following paragraphs.

The right-handmost column in Table 2.2 shows that the weights measured on the NCB device differ from those retrieved by using the CB one by adding an amount of 0.015 g (i.e., 15 mg) as a consequence of the non-correct zeroing of the balance. This kind of error, which can be considered a reproducible inaccuracy consistently occurring always in the same direction, is referred to as “systematic error.” Systematic errors, due to problems persisting throughout the entire experiment performed in a lab, affect experimental reproducibility rather than repeatability. In fact, systematic errors allow comparability among results obtained by the same investigator, which uses instruments that are not correctly calibrated. Conversely, comparing results achieved by researchers operating in different labs is infeasible. The only way to account for systematic errors is regular instrumental maintenance.

2.5 ACCURACY AND PRECISION

Figure 2.4 shows a target with five darts (the black circles) launched. The center of the target represents the true value of any parameter that a scientist wishes to measure, while the darts represent the numerical values that come from the experimental measurements done in five replicates. Due to a combination of systematic and random errors, the five darts (i.e., the five measurements) are randomly spread on the target in Figure 2.4A. This means that all the measurements fail in both accuracy and precision. Figure 2.4B shows that the five darts are in proximity to the center of the target but are spread apart. For this reason, it is possible to argue that the measurements are accurate, but not precise, due to random errors. An increase in precision with a lack of accuracy due to the effect of systematic errors is reported in Figure 2.4C. Here, all the darts have been placed close to each other (that is good precision), while their positions indicate that the measurements are not close to the target’s center (low accuracy).

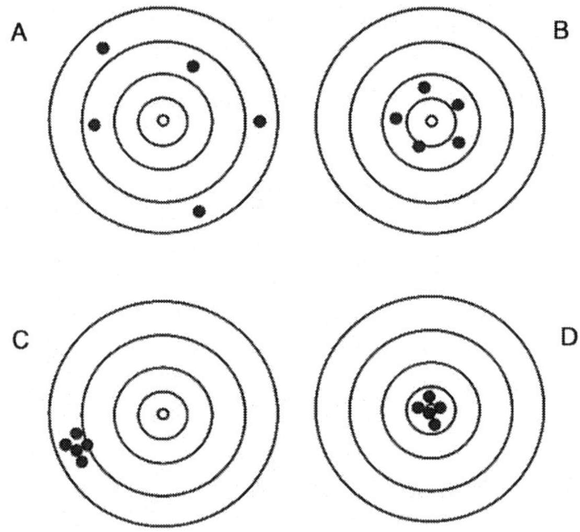

Figure 2.4 (**A**) Measures are imprecise and inaccurate; (**B**) measures are accurate and imprecise; (**C**) measures are precise and inaccurate; (**D**) measures are accurate and precise.

Finally, Figure 2.4D reports a set of experimental measurements that are both accurate and precise. In fact, not only have all the darts been placed in proximity to the target center, but their distance from that center provides quite similar values.

According to the discussion above, accuracy refers to how close the measured values are to the actual (i.e., true) value—the target's center. Precision refers to how close to each other the values lie when the experiment is repeated many times. The closer these values lie to each other, the more precise the results become.

2.6 HOW TO REPORT MEASUREMENTS AND EXPERIMENTAL UNCERTAINTIES

The results achieved by the calibrated balance in Table 2.2 show that the weights range into an interval of values whose limits are 1.950 g and 2.010 g. For this reason, if W is the unknown true value of the sample weight, it is possible to write:

$$1.950 \text{ g} \leq W \leq 2.010 \text{ g} \tag{2.1}$$

The relationship (2.1) can be also re-written as in (2.2):

$$\bar{W} \pm \sigma \text{g} \tag{2.2}$$

where the symbol "±" refers to the fluctuations between the aforementioned upper and lower limits; $\bar{W}$ is indicated as the average (or mean) of the experimental measurements, giving a quantitative idea of the true value of the measure; σ is a value referred to as standard deviation, accounting for the random errors of the experimental measurements.

The average ($\bar{W}$) of a data set made by N different values is given by relation (2.3):

$$\bar{W} = \frac{1}{N}\sum_{i=1}^{N} w_i \tag{2.3}$$

while the random error is calculated by applying relation (2.4):

$$\sigma = \sqrt{\frac{1}{N}\sum_{i=1}^{N}\left(w_i - \bar{W}\right)^2} \tag{2.4}$$

In both relations (2.3) and (2.4), i is the number of the collected experimental data. According to Equation (2.3), the average value obtained using Microsoft Excel® for the seven trials reported in the left-side column of Table 2.2 is $\bar{W} = 1.992428571$ g, while the standard deviation is $\sigma = 0.019898552$ g. Let us express $\bar{W} \pm \sigma$ as shown in Figure 2.5A.

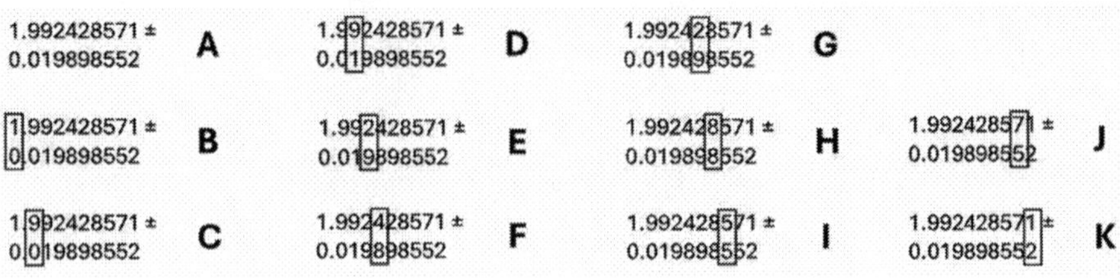

Figure 2.5 The average value and standard deviation are obtained by applying Microsoft Excel® to the data on the left-hand side in Table 2.2. The squares from B to K show the error in each of the different places of the average value returned by the simple mathematical calculation.

Each decimal place in $\bar{W}$ is subject to some degree of uncertainty, as illustrated in Figures 2.5B through 5K. Specifically, the first decimal place (Figure 2.5B) has no associated error, meaning it remains exact. The same applies to the first decimal place (Figure 2.5C). However, starting from the second decimal place (Figure 2.5D), an error of 1 appears, and all subsequent decimal places (Figures 2.5E to 2.5K) are also affected by nonzero errors. Based on this, we consider all the digits up to and including the first place affected by a nonzero error to be significant figures. Any additional digits beyond this point are regarded as non-significant figures. Thus, $\bar{W} \pm \sigma$ must be expressed as 1.99 ± 0.02 g, where $\bar{W}$ has three significant figures, and σ has only one.

It is important to note that the experimental data in the left-hand column of Table 2.2 contain four significant figures. However, due to random fluctuations during the measurements, the average $\bar{W}$ value retains only three. To achieve a more accurate evaluation of $\bar{W}$, all factors contributing to random fluctuations (see above) must be identified and eliminated.

When writing, the standard deviation indicated a few lines above 0.019898552 was rounded to 0.02. This is because specific rules must be followed when reporting significant figures. These are listed below:

1. If the first non-significant figure is less than 5, the last significant figure remains unchanged.

   Example: 11.23**4** ± 0.01[2] must be written as 1.23 ± 0.01 since 4 < 5.

2. If the first non-significant figure is greater than 5, the last significant figure is rounded up

   Example: 1.23**6** ± 0.01 must be written as 1.24 ± 0.01, since 6 > 5

3. If the first non-significant figure is exactly 5, the rounding depends on whether the last significant figure is even or odd:

   — If the last significant figure is even, it remains unchanged.

   Example: 1.225 ± 0.01 must be written as 1.22 ± 0.01 since 2 is even.

   — If the last significant figure is odd, it is rounded up.

   Example: 1.235 ± 0.01 must be written as 1.24 ± 0.01 since 3 is odd.

4. If $\overline{W} \pm \sigma$ is given by 1.2451 ± 0.01, then $\overline{W}$ rounds to 1.25, because the fourth decimal place is nonzero.

2.7 HOW TO WRITE SIGNIFICATIVE FIGURES BY USING SCIENTIFIC NOTATION

With the widespread adoption of computers, most people have become familiar with the so-called "scientific notation." This term refers to a numerical representation in which a number is expressed as a power of 10, consisting of a coefficient (between 1 and 9) and an exponent (i.e., 10 raised to a power). The exponent is also referred to as the "order" or "order of magnitude" (Figure 2.6).

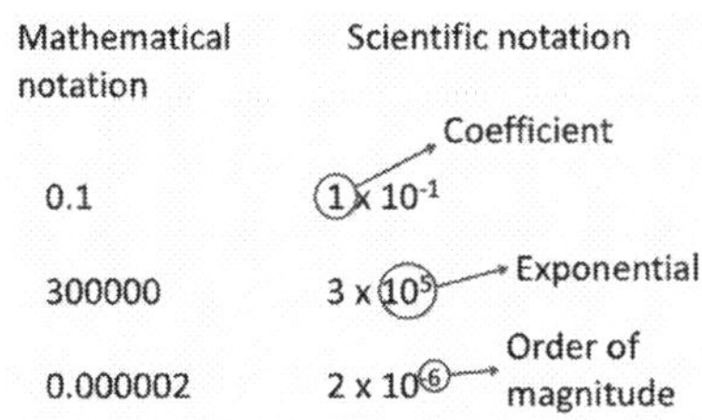

Figure 2.6 Comparison between mathematical notation (left) and scientific notation (right). The coefficient, exponent, and order of magnitude are highlighted in the scientific notation.

[2] To ensure clarity and precision, it is important to note that some authors suggest considering an additional significant figure when the error margin is on the order of 0.1 (e.g., 0.01, 0.001, etc.). For instance, a measurement of 0.234235 ± 0.001234 can be appropriately expressed as 0.2342 ± 0.0012. However, this does not alter the standard rules for rounding and reporting significant figures.

The importance of scientific notation is due to the possibility of handling numbers that are either too small or too large, such as in the case of the electron charge value and the distance traveled by light in one year. The former is 0.00000000000000000016 Coulomb (or C), while the latter is 9460000000000000 meters (or m). It appears clear that such small/large values disturb the reading while burdening calculations. By applying the scientific notation, the electron charge is written as 1.6×10^{-19} C, while the distance traveled by the light in one year is 9.46×10^{15} m. When an experimental measurement is expressed in scientific notation, the following rules apply regarding significant figures:

1. All the non-zero digits are significative.
2. Any zero between two non-zero digits is significant.
3. Trailing zeros (zeros following a nonzero digit) are significant.
4. Leading zeros (zeros at the beginning of a number) are not significant.
5. Zeros in the exponent are not significant.

To clarify the aforementioned points, let us determine the number of significant figures in the following numerical values:

(a) 10
(b) 1.200
(c) 1×10^3
(d) 0.000001

— Case (a): 10 has two significant figures.
— Case (b): 1.200 has four significant figures.
— Case (c): 1×10^3 has one significant figure.
— Case (d): 0.000001 has one significant figure.

2.8 THE OPERATIONS WITH THE SIGNIFICATIVE FIGURES AND THE PROPAGATION OF THE ERROR

Based on the previous discussion, the maximum number of digits that can be reliably read from an instrument is referred to as the "number of significant figures." It is important to note that experimental numerical values and their standard deviations can be combined to derive new quantities, for which new standard deviations must also be considered. For example, molar concentration is determined by combining two measured quantities—the mass of a compound and the volume of the solvent in which it is dissolved—each with its standard deviation. As a result of this combination, the newly derived quantity is inevitably affected by a random error directly related to the standard deviations of the original quantities.

When performing mathematical operations on different numerical values, it is essential to ensure that all values are expressed consistently. Additionally, when working with numbers in scientific notation, calculations must be applied to both the coefficient and the exponent, as demonstrated in the following examples:

$$(2.3 \times 10^{-1}) \times (1.2 \times 10^{-5}) = (2.3 \times 1.2) \times (10^{-1} \times 10^{-5}) = 2.76 \times 10^{-6}$$

$$\frac{2.3 \times 10^{-1}}{1.2 \times 10^{-5}} = \frac{2.3}{1.2} \times \frac{10^{-1}}{10^{-5}} = 1.91\bar{6} \times 10^{4}$$

$$(2.3 \times 10^{-1}) + (1.2 \times 10^{-1}) = (2.3 + 1.2) \times 10^{-1} = 3.5 \times 10^{-1}$$

$$\begin{aligned}(2.3 \times 10^{-1}) + (1.2 \times 10^{-2}) &= (2.3 \times 10^{-1}) + (0.12 \times 10^{-1}) \\ &= (2.3 + 0.12) \times 10^{-1} = 2.42 \times 10^{-1}\end{aligned}$$

$$\begin{aligned}(2.3 \times 10^{-1}) - (1.2 \times 10^{-2}) &= (2.3 \times 10^{-1}) - (0.12 \times 10^{-1}) \\ &= (2.3 - 0.12) \times 10^{-1} = 2.18 \times 10^{-1}\end{aligned}$$

It can be observed that, in the sum and difference calculations above, the numbers were made comparable by expressing their exponents with the same power. Additionally, it is important to remember that linear combinations (i.e., sum and difference) of different quantities are only possible when they share the same dimensions (e.g., weight + weight, length – length, etc.).

How can the standard deviation be determined for quantities obtained by combining numerical values measured for different entities? While the mathematical derivations are beyond the scope of this book, the relevant rules are summarized below.

In the case of a linear combination, the total standard deviation is obtained as the sum of the individual standard deviations:

$$\sum_{i=1}^{N}\left(a_i \pm \sigma_{a_i}\right) = \left(\sum_{i=1}^{N} a_i\right) \pm \left(\sum_{i=1}^{N} \sigma_{a_i}\right)$$

In the case of a product or ratio of different experimental values, the standard deviation is determined as a linear combination of the relative errors multiplied by the computed value of the product or ratio:

$$\prod_{i=1}^{N}\left(a_i \pm \sigma_{a_i}\right) = \prod_{i=1}^{N} a_i \pm \left[\left(\prod_{i=1}^{N} a_i\right) \times \left(\sum_{i=1}^{N} \frac{\sigma_{a_i}}{a_i}\right)\right]$$

According to the previous discussion, the numerical values obtained from the aforementioned combinations must be expressed using a number of significant figures that do not exceed the smallest number of significant figures present in any of the original components.

Furthermore, the standard deviation resulting from these combinations cannot be smaller than the largest standard deviation among the individual components used in the calculation.

2.9 THE INTEGRAL ROLE OF STATISTICAL METHODS IN UNDERSTANDING EXPERIMENTAL DATASETS

All that has been reported above is fundamental to understanding how experimental results should be presented to provide a clear description of the systems under investigation. However, nothing has been written about how large datasets can be handled to deliver a comprehensive analysis that accounts for trends, outliers, and patterns in the data. Proper handling of large datasets requires statistical methods, machine learning algorithms, or other data processing techniques, which are essential for extracting meaningful conclusions and ensuring reproducibility. Without these methods, the risk of misinterpretation increases, particularly when dealing with complex systems or multivariable experiments. Therefore, moving forward, the focus should shift to the exploration of key statistical methods applied in chemistry, particularly in soil physical chemistry, with an emphasis on quality control in analytical chemistry and multivariate analysis techniques such as Principal Component Analysis (PCA), Analysis of Variance (ANOVA), and the emerging field of machine learning. These tools not only enhance current chemical research but also open new fields for future exploration.

2.9.1 Normal Distribution

The normal distribution, also known as the bell curve, is a fundamental concept in experimental science. The bell curve describes how data tends to cluster around an average value, with fewer instances occurring as one moves further from the center. This pattern shows up naturally in many experiments because of the Central Limit Theorem, which basically says that when one adds up a lot of independent random factors, the result tends to form a normal distribution, no matter what the original factors looked like.

As reported in the previous paragraphs, many small, random errors or fluctuations can affect measurements during an experiment. These random errors often add up and are independent of each other, leading to results that follow a normal distribution. By understanding this, researchers can make better guesses about the whole population based on sample data, check how reliable their measurements are, and figure out the involved uncertainties.

The normal distribution is defined by two key numbers: the mean (that in the paragraphs above has been indicated as $\overline{W}$), and the standard deviation (σ), describing how to spread the data are around the mean.

An important use of the normal distribution relies on testing hypotheses and building confidence intervals. It helps determine how likely it is to observe a particular result under a certain assumption, which is crucial for deciding whether to accept or reject hypotheses.

In fields like physics, chemistry, and biology, the normal distribution helps to model natural phenomena and to interpret experimental results. For example, it aids in understanding measurement errors, reaction times, and variations among living organisms. By spotting patterns that fit a normal distribution, scientists can use powerful statistical tools to analyze data more effectively.

To understand the meaning of a bell-shaped curve, the reader should examine the graphs presented in Figure 2.7. This figure reports synthetic data on how long plant roots grow following soil amendment with biochar. The histograms illustrate the frequency of different root growth measurements obtained from experimental data. Each bar corresponds to a specific range of growth values, and the height of the bar indicates how often those values occurred within the sample population.

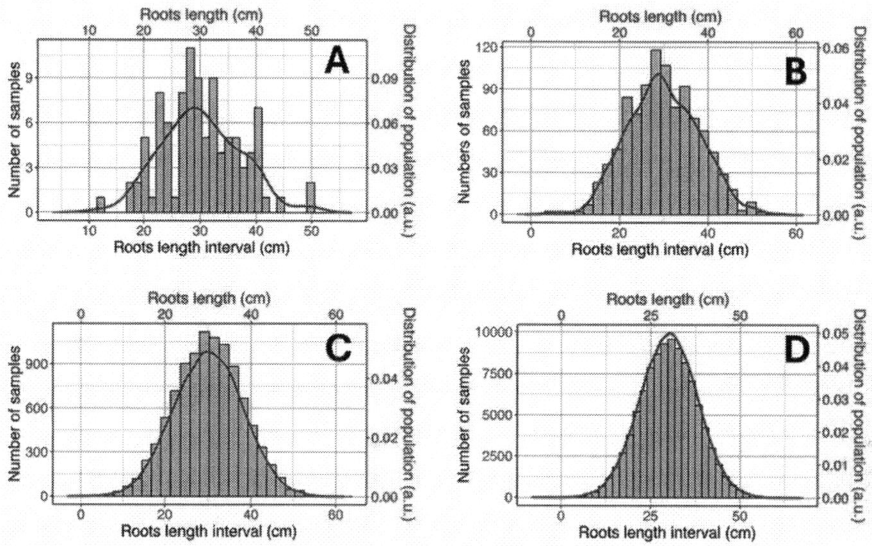

Figure 2.7 Theoretical exercise depicting the root growth of plants in biochar-amended soil. The number of plant samples increases from 100 (**A**) to 1,000 (**B**), 10,000 (**C**), and 100,000 (**D**). As the number of samples increases, the bell-shaped curve becomes more symmetrical. The reader is referred to the main text for a more detailed description of the meaning of this figure.

The histograms clearly show that the majority of data points are concentrated around the mean value, suggesting that most plants experienced a similar rate of root growth following the biochar treatment. This clustering of values forms a bell-shaped curve characteristic of a normal distribution. In other words, the data are symmetrically distributed, with a balanced spread of values on both sides of the mean.

A curve representing the theoretical probability density function of a normal distribution is superimposed on each histogram. The upper and right-hand axes correspond to this curve. This curve is a visual guide to assess how closely the experimental data align with the expected normal distribution. The symmetry of the curve—which becomes more pronounced as the number of samples increases, suggesting that the larger the number of samples, the better the quality of the experimental data—indicates that there are relatively few instances of extreme root growth values, either significantly higher or lower than the average.

It must be emphasized once again that the data reported in Figure 2.7 are synthetic. This means they were obtained as a mathematical exercise to illustrate the relationship between histograms and bell-shaped curves. If the data were real, they would suggest that the close alignment between the histograms and the normal distribution curves indicates that adding biochar consistently affects root growth across the entire plant population. Therefore, as a purely theoretical exercise, one can conclude that such consistency is valuable in agricultural practices because it implies predictability in biochar's influence on plant development. Farmers and agronomists could expect similar outcomes when applying biochar under comparable conditions.

Furthermore, the normal distribution highlighted in Figure 2.7 also implies that the results are not heavily influenced by outliers or anomalous data points. This enhances the reliability and validity of the collected data, reinforcing the scientific credibility of biochar as an effective soil amendment.

In conclusion, the theoretical exercise reported in Figure 2.7 reveals that plant root growth follows a normal distribution after applying biochar. This should underscore the efficacy and consistency of biochar in enhancing root development. The findings should have positive implications for sustainable agriculture and soil management practices, suggesting that biochar could be a reliable means of improving plant growth and overall crop yields.

2.9.2 The Non-Normal Distribution

Non-normal distributions are probability distributions that deviate from the classic bell-shaped curve of the normal distribution. Unlike normal distributions, which are symmetrical around the mean and have specific properties like fixed standard deviations, non-normal distributions can be asymmetrical, have heavier or lighter tails, or even display multiple peaks. The main differences compared to normal distributions are related to skewness, kurtosis, and multimodality.

As mentioned above, while normal distributions are symmetrical, non-normal ones can be skewed to the left (negative skew) or the right (positive skew). This means the data tails off more on one side than the other. For example, income levels in a population are often right-skewed, with a majority earning less and a few earning significantly more.

Kurtosis refers to the "tailedness" of a distribution. Non-normal distributions can have heavier (leptokurtic) or lighter (platykurtic) tails than normal distributions. Heavy tails indicate a higher likelihood of extreme values.

Finally, while normal distributions show only a single peak (mode), non-normal distributions can be bimodal or multimodal, indicating the presence of multiple subgroups within the data.

Non-normal distributions may occur in income and wealth data, where a right-skewed shape is usually observed. They are also typical when dealing with environmental data, such as measurements of pollutant concentrations, which show skewed distributions because of sporadic extreme events. Finally, biological and medical data may exhibit non-normal distributions due to factors like incubation periods or time-dependent concentrations of metabolites in the body.

Figure 2.8 presents a theoretical exercise obtained by simulating soil contamination by heavy metals using *R* software. The figure was generated by modeling a gamma distribution, which is inherently right-skewed. Each histogram displays the frequency of observed heavy metal concentrations across all hypothetical soil samples. The width of each bar represents a range of concentration levels, and the height of the bar indicates the number of samples falling within that range (that is, the bottom and left-hand axes of each graph).

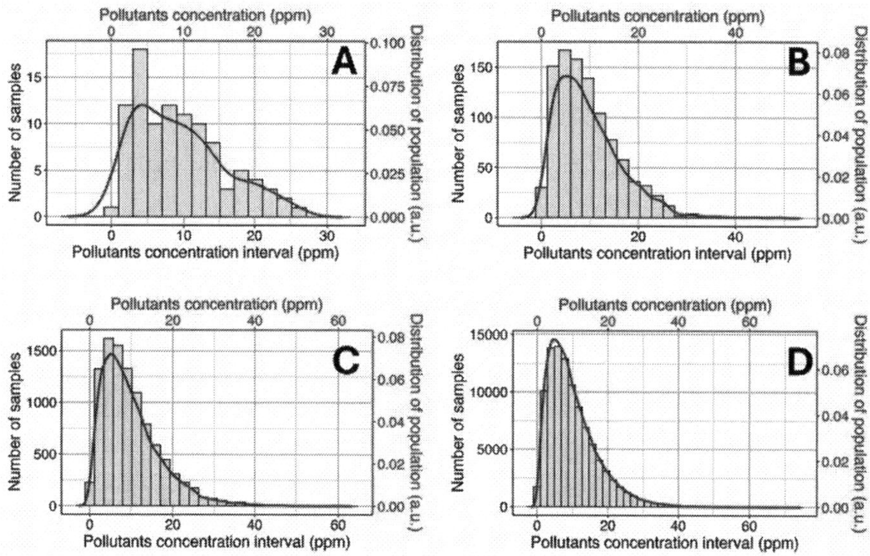

Figure 2.8 Theoretical datasets show soils contaminated by heavy metals. The data were obtained with the *R* software. The number of soil samples accounted for are 100 (**A**), 1,000 (**B**), 10,000 (**C**), and 100,000 (**D**). The larger the number of soil samples, the smoother the gamma distribution shape used to model the data.

The majority of the concentrations are relatively low, suggesting that most soil samples have minimal heavy metal contamination. This is reflected in the higher bars on the left side of the histogram. However, the histogram exhibits a long right tail, indicating the presence of a few samples with exceptionally high concentrations of heavy metals. This pattern is characteristic of environmental data where localized contamination occurs, possibly due to specific industrial activities, improper waste disposal, or natural mineral deposits.

A curve representing the theoretical probability density function of the gamma distribution[3] used to model data is superimposed on the histograms. This curve indicates the expected behavior of the distribution based on the gamma model parameters. The gamma distribution is appropriate here because it models continuous positive data that are skewed to the right, which aligns with the nature of heavy metal concentrations in soils. As with the normal distribution presented previously in Figure 2.7, here, too, the larger the number of samples (in this case, soil samples), the smoother the gamma distribution becomes.

[3] An accurate description of the gamma distribution is out of the aim of the present book. However, it is possible to state that in summary, it is a statistical tool suited for modeling right-skewed, positive data such as heavy metal concentrations in soils. Its use in environmental studies helps scientists and policymakers understand contamination patterns and make informed decisions.

The close alignment between the histogram and the gamma distribution curve suggests that it is a suitable model for these data. This modeling approach allows for better statistical analysis and interpretation of the contamination levels. It aids in estimating the probability of encountering certain concentration levels and helps in risk assessment by identifying the likelihood of extreme contamination events.

Understanding this distribution is crucial for environmental scientists and policymakers. It enables them to identify contamination hotspots, assess the overall health of the soil environment, and develop targeted remediation strategies. By recognizing that a few samples have very high contamination levels, efforts can be concentrated on those areas to mitigate environmental and public health risks.

2.9.3 Quality Control in Analytical Chemistry: Statistical Foundations for Precision

Quality control in analytical chemistry is a critical area where statistics ensures that measurements are both precise and accurate. Analytical methods used in pharmaceuticals, environmental testing, and materials science must undergo rigorous validation processes to confirm that they meet established standards.

Control charts are a primary statistical tool used to monitor the performance of an analytical method over time. By plotting measurements against statistically determined control limits, chemists can identify deviations from expected performance early on, which might indicate a potential failure in the process. For example, in environmental chemistry, control charts can be used to monitor the concentrations of pollutants in water bodies. If pollutant levels exceed predefined thresholds, corrective actions can be triggered to address contamination sources.

Another important aspect of quality control is the validation of analytical methods, which involves rigorous statistical analysis to ensure that methods are reproducible and reliable. This process includes assessing linearity, precision, accuracy, and sensitivity, which comprehensively evaluate a method's performance. Statistical hypothesis testing, such as *t*-tests and ANOVA, is widely used to compare methods and assess whether their differences are statistically significant.

2.9.4 Understanding the *t*-Test in Statistical Analysis

The *t*-test is a widely used statistical method that helps researchers determine whether the means of two groups are significantly different from each other. This test is particularly useful when comparing small sample sizes, making it a fundamental tool in various fields, including chemistry, biology, social sciences, and medicine. At its core, the *t*-test evaluates whether the observed differences between groups are likely due to chance or if they reflect true underlying differences in the population from which the samples were drawn.

There are several variations of the *t*-test, the most common being the independent samples *t*-test, the paired samples *t*-test, and the one-sample *t*-test. Each variation addresses different research questions, but all share the same basic principle: they compare means and account for variability within the data.

The independent samples *t*-test, or the two-sample *t*-test, compares the means of two unrelated groups. For instance, in environmental chemistry, this test might be used to compare the concentration of pollutants in two different rivers. The null hypothesis (H_0) for an independent *t*-test is that the means of the two groups are equal, implying no significant difference between the pollutant concentrations. The test calculates a *t*-statistic based on the difference between the two sample means, the variability within each group, and the size of the samples. If the *t*-statistic is large enough (relative to the variability in the data), the null hypothesis is rejected, suggesting a statistically significant difference between the two groups.

The paired samples *t*-test is applied when the two groups being compared are related or matched somehow. This test is frequently used in before-and-after studies or experiments where subjects are measured under two different conditions. For example, in a study assessing the effectiveness of a new water purification technique, a researcher might collect water samples from the same location before and after applying the technique. The paired *t*-test would then compare the mean pollutant levels before and after treatment. Here, the null hypothesis states that there is no difference in the means, while the alternative hypothesis suggests that the treatment led to a significant change in pollutant levels. Because the paired *t*-test accounts for the two sets of related observations, it is more sensitive to detecting small differences than the independent *t*-test.

The one-sample *t*-test is used when comparing the mean of a single group to a known value or theoretical expectation. This is useful when a researcher wants to determine if a sample differs from a population mean. For instance, a chemist might want to compare the concentration of a certain chemical in a water sample to a known safe threshold established by environmental regulations. The null hypothesis would be that the sample mean equals the regulatory threshold, while the alternative hypothesis would propose that the sample mean is significantly different.

To perform any *t*-test, several assumptions must be met. The most important thing is that the data follows a normal distribution, especially as the sample size decreases. The test also assumes that the data points are independent of one another, and in the case of the independent *t*-test, the two groups should have approximately equal variances. If these assumptions are violated, the *t*-test results may not be reliable, and alternative non-parametric tests like the Mann-Whitney *U* test or Wilcoxon signed-rank test may be more appropriate.

The Mann-Whitney *U* test is a non-parametric statistical test used to compare the differences between two independent groups when the assumption of normality is unmet. Unlike the *t*-test, it does not rely on the assumption that the data follows a normal distribution, making it ideal for ordinal data or non-normally distributed continuous data. The test ranks all the data points from both groups and then compares the sum of ranks to assess whether one group tends to have higher values than the other. It is often used as an alternative to the independent samples *t*-test.

The Wilcoxon signed-rank test is a non-parametric test used to compare two related or paired groups when the data does not meet the assumptions of normality. It is commonly applied in cases where a paired *t*-test would be inappropriate. The test works by ranking the differences between paired observations, ignoring the sign, and then comparing the ranks of positive and negative differences. It evaluates whether the two data sets have a significant median difference, making it ideal for analyzing before-and-after scenarios or matched pairs with non-normal distributions.

The main difference between the Mann-Whitney U test and the Wilcoxon signed-rank test lies in the type of data and research design they are used for:

The Mann-Whitney U test is applied to compare two independent groups. It is a non-parametric alternative to the independent samples t-test and is used when you have two unrelated groups (e.g., two different populations) and want to compare their distributions.

The Wilcoxon signed-rank test is used to compare two related or paired groups. It is the non-parametric equivalent of the paired t-test and is applied when you have repeated measures or matched pairs, such as "before and after" data for the same subjects.

In summary, the Mann-Whitney U test compares independent samples, while the Wilcoxon signed-rank test compares dependent or paired samples.

Overall, the t-test is a versatile and essential tool for hypothesis testing in many scientific disciplines. It allows researchers to make inferences about population means based on sample data, helping to confirm or refute claims about the significance of observed differences.

2.9.5 Multivariate Data Interpretation: Principal Component Analysis (PCA)

With the advent of high-throughput techniques in chemistry, datasets are becoming increasingly large and complex, often comprising thousands of variables and data points. Multivariate statistical techniques, particularly Principal Component Analysis (PCA), have become essential tools for reducing this complexity and extracting meaningful patterns from data.

PCA is a dimensionality-reduction technique that transforms a large set of correlated variables into a smaller set of uncorrelated variables known as principal components. These components capture the maximum variance in the data, allowing chemists to focus on the most significant patterns. In fields like spectroscopy, where datasets are often dense and difficult to interpret visually, PCA simplifies the dataset by identifying key factors that drive variability while preserving as much of the original information as possible[4].

[4] Some examples of PCA application in spectroscopy are in the following references:

- Bambina, P., et al. 2023. ^{1}H NMR-based metabolomics to assess the impact of soil type on the chemical composition of Nero d'Avola red wines. Journal of Agricultural and Food Chemistry. 71: 5823–5835.
- Bambina, P., et al. 2024. ^{1}H-NMR spectroscopy coupled with chemometrics to classify wines according to different grape varieties and different terroirs. Agriculture. 14: 749.
- Bambina, P. and P. Conte. 2024. HRMAS ^{1}H NMR and CPMAS ^{13}C NMR spectroscopies coupled with chemometrics for the metabolomic investigation of commercial teas. Food Chemistry. 461: 140816.

In environmental chemistry, PCA is frequently applied to analyze water quality data. For example, in a study of river pollution, water samples might be analyzed for various parameters, such as pH, heavy metal concentrations (e.g., lead, mercury, and cadmium), and nutrient levels (e.g., nitrates and phosphates). With so many variables, it can be challenging to determine the key contributors to water quality deterioration. PCA can reduce the dimensionality of the dataset, highlighting which factors are the most important in driving water pollution. For example, one principal component might reveal that high levels of nitrates and phosphates are the primary contributors to pollution in agricultural runoff, while another component might highlight industrial heavy metal contamination.

In air quality monitoring, PCA has been applied to analyze large datasets of airborne pollutants measured across different locations and seasons. In a study of urban air pollution, chemists could analyze concentrations of various gases (e.g., CO, NO_x, and SO_2) and particulate matter. PCA can be used to identify major pollution sources by reducing the complexity of the data. For instance, one principal component might suggest that traffic emissions are the primary contributor to CO and NO_x levels, while another component could reveal industrial sources responsible for higher concentrations of SO_2. This enables more targeted policy interventions, such as controlling industrial emissions in areas significantly contributing to poor air quality.

In soil contamination research, PCA is often employed to investigate the presence of heavy metals and organic pollutants across contaminated sites. For example, if multiple soil samples are collected from an industrial site, PCA can help determine which pollutants (e.g., lead, arsenic, and PAHs) are most strongly associated with specific industrial activities. By simplifying the dataset into principal components, researchers can distinguish between different contamination sources, such as historical mining operations, versus more recent agricultural chemical use.

Figure 2.9 presents a theoretical exercise with synthetic data showing different pollutants hypothetically found in contaminated soils. The graph is a biplot resulting from a Principal Component Analysis (PCA) of soil contamination by heavy metals and organic contaminants. In particular, the horizontal and vertical axes represent the first two principal components identified by the PCA. These components capture the majority of the variability in the data, effectively summarizing complex information into two dimensions. Each grey point corresponds to a soil sample, positioned based on its relative chemical composition concerning the principal components. Samples that are close together on the plot have similar contaminant profiles. The black arrows represent variables, namely the heavy metals and the organic contaminants. The direction of each arrow indicates the correlation of that variable with the principal components, while the length reflects its influence on the data variability. Longer arrows mean variables with a larger impact on the variability captured by the components.

This visualization enables the identification of relationships between the samples and the analyzed variables, thereby facilitating the interpretation of data on soil contamination. By examining the distribution of the samples (grey points) and the variables (black arrows), patterns of contamination can be discerned. For instance, samples located in the direction of a specific arrow are likely influenced

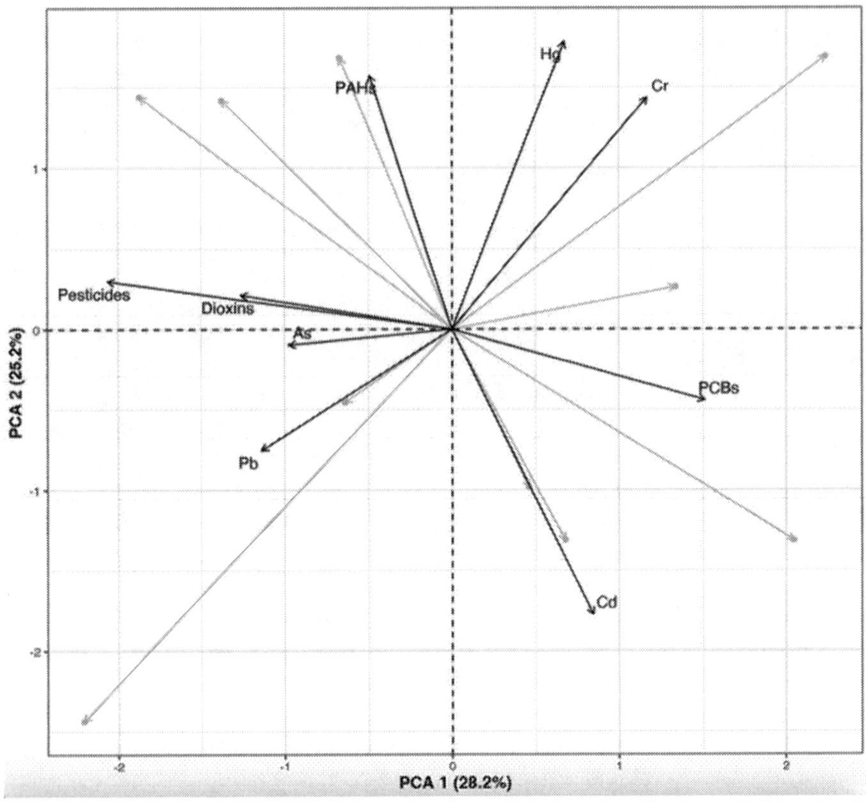

Figure 2.9 Biplot graph resulting from a Principal Component Analysis (PCA) of soils contaminated by organic and inorganic pollutants. PCBs stand for polychlorobiphenyls, while PAHs stand for polycyclic aromatic hydrocarbons. This graph does not present real data; it was obtained by simulating contamination using the R software. Therefore, it should be considered only as an exercise to illustrate how PCA graphs appear.

by that particular contaminant. This information is crucial for understanding the underlying factors affecting soil quality and can inform decisions on soil remediation and management.

Future research could explore real-time PCA algorithms for dynamic environmental systems, enabling immediate analysis of high-throughput monitoring data, such as continuous water or air quality assessments. Integrating PCA with machine learning could also improve predictive models for understanding long-term environmental changes, offering more accurate insights into pollution trends and their impact on ecosystems.

2.9.6 Statistical Hypothesis Testing: Analysis of Variance (ANOVA)

The complexity of chemical processes often necessitates the simultaneous comparison of multiple variables, especially when optimizing reaction conditions or validating analytical methods. Analysis of Variance (ANOVA) is a powerful statistical technique that allows chemists to assess whether differences between group means are statistically significant. Unlike simpler tests like the *t*-test, which compares two means, ANOVA can handle multiple groups or conditions simultaneously, making it ideal for complex experimental designs.

In ANOVA, a *p*-value is accounted for to determine whether the differences between the means of the groups being compared are statistically significant. Specifically, it represents the probability of observing the data or something more extreme, assuming that the null hypothesis is true. The null hypothesis in ANOVA typically states that there are no differences between the group means.

— If the p-value is small (typically less than a significance level like 0.05), it indicates that the observed differences between group means are unlikely to have occurred by chance, leading to the rejection of the null hypothesis.
— A large *p*-value suggests that the differences between the group means could have arisen by random variation, and thus, the null hypothesis cannot be rejected.

In environmental chemistry, factorial ANOVA[5] is often used to optimize water treatment processes by analyzing multiple treatment factors. For example, a chemist might investigate how different factors, such as pH levels, coagulant dosage, and settling time, impact the removal efficiency of contaminants from wastewater. By applying factorial ANOVA, the chemist can determine whether individual factors (e.g., increasing coagulant dosage) significantly affect contaminant removal and whether interactions between factors (e.g., pH and coagulant dosage) lead to improved or diminished treatment efficiency. Understanding these interactions is crucial for designing the most effective and cost-efficient water treatment processes.

One-way ANOVA[6] can be used to compare the effects of different pollutants on the health of aquatic organisms in ecotoxicology studies. For instance, a researcher may expose fish or invertebrates to varying concentrations of heavy metals (e.g., copper, lead, zinc) in a controlled laboratory environment

[5] Factorial ANOVA is a statistical method used to analyze the effects of two or more independent variables (also called "factors") on a dependent variable simultaneously. Unlike one-way ANOVA, which tests the effect of a single factor, factorial ANOVA allows for the investigation of multiple factors and their interaction effects. This makes it especially useful for understanding complex relationships between variables in experimental designs.

[6] In one-way ANOVA, there is only one independent variable (factor) with multiple levels or categories, and the goal is to see how this factor affects the dependent variable. The null hypothesis assumes that all group means are equal, while the alternative hypothesis suggests that at least one group mean is different. For example, in a study examining the effect of different fertilizers (Factor: Fertilizer Type) on plant growth (Dependent Variable: Plant Height), one-way ANOVA would test whether the mean plant height differs significantly among the groups receiving different fertilizers. If the *p*-value from the ANOVA is below a chosen significance level (e.g., 0.05), the null hypothesis is rejected, indicating that there are statistically significant differences among the group means.

and measure their growth or survival rates. One-way ANOVA would allow the chemist to evaluate whether differences in pollutant concentration result in statistically significant effects on the organisms' health. This type of analysis helps determine the threshold concentrations at which pollutants become toxic, informing environmental regulations and standards.

In soil remediation studies, ANOVA can be applied to compare the effectiveness of different remediation techniques. For example, a study might compare methods such as bioremediation, phytoremediation, and chemical leaching to remove heavy metals from contaminated soil. By applying ANOVA, researchers can assess whether the differences in contaminant removal rates between the techniques are statistically significant, helping to identify the most effective approach for a specific contamination site. Further, interaction terms can be included in the analysis to determine if combining methods (e.g., bioremediation followed by chemical leaching) results in improved outcomes compared to using a single method alone.

Future research could extend ANOVA applications in environmental chemistry to longitudinal studies, where environmental changes are monitored over time. Repeated measures ANOVA could provide insights into how pollution levels evolve in response to remediation efforts or changes in industrial activity. Additionally, Bayesian[7] ANOVA[8] could offer probabilistic interpretations of complex environmental data, enabling chemists to incorporate prior knowledge into their analyses, such as historical pollution data, to improve predictions of future environmental impacts.

---

[7] Bayesian principles are rooted in Bayesian probability, which interprets probability as a measure of belief or certainty about an event, rather than just a long-run frequency of occurrence. These principles are based on Bayes' Theorem, which updates the probability of a hypothesis based on new evidence. The key components of Bayesian analysis are: (1) Prior probability: This reflects the initial belief about a parameter or hypothesis before observing any data. It can be based on prior knowledge, experience, or assumptions. (2) Likelihood: This represents the probability of observing the current data given a specific hypothesis or parameter value. (3) Posterior probability: This is the updated belief about the hypothesis or parameter after considering the new data. It is calculated by combining the prior probability and the likelihood using Bayes' Theorem. Bayesian analysis allows continuous updating of beliefs as more evidence becomes available, providing a flexible and probabilistic interpretation of uncertainty. This framework is particularly useful when dealing with complex models, small sample sizes, or when incorporating prior knowledge into the analysis.

[8] Bayesian ANOVA is a statistical method that applies Bayesian principles to the traditional Analysis of Variance (ANOVA). Unlike classical ANOVA, which relies on frequentist p-values to test for differences between group means, Bayesian ANOVA estimates the posterior probability distributions of model parameters. This approach allows researchers to directly assess the probability of hypotheses given the data, rather than relying on p-values. Bayesian ANOVA also accommodates prior information, enabling the integration of previous knowledge into the analysis. It is particularly useful when sample sizes are small or when researchers want to quantify uncertainty in a more flexible, probabilistic framework.

2.10 MACHINE LEARNING IN CHEMISTRY: NEW FRONTIERS IN PREDICTIVE MODELING

As chemical datasets grow in size and complexity, machine learning (ML) has emerged as a transformative tool, allowing chemists to develop predictive models, optimize synthetic pathways, and discover new materials. Machine learning algorithms can analyze vast amounts of data, identifying patterns and relationships that may be impossible to discern using traditional methods. By training these models on existing data, chemists can predict the outcomes of untested reactions, reducing the need for costly and time-consuming experiments.

Machine learning is increasingly applied in environmental chemistry to predict pollution levels based on historical data and real-time monitoring. For example, machine learning models can be trained on data from sensors measuring NO_x, CO, $PM_{2.5}$, and meteorological conditions (e.g., wind speed and temperature) in a study predicting air pollution in urban areas. These models can then predict pollution levels for the following day or week, enabling local authorities to implement timely measures to reduce exposure to harmful pollutants. Moreover, machine learning can be used to identify patterns in how traffic, industrial activity, and weather conditions interact to influence pollution levels, offering valuable insights for urban planning and environmental policy.

In groundwater studies, machine learning can be used to predict contamination risks from industrial activities, agricultural runoff, or leaching from landfills. By training models on data from contaminated sites, machine learning algorithms can identify key factors contributing to groundwater pollution, such as proximity to waste disposal sites, soil type, and rainfall patterns. These models can then be used to predict future contamination risks in unmonitored areas, enabling better resource allocation for groundwater protection efforts.

Looking to the future, integrating machine learning with remote sensing technologies could offer real-time pollution monitoring over large geographical areas. For example, satellite data on land use, industrial activity, and vegetation cover could be combined with machine learning models to predict pollution dispersion patterns and inform remediation strategies.

EXERCISES

1. A carpenter measured the height of a door. It was 210 cm with a confidence interval between 205 cm and 215 cm. Write the height of the door in the form $a \pm \sigma_a$.
2. The density (in kg m^{-3}) of an extra-virgin olive oil was measured at 22°C. The table reports the measurements repeated 10 times. Calculate the average value and the standard deviation for oil density. Write the result with the right number of significative figures.

913.13	913.04
914.00	913.93
913.00	913.16
913.18	913.20
912.98	913.09

3. The water holding capacity (WHC) of soil (in g per 100 g of soil) was measured thrice. It resulted in 25.30, 26.68, and 26.80. Calculate WHC average and standard deviation. Write the result with the correct number of significative figures.
4. A 3 M NaCl water solution must be prepared. A student dissolves 87.6613 ± 0.0001 g of sodium chloride in a 500.0 ± 0.1 mL volumetric flask containing deionized water. Write the right concentration of the NaCl solution. (A.W. Na, 22.990 ± 0.001 g mol^{-1}; A.W. Cl, 35.45 ± 0.01 g mol^{-1})
5. Two NaCl solutions have been prepared (Solution 1: 3.00 ± 0.01 g mol^{-1}; Solution 2: 2.00 ± 0.01 g mol^{-1}). 500.0 ± 0.1 mL of each solution have been sampled and mixed to obtain single NaCl solution. Calculate the exact concentration of the resulting NaCl solution by writing the right number of significative figures.
6. Define the normal distribution and explain its importance in analyzing experimental data.
7. What is a p-value, and how is it used to determine statistical significance?
8. Describe the principles and applications of principal component analysis (PCA) in interpreting multivariate datasets.
9. Explain the difference between factorial ANOVA and one-way ANOVA. Provide an example of how each type of ANOVA can be applied in environmental chemical research.
10. Discuss the benefits of using machine learning techniques in environmental data analysis and provide examples of potential applications.

FURTHER READING

Dickinson, F. and A. McKinley. 2021. Introduction to contextual maths in chemistry. Royal Society of Chemistry, Croydon, UK. ISBN: 9781839164163

Launer, R.L. and G.N. Wilkinson (eds). 1979. Robustness in Statistics: Proceedings of a Workshop. Academic Press, New York (USA). 201–236. ISBN: 9780124381506

Taylor, J.R. 1997. An introduction in error analysis. The Study of Uncertainties in Physical Measurements, 2nd Ed. University Science Books. ISBN: 0935702423

Chapter 3

What is the Soil

3.1 INTRODUCTION

The first question that arises in young people when they approach soil science is: "What is a soil?" This appears quite a trivial question because, apparently, all of us identify "soil" with the ground where we do all our activities. This is not enough to understand what a soil is. If we go around the globe—not physically, but by using the powerful tools offered by our technology such as an internet connection—we can realize that our activities are spread worldwide: we find humans acting not only in the temperate regions of the Earth but also in the Amazon rainforest, the Sahara Desert, Antarctica, in the surroundings of the North Pole, in Earth's profundities, and so on. Based on this, is the sand of the Sahara Desert soil? Is the ground of the Amazonian Forest soil? Or do we need to consider soil only in the most temperate zones of the Earth? Where does soil start, and where does soil end? Can we consider soil what we find in the profundities of mines?

3.2 EARTH AS AN APPLE AND SOIL AS A THIN APPLE PEEL

Various books and websites provide insightful and educational descriptions of what constitutes soil and the depth required to excavate to reach what is formally defined as "soil." One of the most illustrative comparisons is to imagine Earth as an apple (Figure 3.1). First, cut the apple into four equal parts. Three of these four slices represent Earth's water—oceans, seas, lakes, and rivers—while only one-quarter represents land. Next, cut this land portion in half. One of these halves (equivalent to 1/8 of Earth) consists of swamps, deserts, ice, and mountains, areas generally unsuitable for human habitation. The remaining half (another 1/8 of the apple) represents the remaining land. Now, divide this second half of

DOI: 10.1201/9781003780212-3

the 1/8 portion into four equal parts. Three of these four parts (i.e., 3/32 of the apple) consist of regions that are too rocky, too wet, too hot, or too cold, as well as urbanized areas such as cities and highways. The final 1/32 of the apple represents the remaining habitable land. Finally, peel this last slice. The thin layer of peel represents what can be considered "soil"—the crucial layer supporting life.

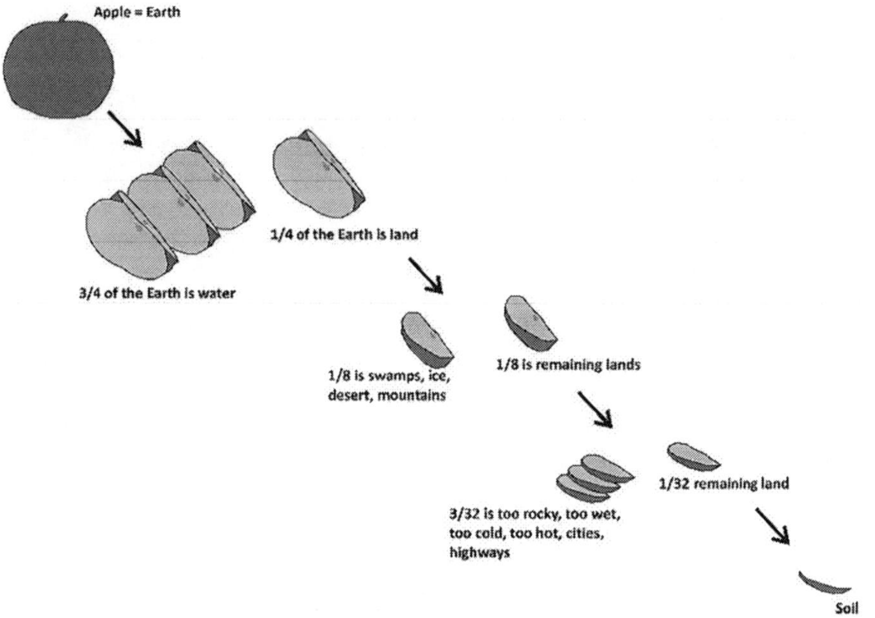

Figure 3.1 Representation of Earth as an apple. Three-quarters of the planet is covered by water, while the remaining quarter represents land. Half of this land is unusable (e.g., deserts, mountains, ice), leaving only 1/8 of Earth as potentially habitable land. Further division shows that just 1/32 of the planet consists of usable land. The thin peel of this final section represents the soil that sustains global food and textile production.

3.3 SOIL POLY-FUNCTIONALITY

From the example in the previous paragraph, it becomes evident that soil can be regarded as the thin surface layer of the Earth's crust where food and textiles are produced. However, while this description is compelling, it fails to convey just how "thin" soil truly is. How deep do we need to define soil boundaries? From the top downward, does soil begin with the undecomposed litter layer, or does it start where litter is absent and decomposed organic matter predominates? Does soil end at the deepest point where plant roots can reach or where all signs of biological activity cease? Furthermore, the "soil/apple-peel" analogy mentioned earlier reflects only an agronomic perspective, emphasizing soil's role in crop production. From an engineering viewpoint, soil is a structural foundation for buildings, bridges, and roads. To geologists, soil is a component of the broader rock cycle. Quarry workers may view soil as an obstacle to be removed for accessing lithoid materials, while ecologists see soil as a system of properties essential for supporting plant and animal life. For the energy industry, the soil is often reduced to a substrate for biomass production. Many other soil functions exist but are not listed here, as the intent is not to provide an exhaustive catalog of soil uses, which can be found elsewhere.

The complexity and poly-functionality of soil also highlight that defining its boundaries is inherently arbitrary, dependent entirely on the observer's perspective. For instance, agronomists typically focus on the surface layer, extending only 5–10 cm deep, while pedologists examine soil to depths of several meters, and geologists may consider much deeper layers.

Clearly, all these perspectives are subjective, each addressing a particular soil function or human interest. As Robert White has argued (see the Further Reading), soil's poly-functionality makes it impossible to establish an exact and universally rigorous definition. Consequently, the quest to define soil becomes an ultimately futile exercise.

3.4 AN OBJECTIVE DEFINITION OF SOIL

Notwithstanding what has been reported in the previous paragraphs, an objective definition of soil exists: a definition that does not account for a specific soil function or human interest. It has been given by pedologists who consider soil as a complex open system where the atmosphere, hydrosphere, lithosphere, and biosphere intersect (Figure 3.2).

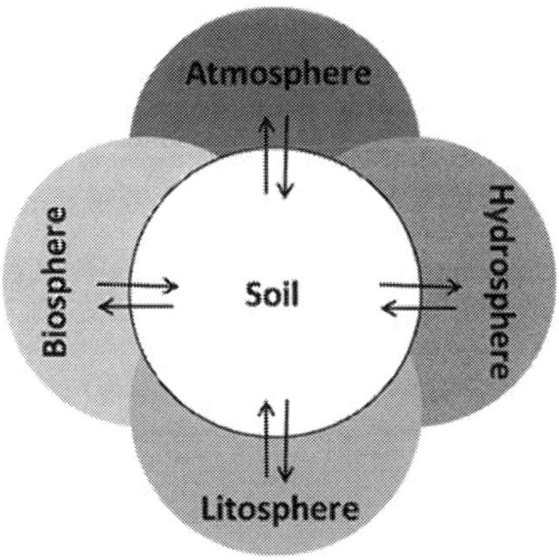

Figure 3.2 Soil is considered the intersection point between the atmosphere, hydrosphere, lithosphere, and biosphere. A couple of arrows indicate the exchanges among soil and the other environmental compartments.

Here, the atmosphere is considered as the ensemble of materials (such as vapor/gases and light solid particles) that are retained around the planet by the gravitational force. These materials are also subjected to dynamics, which are affected by air temperature and pressure. The hydrosphere refers to all types of water present on the planet Earth in the liquid phase. The biosphere is the ensemble of all the living organisms. The lithosphere refers to the solid and most rigid part of the planet Earth, which is commonly indicated as rocks. In particular, the soil is vertically delimited, on the top by the atmosphere (that means that the undecomposed litter is part of the soil), while on the bottom by the lithological matrix (i.e., the undecomposed rock). Waters (such as oceans, rivers, and lakes) represent the horizontal borders within which soil is confined.

According to the aforementioned definition, a ground where living organisms are not found cannot be defined as soil. For example, the Sahara Desert, the frozen ground in Antarctica, the forest ground in Amazonia, and the ground found at very high temperatures in volcanic systems[1] are all considered soils, as all of them can support forms of life. Certainly, life forms are adapted to the environmental conditions where they grow; thus, we cannot expect to find the same living organisms in all the aforementioned soils. The soils that are of the greatest importance to us as humans are those supporting plants that can be used for food[2] or textile production.

Figure 3.2 shows that soil—as an open system—exchanges matter and energy (i.e., different forms of energy) with the atmosphere, hydrosphere, lithosphere, and biosphere[3]. As an example, soil releases gases and radiation (i.e., heat) to the atmosphere, while it can receive water, aerial nanoparticles, and organic and inorganic substances through rains and Aeolian transport from the atmosphere; soil exchanges heat with the hydrosphere while releasing leachable matter and receiving organic and inorganic chemicals through the phenomena of deposition and erosion; soil weathering provides materials to the lithosphere, while diagenesis[4] allows formation of the particles which, all together, generate soil characteristics.

As will become clearer throughout this book, the equilibria governing exchanges among the different environmental compartments make the soil a continuously evolving medium with a structured organization.

---

[1] Examples are the volcanic soils in New Zealand as reported in the following link: https://www.newscientist.com/article/2134206-extreme-plants-thrive-at-72c-in-new-zealands-hot-volcanic-soil/

[2] Nowadays, unmanned trips in search of forms of life on the planets of our solar system are organized (as an example, the reader can be addressed to the following link: https://www.nasa.gov/mission_pages/msl/index.html). Till now, no traces of living organisms have been found. For this reason, according to the definition we are going to apply in the present book, the ground of the extra-terrestrial planets cannot be considered as "soils," although many books and websites make use of such a word.

[3] The atmosphere, hydrosphere, lithosphere, and biosphere are commonly referred to as environmental compartments. Soil is also considered an environmental compartment, as it serves as an interface between these spheres and plays a crucial role in various ecological and geochemical processes.

[4] Diagenesis refers to all the physical and chemical processes involved in the formation (i.e., synthesis) of minerals, that is, the inorganic part of the soils.

3.5 PROFILES AND HORIZONS

To understand the terms "profile" and "horizon," used in soil science, the concept of the pedon[5] was introduced. This is the smallest three-dimensional unit of soil at the Earth's surface, typically conceptualized as a hexagonal prism (Figure 3.3). A group of contiguous pedons is referred to as a polypedon (Figure 3.3), while the exposed vertical section of a pedon, used to observe soil variability, is known as a soil profile (Figure 3.3). A profile is a vertical section extending from the atmosphere at the surface to the underlying lithological matrix (Figure 3.3). The distinct, morphologically homogeneous layers within the profile are called soil horizons (Figure 3.3).

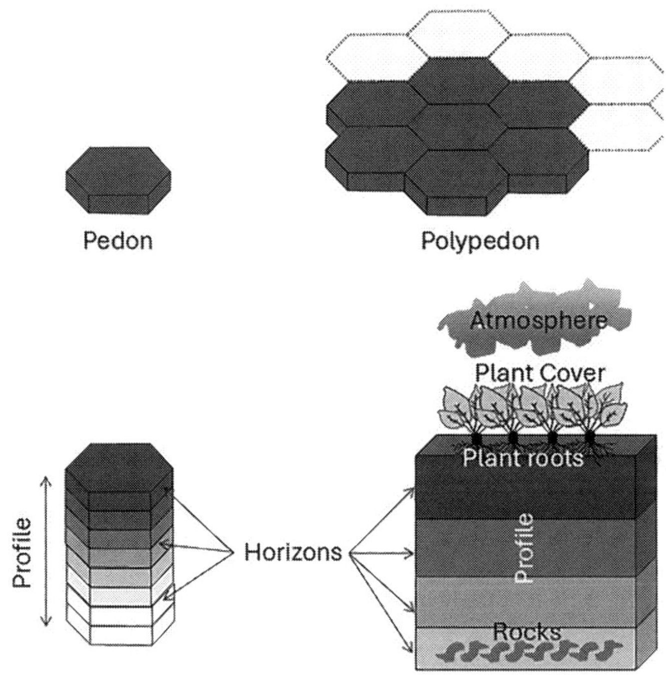

Figure 3.3 Schematic representation of soil units. The pedon is the smallest three-dimensional unit of soil, typically conceptualized as a hexagonal prism. A group of contiguous pedons forms a polypedon. The soil profile is the vertical section of a pedon, extending from the atmosphere above to the underlying lithological matrix (rock). Within the profile, distinct layers called horizons exhibit variations in soil composition and structure.

Table 3.1 presents examples of horizon notation in a soil profile. Horizons are labeled with capital letters (O, A, E, B, C, and R) from top to bottom. The uppermost layer, O (organic), contrasts with the deepest, R (rock). Intermediate horizons (A, E, B, and C) follow sequentially. Combinations of these letters denote

[5] This term comes from the Greek πέδον, which means soil.

transition horizons. Each capital letter may be followed by one or more lowercase letters (Table 3.1 includes one), indicating specific chemical characteristics. Numbers may also appear to mark lithological discontinuities. A soil profile does not always contain the full sequence of horizons. Accurate identification relies on expert analysis and laboratory testing. Studying profiles and horizons is crucial for understanding pedogenesis—the physical and chemical processes involved in soil formation[6].

Table 3.1 Some of the notations are used to identify horizons along a soil profile[7]

Type of Horizon	Brief Description
	O – Organic
O_f	It consists largely of fibrous materials whose botanical origin is readily recognizable
O_m	It consists of mesic material (i.e., material with a moderate amount of moisture), which is at a stage of decomposition intermediate between fibrous and humic materials
O_h	It consists of humic material, which is at an advanced stage of decomposition in uncultivated soils
O_p	It consists of humic material, which is at an advanced stage of decomposition in cultivated soils
A – Near the surface, with well-mixed humified organic material either by natural or anthropic processes	
A_h	It is rich in organic matter being the latter >1% (w w^{-1})
A_p	It is disturbed by anthropic activities such as cultivation and logging
E – Eluvial, subsurface horizon with low organic content, having lost material towards lower horizons	
E_a	It consists of lacking mottled, bleached particles due to the removal of organic and sesquioxide[8] coatings
E_b	brownish color due to homogeneously distributed iron oxide
B – Subsurface horizon without rock structure. It shows characteristic structure, texture, and color due to illuviation material from the above and/or weathering of parent material	
B_f	It is enriched with amorphous material, mainly compounds containing Al and Fe, which are also combined with organic matter that ranges in the interval 0.5–5% (w w^{-1})
B_h	It contains more than 1% organic C, less than 0.3% pyrophosphate-extractable Fe, and shows a ratio of organic C-to-pyrophosphate-extractable Fe of 20 or more

[6] The term "pedogenesis" comes from the combination of two Greek words: *πέδον*, which means soil, and *γένεσις*, which means birth. For this reason, "pedogenesis" means literally "soil formation."

[7] Refer to the book by White, R. 2010. reported in the Further Reading section.

[8] The term "sesquioxide" is an old chemical word referring to compounds where the amount of oxygen is in the ratio 3:2 with the metal. As an example, the aluminum sesquioxide is Al_2O_3, while iron sesquioxide is Fe_2O_3.

Type of Horizon	Brief Description
B_s	It is rich in sesquioxides by illuviation; it shows an orange-to-red color
B_t	It is rich in alluvial clays
C – Unconsolidated or weakly consolidated mineral material retaining rock structure	
C_k	Contains more than 1% (w w^{-1}) $CaCO_3$ as concretion or coatings
C_m	Continuously cemented
C_r	It is weakly consolidated but dense enough to prevent root ingress except along cracks
R	Consolidated rock layer

3.6 THE BEGINNING: WE ALL ARE STARS' CHILDREN

Soils are incredibly complex systems containing nearly all the elements of the periodic table, although their concentrations vary widely depending on the type of parent material from which they originate. These elements form the foundation of the chemical composition of soils, influencing their properties and functions.

For example, Figure 3.4[9] illustrates the average composition of soils worldwide, highlighting that oxygen is one of the most abundant elements, alongside silicon, barium, and strontium. These elements play critical roles in soil structure, nutrient availability, and plant growth, reflecting the diverse pathways through which they were formed and distributed across the Earth.

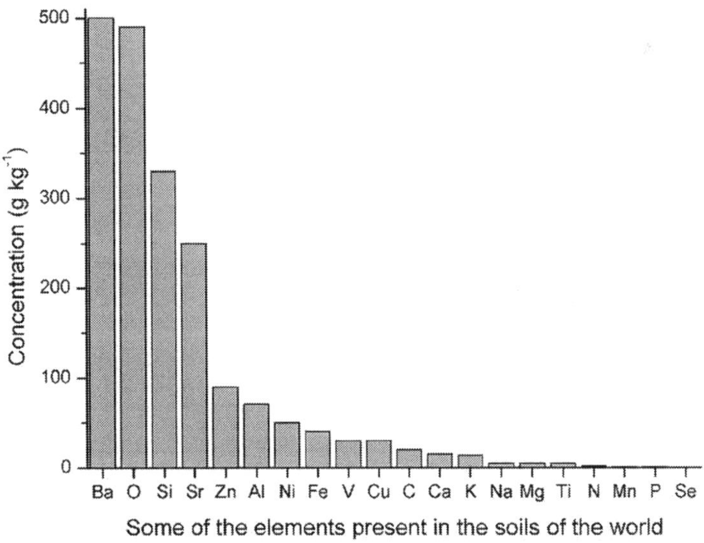

Figure 3.4 Average composition of the soils of the world. Only 22 elements are shown[25].

[9] This figure is based on the data reported in J. Yanai, et al. 2012. Elemental composition of agricultural soils in Japan in relation to soil type, land use, and region. Soil Science and Plant Nutrition. 58: 1–10.

3.6.1 The Origins of the Elements

To understand where these elements come from, we must travel back around 14 billion years to the universe's beginning. At that time, the universe was an incomprehensibly dense and hot point—a "space-time singularity." Physicists describe this singularity as a state where space, time, and all the laws of physics as we know them are indistinguishable. Then, a sudden and monumental event known as the Big Bang occurred, marking the beginning of the universe's expansion. This expansion initially released enormous amounts of energy through intense heat and electromagnetic radiation (photons). Subatomic particles, such as electrons and quarks, were also generated, but the extreme temperatures—reaching trillions of Kelvin—prevented these particles from combining into stable structures. At this stage, the universe was a seething, chaotic soup of energy and particles. Over the first fractions of a second, as the universe continued to expand, its temperature began to decrease, making it possible for these particles to begin interacting and forming the first building blocks of matter.

3.6.2 Formation of Atoms and Stars

Roughly one second after the Big Bang, the universe cooled enough for protons (the nuclei of hydrogen atoms) to form. Over the next few minutes, some protons merged to form helium nuclei and small amounts of heavier elements like lithium and beryllium. This process, called nucleosynthesis, was driven by the high-energy collisions of particles but stopped as the universe expanded and cooled further, leaving most of the matter in the form of hydrogen and helium.

It was not until several hundred thousand years later when the temperature dropped to around 3,000 K, that electrons could combine with these nuclei to form the first neutral atoms. At this point, the universe became transparent, allowing light to travel freely for the first time—a moment captured in the cosmic microwave background radiation we observe today. Under the influence of gravity, these hydrogen and helium atoms began to clump together into vast clouds, eventually collapsing to form the first stars.

Inside these early stars, intense gravitational pressure and heat initiated nuclear fusion. Hydrogen atoms fused into helium, releasing enormous amounts of energy that balanced the inward pull of gravity. As stars aged, their cores became hot and dense enough to fuse helium into heavier elements such as carbon, oxygen, and silicon. This process of fusion, layered by the density of elements, is depicted in Figure 3.5, which shows how lighter elements like hydrogen accumulate near the surface, while heavier elements like iron are found deeper within the star.

3.6.3 Supernovae and the Distribution of Elements

As shown in Figure 3.6, stars maintain equilibrium by balancing gravitational forces pulling inward (G) and the outward pressure generated by thermal energy (H) released from nuclear fusion. This delicate equilibrium allows stars to sustain their structure while continuously producing heavier elements through fusion.

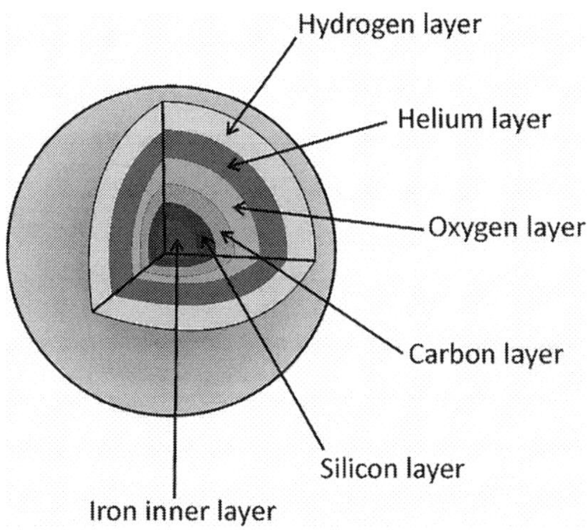

Figure 3.5 Different layers in a star. As a heavier element is synthesized by the reactions in the more surficial layers, it goes towards the center of the star because of the gravitational force. This is the reason why the star's surface contains mainly the lightest hydrogen, while the heaviest iron is accumulated in the innermost core of the star.

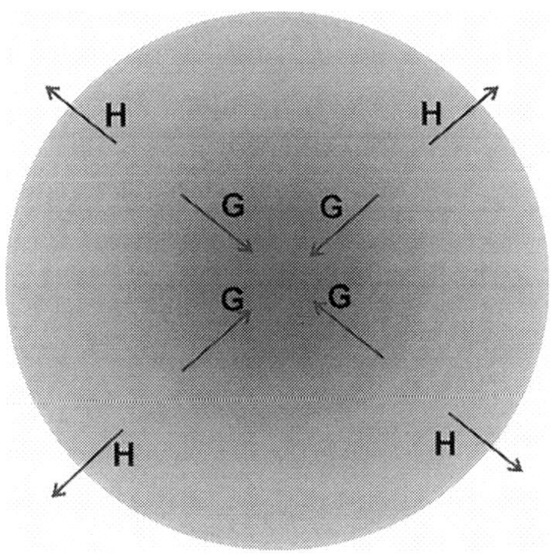

Figure 3.6 Equilibrium conditions in a star. Gravitation (G) is balanced by the thermal energy (H) produced due to the fusion of hydrogen atoms.

However, nuclear fusion halts when a star core accumulates too much iron because, unlike lighter elements, iron cannot release energy through fusion, thereby stopping the energy production that counterbalances gravity. Without the outward thermal energy to counteract gravity, the star collapses under its weight in a dramatic event known as a supernova. The immense energy released in a supernova allows for the formation of elements heavier than iron, such as gold and uranium, through a process called neutron capture. Star explosion scatters material into the surrounding space, enriching the interstellar medium with the raw ingredients for new stars, planets, and eventually life. Small amounts of molecules like water (H_2O), carbon dioxide (CO_2), and methane (CH_4) also form during these explosions.

3.6.4 Planet Formation and Soil Origins

The debris from supernova explosions drifts through space, forming clouds of gas and dust. Over millions of years, these materials clump together under the influence of gravity, eventually forming the protoplanetary disks that give rise to planets. Although the exact mechanisms of planet formation remain a subject of scientific investigation, it is understood that collisions between particles in these disks create larger and larger bodies, culminating in fully formed planets.

On Earth, the elements produced during stellar nucleosynthesis and scattered by supernovae eventually became part of the planet's rocky crust through planetary accretion and differentiation. Over billions of years, geological processes such as weathering and erosion broke down these rocks, releasing the elements into the environment and initiating soil formation (pedogenesis). This long chain of processes connects the distant history of stars to the soil beneath our feet.

3.7 THE LITHOGENESIS

What we call "soil," according to the definition given above, comes from the degradation and the transformation of a parent material containing all the elements produced during the star synthesis and the formation of the planet. This parent material is made of rocks.

A scientific definition of "rock" was provided by Geoffrey Mibei[10], who stated that "a rock is a naturally occurring, solid, cohesive aggregate of one or more minerals or mineral-like materials." In this context, a mineral is defined as a naturally occurring, solid substance with a specific chemical composition and an ordered atomic structure. While most minerals are inorganic, some, such as carbonate minerals (e.g., calcite and dolomite) and certain elemental forms like graphite, contain carbon.

Different categories of rocks can be distinguished. The igneous rocks are formed during the cooling and solidification of the magma, which is a molten mixture of different minerals beneath or inside Earth's crust. Igneous rocks are differentiated into plutonic[11] (also referred to as intrusive) and volcanic (also referred to as extrusive). The former is obtained by the slow cooling inside Earth's crust, while the latter by the fast cooling following volcanic eruptions. Due to the slow cooling, plutonic rocks contain well-defined crystals such as those

[10] Mibei, G. 2014. Introduction to types and classification of rocks, presentation at the Short Course IX on Exploration for Geothermal Resources, Organized by UNU-GTP, GDC and KenGen, at Lake Bogoria and Lake Naivasha, Kenya, 2–24. 2014. Available at: https://orkustofnun.is/gogn/unu-gtp-sc/UNU-GTP-SC-19-0302.pdf

[11] The term "plutonic" comes from Pluto, the god of the underworld, for the ancient Greeks.

present in granite[12]. Conversely, the quick cooling rate of the volcanic magma and fragmental ejecta address the formation of pumice[13] and basalt[14].

The different rates at which magma can be cooled trigger a sequence of reactions known as Bowen's series, explaining the crystallization of the common silicate minerals from magma (Figure 3.7)[15]. In particular, Bowen, a Canadian petrologist, performed a series of experiments at the beginning of the 20th century by using different mineral powders. He was able to recognize two different pathways for the formation of igneous rocks. On the one hand, a discontinuous branch (Figure 3.7) where the transformation from olivine to biotite was identified. In particular, as temperature reduces, olivine is formed. Upon progressive temperature reduction, olivine reacts with the existing magma to form pyroxene. The latter transforms into amphibole and then biotite by reacting with magma, which cools it down to progressively lower temperatures.

On the other hand, the continuous branch (Figure 3.7) reports the evolution of the plagioclase feldspars from calcium-rich to sodium-rich systems. Finally, both branches address the same pathway for synthesizing, e.g., orthoclase, muscovite, and quartz. Although Bowen's reaction series is not always valid[16], it is a good explanation for understanding how igneous rocks are formed.

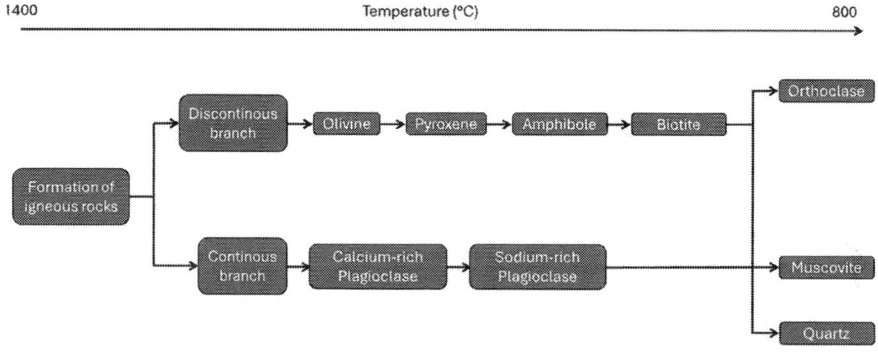

Figure 3.7 Bowen's sequence of mineral formation following magna cooling.

[12] Granite is a mixture of different minerals, such as quartz, feldspars, among which orthoclase is the most abundant, and micas that belong to the class of silicates.

[13] Pumice is a light-colored and highly porous rock formed during volcanic eruptions.

[14] Basalt is a dark rock made mainly of plagioclase and pyroxene, formed during the cooling of the magma coming from volcanic eruptions.

[15] Bowen, N.L. 1915. The later stages of the evolution of igneous rocks. The Journal of Geology. 23: 1–91.

[16] The limits of Bowen's reaction series are documented at the following link: https://geologyscience.com/geology/bowens-reaction-series/ where it is clearly stated that "The entire series is not found in nature, but many magmatic rocks show parts of the sequence. The main limitations are the state of the liquid, the rate of cooling and the tendency of mineral crystals to settle under gravity. If the liquid is depleted from an element needed for a particular mineral, the series containing that mineral is discontinued. If the magma cools faster than the reaction can continue, the early minerals may remain in partially absorbed form. This changes the evolution of magma. If the crystals can rise or sink, they stop reacting with the liquid and settle elsewhere".

According to the United States Geological Survey (USGS)[17], "metamorphic rocks started out as some other type of rock, but have been substantially changed from their original igneous, sedimentary, or earlier metamorphic form. Metamorphic rocks form when rocks are subjected to high heat, high pressure, hot mineral-rich fluids or, more commonly, some combination of these factors. Conditions like these are found deep within the Earth or where tectonic plates meet."[18] Therefore, metamorphism arises when minerals in pre-existing rocks change their nature following pressure and temperature changes by either physical or chemical re-arrangements or by reactions with hot fluids entering the rocks.

Figure 3.8 shows a typical phase diagram where three different polymorphs[19] of the $Al_2(SiO_4)O$ mineral are reported. The lines inside the figure represent

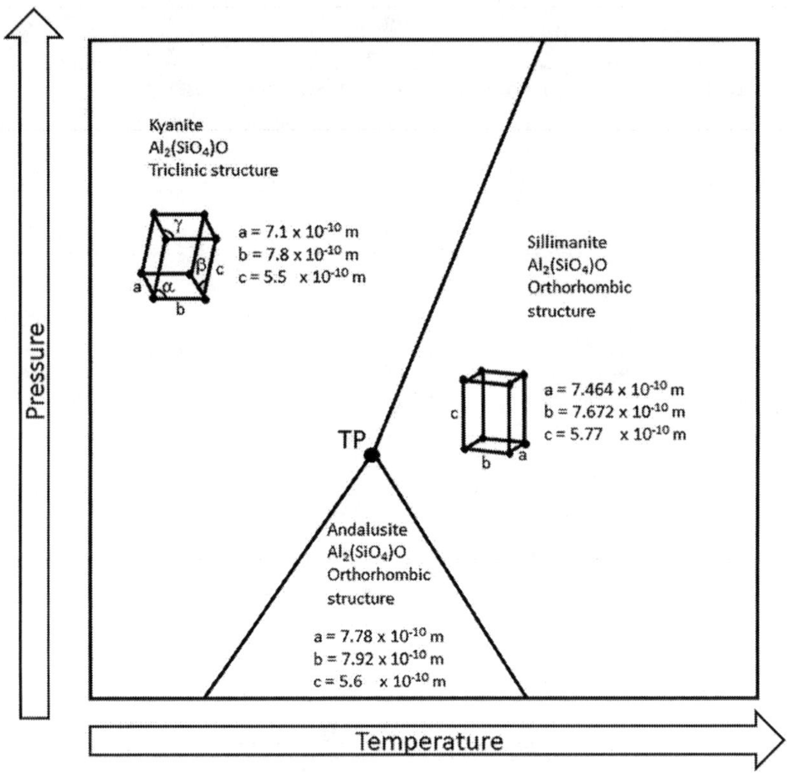

Figure 3.8 Phase diagram of the three polymorphic structures of the $Al_2(SiO_4)O$ mineral. They all co-exist at the equilibrium in the triple point indicated as TP[20].

[17] USGS webpage is at the link https://www.usgs.gov/.

[18] This definition can be found at the following link: https://www.usgs.gov/faqs/what-are-metamorphic-rocks#:~:text=Metamorphic%20rocks%20started%20out%20as,some%20combination%20of%20these%20factors.

[19] Polymorphism refers to the existence of solid materials in different crystalline forms that can switch from one to another when temperature and pressure are changed.

[20] In the triclinic structure, $a \neq b \neq c$, while all the angles are not 90°. In the orthorhombic system, $a \neq b \neq c$, while $\alpha = \beta = \gamma = 90°$.

the equilibria between the different phases. The intersection point indicated as *TP* represents the triple point where all the phases are simultaneously in equilibrium. At high pressure (>0.4 GPa) and low temperature (<600°C), kyanite predominates; above 600°C sillimanite is the main polymorph; finally, when the pressure is <0.4 GPa and the temperature is in the range of 200–800°C, andalusite is the main form of the aforementioned aluminum silicate. The equilibria in the phase diagram reported in Figure 3.8 are also referred to as polymorphic reactions and belong to the reaction group indicated as "solid-solid reactions."

Apart from those, other typical reactions involved in the formation of metamorphic rocks are the ion-exchange ones, such as in the equilibrium:

$$Fe_3Al_2(SiO_4)_3 + KMg_3(Si_3Al)O_{10}(OH)_2 \leftrightarrows Mg_3Al_2(SiO_4)_3 + KFe_3(Si_3Al)O_{10}(OH)_2$$

Here, the potassium in phlogopite (contained in biotite) replaces the iron in almandine (contained in garnet) to form pyrope (in garnet) and annite (in biotite). The reaction rate of the exchange reactions is affected by the diffusion rate of the ions involved in the redistribution among minerals. This, in turn, is strongly affected by the temperature rather than pressure[21]. Other examples of metamorphic reactions are dehydration, such as in equation (3.1), where kaolinite reacts with quartz to produce pyrophyllite and water; the decarbonations, such as in equation (3.2), where carbonate and quartz react to wollastonite and carbon dioxide, and dehydration-decarbonation reactions, as in equation (3.3), where a combination of dehydration and decarbonation allows transformation of dolomite and tremolite in calcite and forsterite.

$$Al_2Si_2O_5(OH)_4 + 2SiO_2 \rightarrow Al_2Si_4O_{10}(OH)_2 + H_2O \tag{3.1}$$

$$CaCO_3 + SiO_2 \rightarrow CaSiO_3 + CO_2 \tag{3.2}$$

$$11CaMg(CO_3)_2 + Ca_2Mg_5Si_8O_{22})(OH)_2 \rightarrow 13CaCO_3 + 8Mg_2SiO_4 + 9CO_2 + H_2O \tag{3.3}$$

The reader must consider that this paragraph is not intended to be exhaustive about the reactions involved in metamorphic rock formation because this is out of the aim of the book. However, a complete description of the aforementioned reactions can be found in Wei[22].

Sedimentary rocks are obtained by the deposition on Earth's surface and/or in water bodies of small detritus arising from pre-existing rock and/or organic deposits[23]. The latter comes from the accumulation of decomposed dead organisms[24]. The aforementioned small detritus is also referred to as "sediments." A sediment is any solid material that can be carried from one location to another under the action of wind, water, or ice. The frictional forces at the interface

[21] A good excursus on the metamorphic reactions is at the following link: https://www.britannica.com/science/metamorphic-rock/Isograds#ref80316

[22] Wei, C. 2021. Types of metamorphic reactions. *In*: Encyclopedia of Geology, 2nd Ed., D. Alderton and S.A. Elias (eds). 397–411. Academic Press. ISBN 9780081029091.

[23] https://www.usgs.gov/faqs/what-are-sedimentary-rocks;

[24] https://landscape.soilweb.ca/organic-environment/

between air, water, ice, and the solid surface of pre-existing organic/inorganic deposits are mainly responsible for sediment formation[25]. When the accumulated sediments are deeply buried, the high pressures generated by the overlying layers allow compaction and cementing, thereby forming sedimentary rocks. Instead of "deposition," the term "sedimentation" can be also used. It refers to the ensemble of processes (both chemical and physical) enabling small particles to settle (for the organic materials), precipitate (for the inorganic systems), and accumulate in the environment. The special mechanisms involved in sedimentation provide the typical sedimentary rock layering or bedding (Figure 3.9).

Figure 3.9 Sedimentary rocks in Middle Triassic marginal marine sequence, southwestern Utah. This photo was taken by Mark A. Wilson (Department of Geology, The College of Wooster), and it has been released under the public domain of the Wikimedia Commons (https://commons.wikimedia.org/wiki/File:Triassic_Utah.JPG).

Common sedimentary rocks arise when sediments are carried in rivers and settled in lakes and oceans (e.g., sandstone, limestone, and shale). When buried, sediments lose water and become cemented, thereby forming sedimentary rocks.

Following weathering, transportation, and entrapment into terrestrial basins and depressions, sediments made by clasts of pre-existing rocks can accumulate. When the clast heap is subjected to high pressions because of burying, compaction and cementing occur, forming the clastic sedimentary rocks. Clastic sedimentary rocks may have particles ranging in size from microscopic clay to huge boulders[26] (Figure 3.10).

When living organisms die, their residues can pile up, be compressed, and cemented, thereby providing the so-called biologic sedimentary rocks (Figure 3.11).

[25] https://education.nationalgeographic.org/resource/sediment

[26] https://www.usgs.gov/faqs/what-are-sedimentary-rocks

Figure 3.10 Clasts of siliciclastic sedimentary rock in conglomerate[27] in the Pennsylvanian of Ohio, USA. (geology hammer for scale). This picture was taken by James St. John (Ohio State University) under the Creative Commons Attribution 2.0 Generic license (from Wikimedia Commons) (https://commons.wikimedia.org/wiki/File:Clasts_of_siliciclastic_sedimentary_rock_in_quartz-pebble_conglomerate_(Sharon_Conglomerate,_Lower_Pennsylvanian;_Jackson_North_roadcut,_Ohio,_USA)_8_(37034531854).jpg).

Figure 3.11 Coquina is an example of biological sedimentary rock being made of finely busted-up seashells. This image was taken by James St. John (Ohio State University) under the Creative Commons Attribution 2.0 Generic license (from Wikimedia Commons) (https://commons.wikimedia.org/wiki/File:Coquina_2.jpg)

[27] A conglomerate is a mixture of heterogeneous sized sediments cemented all together.

3.8 PEDOGENESIS: A MATHEMATICAL APPROACH

Pedogenesis is a branch of soil science dealing with all the factors and mechanisms (either physical or chemical) involved in soil formation, as well as evolution. It is quite an old scientific field since one of the first rational approaches was given by Dokuchaev in 1899[28]. This author approached soil formation and evolution in a mathematical way by suggesting the equation given in (3.4).

$$P = f(K, O, G)B \tag{3.4}$$

Here, P is the soil, K is the climate, O lays for "organisms," G indicates the "ground" parent or rocks, and B is the time. The operator f is an undefined function that relates all the aforementioned factors. The idea behind equation (3.4) is that soil formation is mainly affected by climate, rock weathering, and living organisms, while time accounts for soil evolution.

On the basis of equation (3.4), many other equations have been suggested[29], Jenny's equation (3.5) being the most famous one.

$$S = f(cl, o, r, p, t, \ldots) \tag{3.5}$$

Here, S is the soil, cl is the climate, o are the organisms, r is the relief (that is, the topography), p is the parent material such as the rocks, and t is the time accounting for soil transformation and evolution. The ellipsis reported in Jenny's equation is an indication that "in any landscape, one should consider climate, organisms, topography, parent material and time factors, but that there are other factors (the "dot factors") that are critical in some locations and situations, but not everywhere"[30].

It is worth noting that Jenny's intent was not to provide factors describing pedogenetic processes or components of the soil itself. Rather, the abovementioned "factors provide the context and the boundary conditions within which soil formation and development occurs"[31]. Indeed, according to Jenny, "the factors are not formers, or creators, or forces: they are the variables (state factors) that define the state of the soil system"[32]. In other words, the aforementioned factors only describe the status of the soil in a certain space and time. They can, in turn, be considered as the boundary conditions for soil formation/evolution processes.

[28] Dokuchaev, V.V. 1899. Report to the Transcaucasian Statistical Committee on soil taxonomy in general and especially for the Transcaucasia. Horizontal and vertical soil zones. Office Press of the Civilian Affairs Commander-in-Chief in the Caucasus, Tiflis (in Russian)

[29] Stockmann, U., B. Minasny and A.B. McBratney. 2018. Clorpt functions. *In*: McBratney, A., B. Minasny and U. Stockmann. (eds). Pedometrics. Progress in Soil Science. Springer, Cham. https://doi.org/10.1007/978-3-319-63439-5_18

[30] https://geography.as.uky.edu/blogs/jdp/connecting-dot-factors

[31] Phillips, J.D. 1989. An evaluation of the state factor model of soil ecosystems. Ecological Modelling. 45: 165–177.

[32] Jenny, H. 1941 Factors of Soil Formation. A System of Quantitative Pedology. McGraw-Hill, New York.

– H. Jenny. 1961. Derivation of state factor equations of soils and ecosystems. Soil Science Society of America. Proceedings. 25: 385–388.

It is important to note that Jenny's equation and similar formulations are often referred to as "state equations" because they describe the soil condition at a specific point in space and time. However, the term "state equation" is also widely used in physics and chemistry to refer to mathematical relationships among state variables, such as temperature, pressure, volume, internal energy, enthalpy, and entropy. In this context, Jenny's equation treats climate, organisms, topography, parent material, and time as state variables. However, from a mathematical standpoint, considering "time" as a state variable in a time-dependent system presents a conceptual challenge. In physics and chemistry, time is not classified as a state variable, as state variables describe the equilibrium condition of a system at a given moment, independent of its past evolution. In contrast, Jenny's "time" represents age, meaning that a soil can be considered younger or older relative to another. Since youth and maturity are time-dependent characteristics influencing soil development, Jenny's use of "time" as a state variable is justified within the context of soil formation, even though it differs from its treatment in classical physics and chemistry. Jenny's equation and similar mathematical expressions are also referred to as "factorial equations" because they incorporate state factors, as previously described.

As in equation (3.4), the function f in equation (3.5) remains undefined and serves only to establish relationships among the different soil state factors. The main challenge in determining a precise mathematical expression for f lies in the interdependence of these state factors and the difficulty of quantifying their values as state functions or variables. The only way to define f mathematically is through modeling, where all parameters are held constant except for one which is systematically varied. This approach allows for the derivation of five distinct equations, as expressed in equations (3.6) to (3.10):

$$S = f(cl)_{o,r,p,t,\ldots} \tag{3.6}$$

$$S = f(o)_{cl,r,p,t,\ldots} \tag{3.7}$$

$$S = f(r)_{cl,o,p,t,\ldots} \tag{3.8}$$

$$S = f(p)_{cl,o,r,t,\ldots} \tag{3.9}$$

$$S = f(t)_{cl,o,r,p,\ldots} \tag{3.10}$$

The equations from (3.6) to (3.10) are also known as climofunction, biofunction, topofunction, lithofunction, and chronofunction, respectively. All the subscripts in equations (3.6) to (3.10) mean that those state variables are kept constant.

Different attempts have been made to predict the correct form for equations (3.6) to (3.10). As an example, at the end of the eighties of the 20th century, Richardson and Edmonds[33] used the linear approach reported in equation (3.11):

$$S = a + bF \tag{3.11}$$

[33] Richardson, J.L. and W.J. Edmonds. 1987. Linear regression estimations of Jenny's relative effectiveness of state factors equation. Soil Science. 144: 203–208.

Here, S is the soil property under investigation, a and b are the ordinate intercepts and the regression coefficient, respectively, while F is one of the aforementioned state functions. Later in the nineties, Hugget[34] proposed the brash model based on equation (3.12):

$$\frac{d\boldsymbol{x}}{dt} = f(\boldsymbol{x}) + \boldsymbol{z} \tag{3.12}$$

where $\boldsymbol{x}$ is a vector of variables describing the system, f is a transformation matrix defining the interactions between the state variables, and $\boldsymbol{z}$ is a vector of driving variables[35].

Still, at the beginning of the nineties, Willgoose et al.[36] proposed a model for soil evolution based only on the processes of erosion and deposition mainly affected by landscape transformation. The simplest equation is given in (3.13).

$$\frac{\partial z}{\partial t} = -\nabla q_s \tag{3.13}$$

Here, z is the soil elevation, q_s is the material flux and ∇ is called the *nabla* operator given by $\left(\frac{\partial}{\partial x} + \frac{\partial}{\partial y}\right)$. Equation (3.13) has been the base for the landscape evolution models, as reported in equation (3.14)[37].

$$\frac{\partial h}{\partial t} = \frac{\rho_r}{\rho_s}\frac{\partial e}{\partial t} - \nabla q_s \tag{3.14}$$

where h is the soil thickness, e is the boundary between the soil and the bedrock, ρ_r is the rock density, while ρ_s is the soil density. Equation (3.14) accounts for rock weathering and transport of materials. This equation was re-elaborated by Rosembloom et al.[38] to account for sand, silt, clay, and natural organic matter as reported in (3.15):

$$\frac{\partial h_c}{\partial t} = \left(\frac{\partial \kappa_c}{\partial x} \cdot \frac{\partial z_{sfc}}{\partial x}\right) + \left(\frac{\partial \kappa_c}{\partial y} \cdot \frac{\partial z_{sfc}}{\partial y}\right) \tag{3.15}$$

---

[34] Huggett, R.J. 1991. Climate, Earth Processes and Earth History. Springer, Heidelberg. ISBN: 9783642762703.
– R.J. Huggett. 1995. Geoecology. An Evolutionary Approach. Routledge, London. ISBN: 9780415087100.

[35] Minasny, B., et al. 2008. Quantitative models for pedogenesis–A review. Geoderma. 144: 140–157.

[36] Willgoose, G., et al. 1991. A coupled channel network growth and hillslope evolution model. 1. Theory. Water Resources Research. 27: 1671–1684.

[37] Dietrich, W.E., et al. 1995. A process-based model for colluvial soil depth and shallow land sliding using digital elevation data. Hydrological Processes. 9: 383–400.

[38] Rosembloom, N.A., et al. 2001. Geomorphic evolution of soil texture and organic matter in eroding landscapes. Global Biogeochemical Cycles. 15: 365–381.

In equation (3.15), the subscript c refers to the soil textural component (sand, silt, or clay), h_c is the height of the textural component in the 20 cm depth surface horizon, z_{sfc} is given in (3.16), while κ_c is reported in (3.17).

$$z_{sfc} = h_{sand} + h_{silt} + h_{clay} + z_{bedrock} \tag{3.16}$$

$$\kappa_c = f_c F_{depth} F_{SOC} \kappa_c^* \tag{3.17}$$

The terms in equations (3.16) and (3.17) are as follows:

$z_{bedrock}$ is the elevation of the underlying bedrock; f_c is the "concentration of each textural component reflecting its fractional contribution to the 20 cm surface layer"[38]; F_{depth} is a parameter indicating that, as soil depth is depleted, diffusivity[39] goes to zero, thereby ensuring that sediment transport does not exceed the supply of the available loose material; F_{SOC} is a parameter relating the loose of soil material as affected by the presence of soil organic matter. Rosembloom et al.[38] used the numerical relation $F_{SOC}^{clay} = 0.4e^{-25[SOC]} + 0.6$ to indicate that soil organic matter reduces clay diffusivity by the action of the interactions that will be described in the next chapters.

Finally, κ_c^* represents the initial diffusivity of each textural soil component.

This paragraph is not intended to provide an exhaustive description of all existing pedological mathematical models. More detailed information can be found in specific publications (for example, the Further Readings). Here, only a general idea of the complexity involved in developing pedogenetic mathematical models has been presented.

3.9 PEDOGENESIS: A QUALITATIVE PHYSICAL AND CHEMICAL UNDERSTANDING

It is clear from the discussion above that all the mathematical models can predict to some extent how soil depth can change because of particle translocation, hillslopes, amount of soil organic matter, and so on. However, none of the models account for the mechanisms involved in soil formation at a molecular level. Therefore, a question arises: what happens from a physical and a chemical point of view during pedogenesis?

Let us consider, firstly, a bedrock subjected to a heating/cooling cycle (Figure 3.12A) due to the day-night alternance and to the season changes.

The temperature variations due to the heating/cooling allow mechanical stresses as a consequence of the anisotropy of the bedrock thermal expansion coefficient. In other words, the bedrock chemical components are organized in a three-dimensional structure whose physical properties are not the same along the three axes of a Gaussian space. This means that heat diffusion from the surface to the inner bedrock part (and vice versa) occurs at different rates depending

[39] Diffusivity relates to the spread of a sediment or a mass constituent under the action of a concentration gradient.

on the diffusion direction. From a chemical point of view, the aforementioned anisotropy provokes an asymmetrical elongation of the chemical bonds among the different chemical bedrock components. Because of this, bedrock fracturing occurs, thereby leading to the formation of cracks and several small-sized dust particles (Figure 2.12B). The larger the anisotropy of the thermal expansion coefficient, the easier the bedrock disaggregation is.

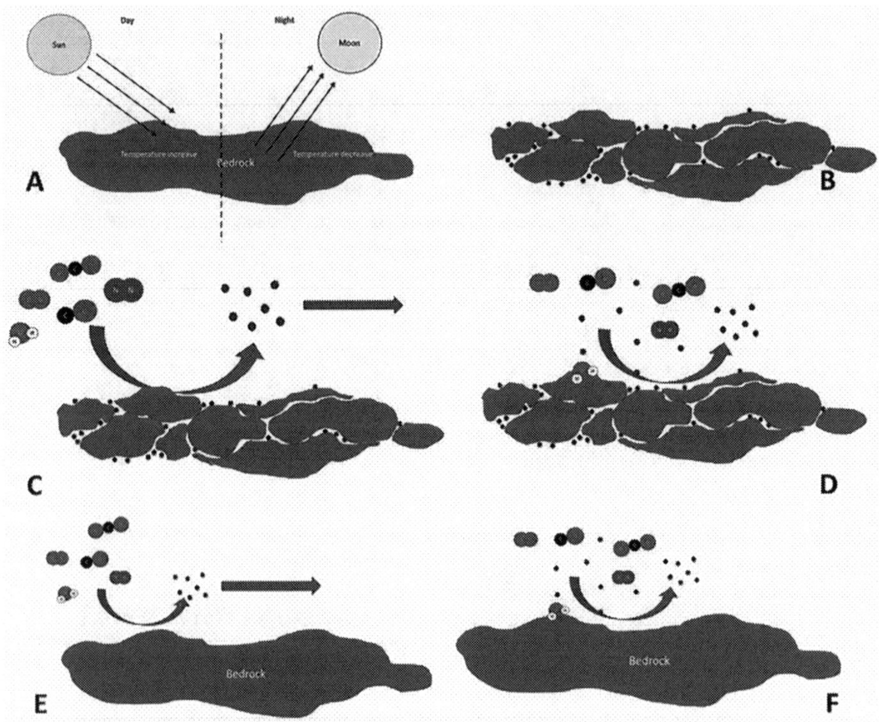

Figure 3.12 Under the action of heating/cooling cycles (**A**), the bedrock cracks and small dust particles form (**B**). Wind "captures" dust powders and allows translocation to a different location (**C**). The frictional forces between the translocating dust particles and the bedrock surface contribute to further disaggregating the bedrock (**D**). The frictional forces also occur between air gas particles and bedrock surface, producing movement of clouds of dust to another place (**E**). The dust produced by the friction of air on the bedrock surface further contributes to the disaggregation of the bedrock (**F**).

It is worth noting that the heating/cooling cycle also acts on water molecules trapped in bedrock. Due to the anomalies of water characteristics (see Chapter 5), this trapped material contributes to bedrock cracks and the formation of small dust particles.

Under the action of the wind, the air constituents can remove the small-sized dust particles and transport them to a different location (Figure 3.12C). Moreover, during the wind-driven translocation, flying dust also contributes to bedrock disaggregation as a consequence of frictional interactions (Figure 3.12D). The frictional forces also act when gas particles in the air hit the bedrock surface, thereby further promoting disaggregation (Figures 3.12E and 3.12F). When wind intensity reduces, dust settles down with the formation of soil layers.

A mechanism based on frictional and tensional forces can also be invoked when one considers the action of water flowing on a bedrock. Indeed, liquid water abrades the bedrock surface, penetrates the cavities, and exerts forces against cavity boundaries, thus inducing bedrock fractionation and powdering. Dust translocation from one place to another occurs along the direction of the water flow.

Finally, when plant roots "are looking for" nutrients, they penetrate the bedrock and favor its physical fracturing with the formation of dust, which is translocated in different places by the flowing water.

It is worth noting that the mechanisms outlined above are referred to as physical bedrock weathering. However, when bedrock is exposed to the atmosphere, it can also chemically react with water, air, and organic acids (e.g., those present in plant exudates), thereby being partly or wholly dissolved. When this happens, we say that chemical bedrock weathering occurs.

An example of a chemical reaction involved in bedrock weathering is given in equation (3.18).

$$CaSO_4 + 2H_2O \rightarrow CaSO_4 \cdot 2H_2O \tag{3.18}$$

where anhydrite ($CaSO_4$) reacts with water to form gypsum ($CaSO_4 \cdot 2H_2O$) in a reaction named hydration. In other words, two water molecules are "trapped" into the crystalline structure of anhydrite (Figure 3.13A) to provide another crystalline structure given by gypsum (Figure 3.13B).

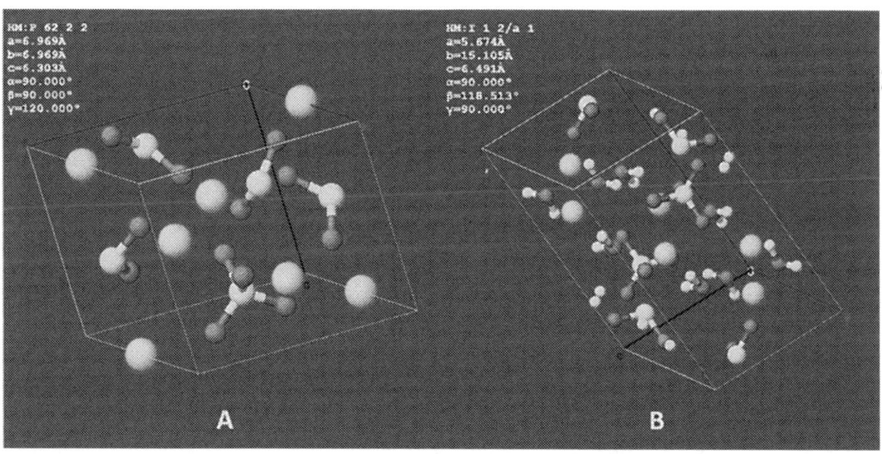

Figure 3.13 Crystal structure of anhydrite (**A**) and gypsum (**B**). These crystal structures have been obtained by the American Mineralogist Crystal Structure Database (http://rruff.geo.arizona.edu/AMS/amcsd.php)

Another example is the transformation of albite (that is, a mineral having the following minimum formula: $NaAlSi_3O_8$) to kaolinite (i.e., $Si_4Al_4O_{10}(OH)_8$) via a reaction that is named "hydrolysis":

$$4NaAlSi_3O_8 + 4H^+ + 18H_2O \rightarrow Si_4Al_4O_{10}(OH)_8 + 8Si(OH)_4 + 4Na^+ \quad (3.19)$$

Hydrolysis also occurs when olivine (Mg_2SiO_4) is transformed into silicic acid (H_4SiO_4):

$$Mg_2SiO_4 + 4H_2O \rightarrow H_4SiO_4 + 2Mg(OH)_2 \quad (3.20)$$

and when biotite (K_2 $[Si_6Al_2]Mg_3Fe_3O_{20}(OH)_4$) transforms into kaolinite and goethite (FeOOH):

$$2K_2[Si_6Al_2]Mg_3Fe_3O_{20}(OH)_4 + 16H^+ + 11H_2O + \frac{3}{2}O_2 \rightarrow Si_4Al_4O_{10}(OH)_8 + 6FeOOH + 8Si(OH)_4 + 4K^+ + 6Mg^{2+} \quad (3.21)$$

Carbonation (3.22), decarbonation (3.23), dissolution (3.24), and redox (3.25) are other examples of chemical reactions involved in rock alteration:

$$CO_2 + H_2O \rightarrow H_2CO_3 \xrightarrow{Ca^{2+}} CaCO_3 \quad (3.22)$$

$$CaCO_3 + 2H^+ \rightarrow CO_2 + Ca^{2+} + H_2O \quad (3.23)$$

$$CaSO_4 + 2H_2O \rightarrow CaSO_4 \cdot 2H_2O \rightarrow Ca^{2+} + SO_4^{2-} + 2H_2O \quad (3.24)$$

$$3Fe_2SiO_4 + 3Mg_2SiO_4 + O_2 \rightarrow 2Fe_3O_4 + 6MgSiO_3 \quad (3.25)$$[40]

EXERCISES

1. Describe nucleosynthesis.
2. Balance and analyze the following reaction:

$$K_2[Si_6Al_2]Mg_3Fe_3O_{20}(OH)_4 + H^+ + H_2O + O_2 \rightarrow Si_4Al_4O_{10}(OH)_8 + FeOOH + Si(OH)_4 + K^+ + Mg^{2+}$$

3. Among the following elements, which one is the most abundant in soils: Al, Si, O, N, P, Ba, Zn, Mn, and P.
4. Describe Bowen's reaction series.
5. Provide three examples of igneous rocks.
6. Describe the factors affecting the physical alteration of the rocks.
7. Provide a list of possible chemical reactions involved in rock weathering.
8. Discuss Dokuchaev's equation.
9. Describe Jenny's equation.

[40] Fe_2SiO_4 is named olivine, Mg_2SiO_4 is forsterite, Fe_3O_4 is magnetite, and $MgSiO_3$ is enstatite.

FURTHER READING

Benbi, D.K. and R. Nieder. (eds). 2003. Handbook of Processes and Modelling in the Soil-Plant System. Food Product Press® and The Haworth Reference Press. ISBN: 1560229152

Hazen, R.M. 2012. The Story of the Earth. The Penguin Books. ISBN: 9780143123644

Jenny, H. 1994. Factors of Soil Formation. A System of Quantitative Pedology. Dover Publications, Inc., New York. ISBN: 9780486681283

Langer, N. 2012. Nucleosynthesis. SS Bonn University (available at: https://astro.uni-bonn.de/~nlanger/siu_web/nucscript/Nucleo.pdf)

Sposito, G. 2008. The Chemistry of Soils, 2nd Ed. Oxford University Press. ISBN: 9780195313697

Tan, K.H. 2014. Principles of Soil Chemistry, 4th Ed. CRC Press. ISBN: 9781439813928

White, R.E. 2006. Principles and Practice of Soil Science. The Soil as a Natural Resource, 4th Ed. Blackwell Publishing. ISBN: 9780632064557

White, R.E. 2015. Understanding Vineyard Soils, 2nd Ed. Oxford University Press. ISBN: 9780199342068

Chapter 4

The Hydrosphere

4.1 INTRODUCTION

This, and the next chapter, could appear strange in a book about soils. Why should we be interested in water when soil properties are our goal? Water is the most abundant Earth's system, and it is involved, directly or indirectly, in many different reactions as well as transport phenomena[1]. As an example, let us consider the reaction reported in (4.1):

$$6CO_2 + 6H_2O \rightarrow C_6H_{12}O_6 + 6O_2 \tag{4.1}$$

This is known as the photosynthesis reaction, where six moles of carbon dioxide formally react with six moles of water in order to provide one mole of glucose and six moles of molecular oxygen. The "invisible" term occurring in equation (4.1) is the sunlight, which is responsible for triggering of the whole photosynthetic process[2].

When dealing with biochemistry, the reader usually pays attention to the carbon-containing molecules. However, water plays an important role in this reaction.

Firstly, the photochemical excitation of chlorophylls occurs, thereby releasing electrons. These electrons are, then, involved in the reduction of nicotinamide adenine dinucleotide phosphate (NADP) to $NADPH_2$ and in the synthesis of the adenosine triphosphate (ATP), while CO_2 is converted to glucose. During the electron transfer, water oxidation occurs, as reported in equation (4.2), which, more correctly, must be written as in (4.3)[3].

$$2H_2O \rightarrow O_2 + 4H^+ + 4e^- \tag{4.2}$$

[1] Felix Franks. 2000. Water, a matrix of life. RSC, 2nd Ed. 1–8. ISBN: 9780854045839

[2] Other missing systems in equation (4.1) are the enzymes without which not only photosynthesis, but also all the biochemical reactions cannot occur.

[3] Felix Franks. 2000. Water, a matrix of life. RSC, 2nd Ed. 143. ISBN: 9780854045839

DOI: 10.1201/9781003780212-4

$$4H_2O \rightarrow O_2 + 4H^+ + 2H_2O + 4e^- \tag{4.3}$$

There is no doubt that, without water, photosynthesis could never have occurred, and life on Earth could not have existed as it is now.

Being ubiquitous, water also plays a vital role in transport phenomena. The term "transport" refers to the physical movement of particles (dissolved, suspended, or dispersed) among different points within the environmental compartments. When dealing with solutions where organic and inorganic species are dissolved, water is crucial for nutrient supply to plants. As the suspended/dispersed particles are transported, water is essential for soil weathering, erosion, and pedogenesis[4].

Water chemistry in the soil plays a fundamental role in various physical, chemical, and biological processes. Soil pH, influenced by dissolved CO_2 and other acids in water, affects nutrient availability by modifying the solubility of essential elements. Water also serves as a medium for ionic interactions and facilitates cation exchange processes that control nutrient availability. In saline or sodic soils, high salt concentration in water can disrupt soil structure and reduce plant water uptake. Furthermore, water acts as a vehicle for redox reactions in waterlogged soils, where limited oxygen reduces iron and manganese, affecting soil color and nutrient mobility. Biological processes, including microbial decomposition of organic matter, rely on water availability, influencing the production of organic acids and nutrient release. Soil buffering capacity, hydration of clay minerals, and dissolution/precipitation reactions are additional processes mediated by water that shape soil chemistry and structure.

Due to the paramount importance of water in the environment, and soils in particular, a detailed analysis of its properties and how they affect soil physical chemistry is crucial. This chapter aims to outline how the big water bodies (BWB) may influence Earth's climate through the so-called water cycle.

4.2 THE CYCLE OF WATER

"Water" is the only word with two different meanings. It relates to both a single molecule (i.e., H_2O) and to a large amount of water molecules, such as those in oceans, seas, rivers, lakes, and clouds. Regardless of its meaning, it is possible to state that water is constantly in motion, not only at a microscopic scale (e.g., molecular vibrations, rotations, and translations) but also at a macroscopic scale. In fact, rivers flow, oceans are subjected to circular motions, and clouds produce rain falling on the Earth's surface due to gravity. Calculations show that every 3,000 years, ocean water evaporates and falls back on Earth by rain. The entire water cycle involving all the water masses on Earth is also known as the hydrogeological cycle. Although some scientists report that the latter is due to the action of the thermal energy coming from the Earth and the Sun, the most recent literature shows that the main reason for the massive water movement is solely

[4]The reader must refer to Chapter 3 to deal with the meaning of pedogenesis.

the Sun. In particular, the Sun removes yearly by evaporation a water volume of around 875 km^3 from oceans and, by evapotranspiration, around 160 km^3 from other land surfaces. The different mechanisms involved in water movement from the different environmental compartments work at different time scales. In fact, rivers' water may evaporate into the atmosphere and return to the land and seas in weeks, while that from oceans returns in only around ten days. Conversely, the iced water in the two terrestrial poles takes thousands of years to complete its cycle. This brief introduction reveals that the hydrogeochemical cycle is directly related to water's capacity to co-exist in three different phases (i.e., solid, liquid, and gas) in the environmental conditions of the planet Earth.

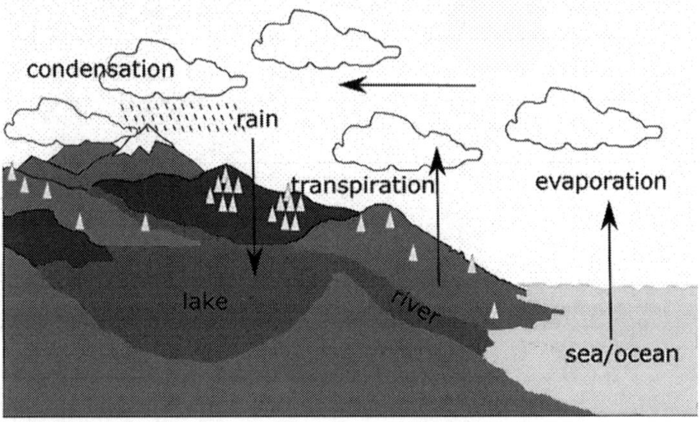

Figure 4.1 Simplicist representation of the cycle of water. Due to sun heating, water evaporates from the Earth to the atmosphere. Here, water moves everywhere due to winds. When the water's hot front meets cold air, water condensates and comes back to the Earth in the form of rain. Once in the soil, water penetrates the soil and can come back to groundwater and other water basins.

Figure 4.1 shows the water hydrogeochemical cycle in a simplistic way. Due to sun heating, water moves away from the Earth's surface (i.e., from seas, lakes, rivers, soils, and the surface of plant leaves) through a mechanism referred to as evapotranspiration. During this process, water is made free of any type of salt and any other system that is dispersed/suspended. Once in the atmosphere, water moves all around due to the wind. Depending on temperature and pressure in the atmosphere, water vapor may condense, thus forming small drops and becoming progressively heavier. When drop sizes are large enough to be subjected to gravity, water returns to the Earth as rain. During condensation and precipitation, water dissolves gases such as oxygen, nitrogen, and other substances present in the atmosphere by both natural origins and anthropogenic activities[5]. Some of the anthropogenic gases are Nitrogen (N_xO_y), and Sulphur oxides (SO_2, and SO_3), ammonia (NH_3), mercaptans (i.e., sulfur-containing organic molecules), and carbon

[5] Volcanic activity is among the natural activities, while agriculture, house heating, industrial activeness, vehicular movements, etc. are considered human activities.

dioxide (CO_2). All of the aforementioned chemicals make rains acidic. In the lower layers of the atmosphere, rainwater is further enriched with dust, soot, bacteria, and spores of microorganisms. For this reason, on arriving on the ground, rainwater is already rich in many dissolved, dispersed, and suspended substances[6].

Once on soils, water can diffuse both horizontally (such as in rivers) and vertically (i.e., movement towards groundwater). Water can still dissolve, disperse, and suspend other substances during the aforementioned movements. As an example, if the soil is rich in gypsum ($CaSO_4 \cdot 2H_2O$), water can be enriched with calcium and sulfate. If magnesite ($MgCO_3$), limestone ($CaCO_3$), and dolomite ($MgCa(CO_3)_2$) are present, high concentrations of magnesium, calcium, and carbonate occur. The dissolution rates of the aforementioned systems depend on rainwater acidity. In fact, acidic rainwater allows carbonate transformation to carbon dioxide, as reported in equations (4.4) and (4.5):

$$CO_3^{2-} + H_3O^+ \rightarrow HCO_3^- + H_2O \tag{4.4}$$

$$HCO_3^- + H_3O^+ \rightarrow CO_2 + 2H_2O \tag{4.5}$$

Even the oxygen dissolved in water may aid rock degradation. As an example, pyrite can be degraded as in equation (4.6)[7]:

$$4FeS_2 + 15O_2 + 14H_2O \rightarrow 4Fe(OH)_{3(solid)} + 16H^+ + 8SO_4^{2-} \tag{4.6}$$

As evidenced in equation (4.6), the production of acidic species can further contribute to the rate increase of the dissolution mechanisms of many other chemicals present in the environment. Besides the outlined chemical reactions, water flow is also responsible for the physical soil weathering, which allows water enrichment with dispersed and suspended materials (see Chapter 8). The hydrogeological cycle described here did not account for the iced water at the Earth's poles. Very briefly, we can say that the vapor from the evaporation of oceans, rivers, lakes, and surficial lands can reach the coldest Earth's regions, where it condensates to form ice. Once iced, water lays in that solid condition for years till a sudden temperature rise favors fusion with a reduction of ice volume and an increase in the amount of liquid water that enters the cycle as already described.

4.3 THE MOVEMENT OF BIG WATER BODIES: THE EFFECT OF WINDS

The water cycle does not end with the little information in the paragraph above. Actually, the movement of oceans, seas, and rivers (i.e., the so-called big water bodies, BWB) plays a very important role in defining Earth's climate and life

[6] A definition of dissolution, dispersion, and suspension is given in this chapter and Chapter 5.

[7] Equation (4.6) is an overall reaction whose details are deeply described in A.H. Kaksonen and J.A. Puhakka. 2007. Sulfate Reduction Based Bioprocesses for the Treatment of Acid Mine Drainage and the Recovery of Metals. Engineering in Life Science. 7(6): 541–564.

sustainability. In fact, the biogeochemical cycles generally induce feedback mechanisms that regulate the factors influencing climate change, including solar radiation intensity and volcanic activity.

All the BWBs are connected to each other. However, all of them are delimited by lands. In particular, oceans are separated by large continental masses whose shapes, depths, and distances affect oceans' connectivity. Starting in the early 2000s, the International Hydrographic Organization identified five different oceanic "basins": Pacific, Atlantic, Indian, Arctic, and Antarctic Ocean (Figure 4.2). All of them are subjected to horizontal and vertical motions. The horizontal movements account only for the surficial part of the oceans, that is, the first hundred meters from the top of the water surface, which is subjected to the direct action of the winds[8].

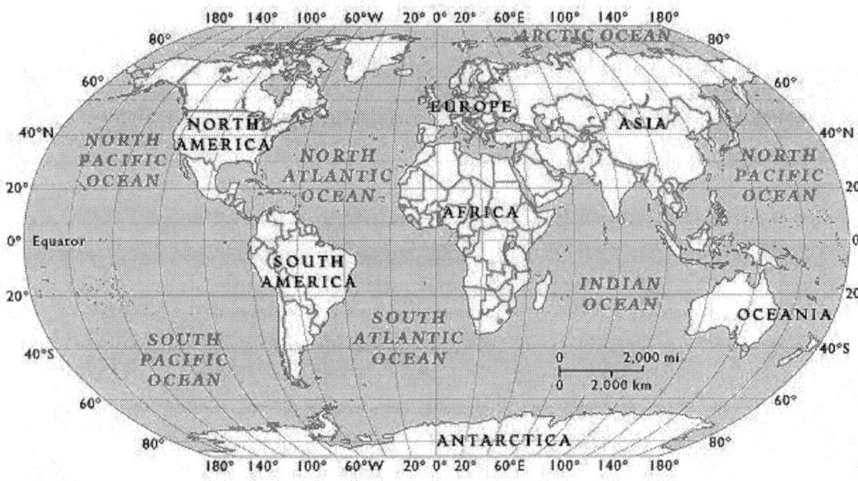

Figure 4.2. Representation of the Earth's globe with an indication of the oceans, latitudes (horizontal lines), and longitudes (vertical curves). Figure created by ChatGPT under the author's instructions.

Western winds, responsible for the Eastern marine currents, are predominant at latitudes between ±30° and ±60°[9]. Around the equator, up to the latitude of ±15°, Eastern Trade winds provoke the Western marine currents. Just around latitude 0° (i.e., at the equator), where winds are poor and mainly Eastern, marine currents are quasi-null, and equatorial calm zones occur.

Continental masses and terrestrial rotation are the two factors affecting marine current flows. When Eastern marine currents meet the continental masses,

[8] Conventionally, wind flows are indicated by considering wind origin, whereas big water bodies' flows are indicated by their point of arrival. Therefore, an Eastern wind flows from East to West, while a marine current flowing from East to West is indicated as a Western current. Both flows (the wind and the marine ones) have the same direction, although their names differ.

[9] The ± sign indicates the same latitude on both the Northern (+) and the Southern (–) Earth's hemispheres.

they deviate towards the North or the South. Because of this, two closed loops, also referred to as vortices, are generated (Figure 4.3). The shape of the vortices is affected by the Coriolis force. The latter can be described as a pseudo or apparent force influencing the motions of objects when they are observed in a reference frame that is moving rotationally in an inertial frame. In particular, when one accounts for a reference frame rotating clockwise, the Coriolis force appears to act to the left of the object's direction of motion. Conversely, this force appears to act to the right in a reference frame rotating anticlockwise (or counterclockwise).

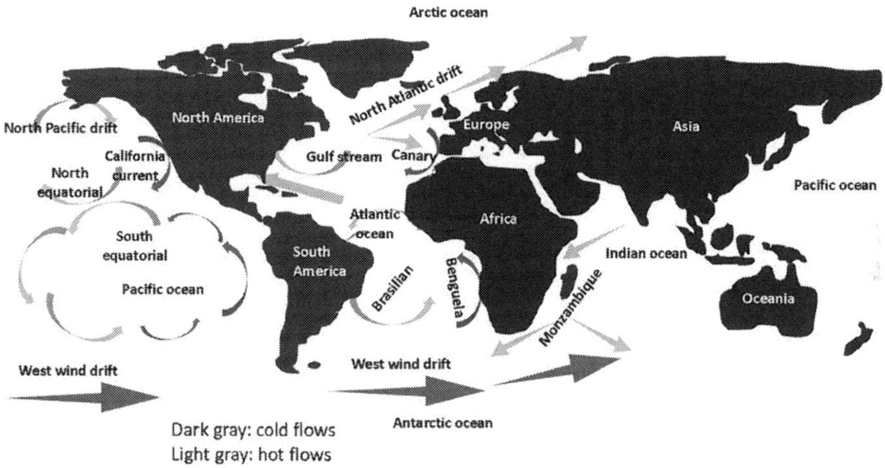

Figure 4.3 Indication of the surface water currents. Because of the presence of continents, the big water bodies' currents deviate from their directions, thereby inducing the formation of vortices.

In order to describe the Coriolis effect on the ocean motions, let us consider, as an example, the vortex in the North Pacific drift (Figure 4.3). The eastern trend winds (the winds moving from East to West) trigger an oceanic current moving towards the West in the bottom part of the vortex at a latitude of around 15° N (the light gray arrow indicated as North equatorial in Figure 4.3). Meanwhile, at around a latitude of 30°N, the same winds allow a water current directed to the East. Due to the Coriolis effect, the aforementioned flows are compressed along the Asiatic coast and stretched while moving towards the North American coasts. In other words, the Coriolis effect induces an intense flow, also referred to as Kuroshio, with a rate of 1 m s^{-1} (that is around 4 km h^{-1}), in the North and a less intense flow in the South around California's coasts. The same phenomenon also occurs in the North Atlantic Ocean, where the intense north-eastern flow coming from the Gulf of Mexico and the Southern part of Florida, along the eastern coast of North America, corresponds to the so-called Gulf Stream. This narrow flow becomes a more widespread northeastern current, the North Atlantic Current, which moves towards the British islands and Norway, carrying warm waters and a milder climate to Western Europe as compared to the conditions at similar latitudes on the North American continent. Mirrored vortices are present in the

Southern Hemisphere, where the Brazilian current in the South Pacific Ocean and the Agulhas near Southeast Africa, near the South Indian Ocean, are generated.

4.4 THE MOVEMENT OF BIG WATER BODIES: THE EFFECTS OF SALT CONCENTRATION AND TEMPERATURE

Up until now, we have focused only on the wind effect on the motion of surface water in oceans. However, water also moves vertically following gradients of temperature and saline concentration. Let us consider Figure 4.4, where the equilibrium between vapor ($H_2O_{(V)}$) and liquid ($H_2O_{(L)}$) water phase is indicated as being affected by exposure to the sun's heat. Evaporation removes water molecules from the big water bodies' surface while the salts remain dissolved. As a consequence, salt concentration near the surface of oceans is larger than that in their depths. In other words, the water solution at the interface with the atmosphere is denser than the solution on the ocean bottom. As all physics and chemistry students know, when two solutions with different densities come into contact, the solution with the higher density moves downward, while the one with the lower density rises. Therefore, a kind of up-and-down water-lift starts. The salt concentration varies also when water colds down. In fact, big water bodies at the two Earth's poles are affected by atmospheric cold temperatures. Because of this, water tends to ice. Dissolved salts are moved away during icing, and big-iced saltless water blocks float on the salt-richer liquid water. As a consequence, the surface water around the ice blocks is denser than that in the bottom. The different salt density triggers the vertical water circulation described above.

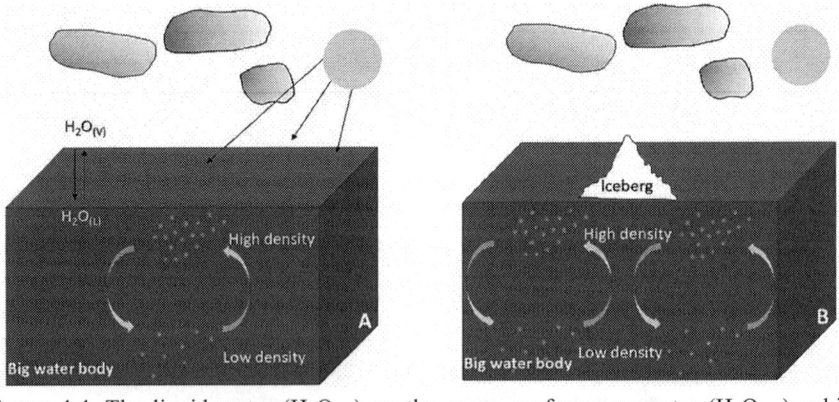

Figure 4.4. The liquid water ($H_2O_{(L)}$) on the ocean surface evaporates ($H_2O_{(V)}$), while the dissolved salts (dots) remain in the liquid phase. Due to this, the density of the solution at the interface with the atmosphere is denser than that at the big water body bottom. Consequently, an up-and-down water-lift starts as described by the arrows (**A**). A similar mechanism can be evoked when water is cooled at the Earth's poles (**B**). The formation of big-iced blocks allows the surrounding water to be salt-richer. Therefore, the density gradient between surface and deep water initiates the aforementioned up-and-down water-lift (**B**).

What has been described above is referred to as "thermohaline circulation"[10] where the composite word contains the terms "thermo" which considers the temperature effect, and "haline," coming from the ancient Greek word *ἁλσ,* which means "salt," thereby accounting for the effect of the salts on the water's vertical movements. The two effects (the temperature and the salt-driven ones) cannot be separated and occur simultaneously.

4.5 THE GLOBAL CONVEYOR BELT

The vertical movement just described in the paragraph above involves an amount of water down to a depth of 5,000 m and more, while the horizontal windy-driven motion, explained previously, covers a depth of around a few hundred meters. In addition, thermohaline circulation is much slower than horizontal circulation. In fact, the former covers a distance of 10–20 km per year, while the latter covers several kilometers per hour. However, the combination of the two different currents, which is also referred to as a global conveyor belt[11], strongly affects Earth's climate, making it suitable for life. In fact, the conveyor belt significantly influences climate regulation by redistributing heat and moisture worldwide. It transports warm water from the tropics to the poles and cold water from the poles to the tropics, thus moderating temperature disparities between these regions. Furthermore, it aids in transferring moisture from oceans to land, which is crucial for cloud formation and precipitation. Global temperatures would soar without the ocean conveyor belt, making the climate markedly drier. Therefore, global climate warming may slow the supply of tropical water towards the coldest Earth's poles and vice versa, thereby reducing the ocean conveyor belt effect on climate with the consequences just described[12]. In terms of consequences on soils, drier and warmer climates can be responsible for soil erosion and loss of cultivable areas, which are essential for food and textile production needed for life maintenance.

EXERCISES

1. List a number of the most important anthropogenic greenhouse gases.
2. Describe the reactions making acid the rains.
3. How many different oceanic basins can be identified?
4. List the oceanic basins.
5. How deep can the winds horizontally move the big water bodies?
6. Define the Coriolis force.

[10] Stommel, H. and A.B. Arons. 1959–1960. On the abyssal circulation of the world ocean–I. Stationary planetary flow patterns on a sphere. Deep Sea Research 6: 140–154.

[11] The definition of global conveyor belt has been taken from: https://oceanservice.noaa.gov/facts/conveyor.html

[12] Ditlevsen P. and S. Ditlevsen. 2023. Warning of a forthcoming collapse of the Atlantic meridional overturning circulation. Nature Communications. 4: 4254.

7. Define the thermohaline circulation.
8. What is the oceanic conveyor belt?
9. Discuss the importance of the oceans in the global carbon cycle.
10. Explain how human activities, such as pollution and climate change, impact the hydrosphere and its interactions with other Earth systems.

FURTHER READING

Ball, P. 1999. H_2O, A Biography of Water, 1st Ed. Weidenfeld and Nicolson. ISBN: 9780753810927.

Franks, F. 2000. Water, A Matrix of Life, 2nd Ed. RSC. ISBN: 9780854045839.

Webb, P. 2023. Introduction to oceanography. creative commons attribution. Downloadable from: https://rwu.pressbooks.pub/webboceanography/

Chapter 5

The Nature and Chemistry of Water

5.1 THE STRUCTURE OF WATER: BASIC THEORY ABOUT ORBITALS

As a molecule, water is described by the very well-known formula H_2O, where two tiny hydrogen atoms are connected via covalent bonds to a 16 times larger oxygen atom. In order to understand water properties, an analysis of the nature of the oxygen-hydrogen bond is needed.

The electron configurations of hydrogen and oxygen are as follows:

$$_{1}H,\ 1s^1$$
$$_{8}O,\ 1s^2,\ 2s^2\ 2p^4$$

Here, the subscript figures on the left side of each atomic symbol refer to the atomic number[1]; the letters *s* and *p* are the atomic orbitals for H and O, respectively; the figures on the left-hand side of the atomic orbitals refer to the principal quantum number (also referred to as *n*); the superscript figures to the right-hand side of each atomic orbital refers to the electrons contained in those atomic orbitals.

Figures 5.1A and B report the shape of the orbitals *s* and *p*, respectively. The *s* orbital is depicted as a sphere whose size depends on the energetic level given by the principal quantum number. In particular, the higher the *n* value, the larger the size of the *s* orbital. The *p* orbital is not "one." It consists of three different orbitals (all called *p*) oriented along the *x*-, *y*-, and *z*-axes of a Cartesian space,

[1] The atomic number is the number of protons in the nucleus of an atom. As the atom is electrically neutral, the number of protons (i.e., positive charges placed in the atomic nucleus) is the same as the electrons. Therefore, the atomic number refers also to the number of electrons in an atom.

DOI: 10.1201/9781003780212-5

each being a kind of "8" where the origin of the Cartesian space is indicated as the node. The node is the place where the atomic nucleus is located, and it is the place where the electrons cannot be present. The three *p* orbitals are, hence, indicated as p_x, p_y, and p_z, according to the axis along which they are aligned[2].

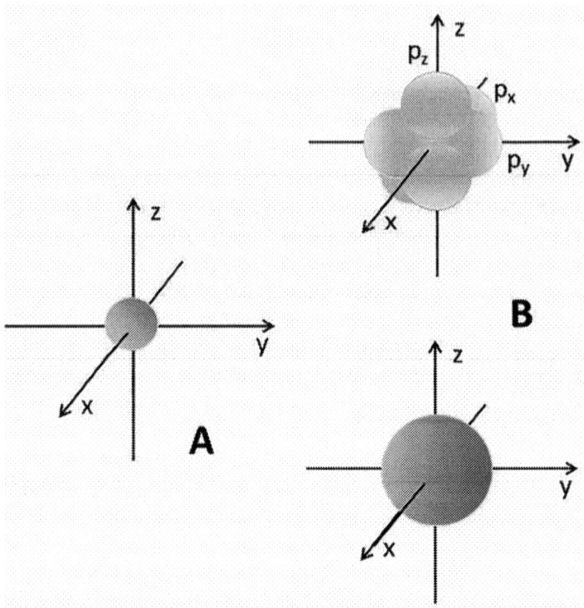

Figure 5.1. Geometric representation of the orbitals of the hydrogen and the oxygen atoms. Hydrogen's 1s orbital is reported in (**A**). Oxygen's 2s (on the bottom) and 2p (on the top) orbitals are shown in (**B**).

It must be kept in mind that orbitals are only the mathematical expressions describing the behavior of the electrons. In other words, the term "orbital" refers to a mathematical function containing all the information about the electrons and derived by solving the very well-known Schrödinger's equation ($\hat{H}\Psi = E\Psi$)[3]. All mathematical functions, including those coming from solving Schrödinger's equation, can be geometrically depicted. Therefore, what is shown in Figure 5.1

[2] By the discussion, it seems that being the *s* orbital a sphere, it should not have nodes. This may lead to the conclusion that electrons in an *s* orbital can also be on the nucleus. This is clearly not true. Quantum mechanics predicts that two different types of nodes are present in an orbital: radial and angular. The angular nodes are flat planes at fixed angles, while the angular ones are spheres at fixed radii. The total number of nodes in an orbital is given by the sum of the radial and angular nodes. The way the number of nodes is calculated is out of the aim of the present book. This short note aims only at the acknowledgment of the fact that the s orbitals also contain nodes. They are only radial and not angular.

[3] A quantum mechanics explanation of the Schrödinger equation and how it can be solved is out of the aim of the present book. The readers are addressed to more specialized literature as reported in the Further Reading.

is a trivial geometric representation of the spaces within which electrons may move. This does not mean that an electron moves only in the spaces represented by the orbitals. It simply means that the probability of finding the electrons in the aforementioned spaces is the largest. An electron can be anywhere around the nucleus. Figure 5.2 shows that the probability density of finding the electron in the *s* orbital, which is represented by the brownest spot in the figure, is the largest inside and in its closest proximity. Conversely, the probability density decreases as far as the distance from the brownest spot increases.

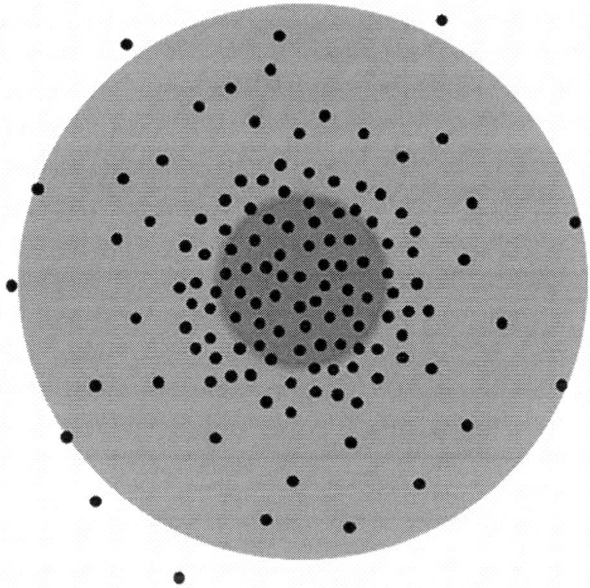

Figure 5.2 Density of probability to find the electron around the nucleus. The brownest spot is the s orbital. The probability density is the highest inside and closest to the s orbital. The probability density decreases as the distance from the s orbital increases.

Quantum mechanics predicts that a covalent bond forms when atomic orbitals overlap. However, the term "overlap" requires clarification. This concept arises from the geometric representation of orbitals.

Imagine two *s* orbitals approaching each other. Since orbitals represent regions of electron probability rather than physical objects, they can overlap, as shown in Figure 5.3A. This depiction may give the false impression that electrons are confined only to the shared space where the two orbitals intersect. However, this interpretation is incorrect.

The correct representation of orbital overlap is shown in Figure 5.3B, where the two *s* orbitals combine to form a single molecular orbital. This new orbital represents the bonding interaction and is delocalized, meaning that electron density is spread throughout the entire region rather than being restricted to the original atomic orbitals. As a result, the molecular orbital differs in shape from the individual *s* orbitals from which it originates. But how is this possible?

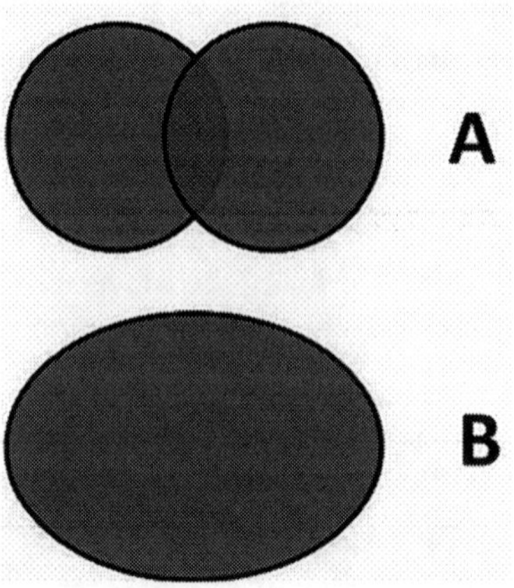

Figure 5.3 (**A**) representation of two overlapping s orbitals. (**B**) The real result of the *s* orbitals overlapping.

As discussed earlier, orbitals are solutions to the Schrödinger equation, representing electron probability distributions rather than physical objects. To describe them mathematically, we denote an atomic orbital as ψ_i, where the subscript $i$indicates a generic orbital. When atomic orbitals combine to form molecular orbitals, their wave functions undergo a linear combination, represented as:

$$\Psi_1 = c_1\psi_1 + c_2\psi_2 + c_3\psi_3 + c_4\psi_4 + etc. \tag{5.1}$$

Where the coefficients c_i (for i = 1, 2, 3, 4, etc.) reflect the contribution of each atomic orbital ψ_i to the final molecular orbital Ψ. This molecular orbital is a quantum state where electron density is delocalized over the entire molecule. However, "linear combination" does not simply refer to an arithmetic sum of wave functions but rather to their superposition, where phase relationships play a crucial role. As a result, various combinations exist, including:

$$\Psi_2 = (c_1\psi_1 + c_2\psi_2) - (c_3\psi_3 + c_4\psi_4) + etc. \tag{5.2}$$

$$\Psi_3 = c_1\psi_1 - c_2\psi_2 - c_3\psi_3 + c_4\psi_4 + etc. \tag{5.3}$$

$$\Psi_4 = c_1\psi_1 - c_2\psi_2 + c_3\psi_3 - c_4\psi_4 + etc. \tag{5.4}$$

and so on.

As dictated by quantum mechanics, the number of possible molecular orbitals is always equal to the number of atomic orbitals involved in the combination. Specifically, if two atomic orbitals combine, only two molecular orbitals are formed: one bonding and one antibonding. These are given by:

$$\Psi_1 = c_1\psi_1 + c_2\psi_2 \tag{5.5}$$

$$\Psi_2 = c_1\psi_1 - c_2\psi_2 \tag{5.6}$$

When two *s* orbitals combine, the resulting molecular orbitals are shown in Figure 5.4A. Similarly, when two *p* orbitals interact, they form bonding and antibonding π orbitals, as illustrated in Figure 5.4B.

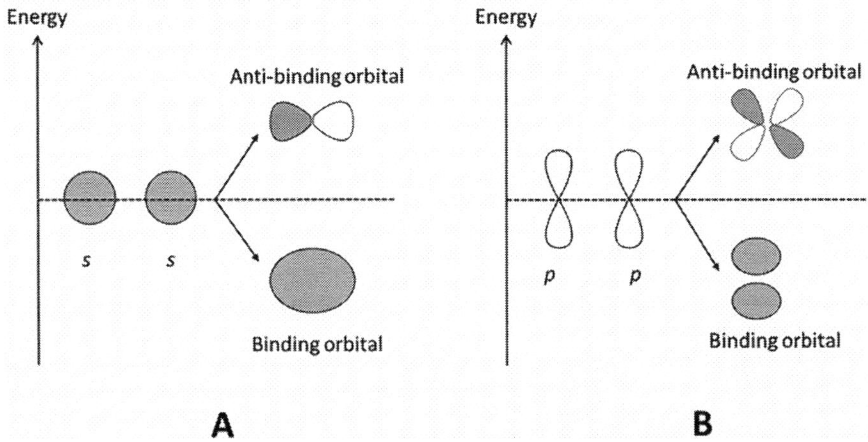

Figure 5.4 The linear combination of two *s* (A) and *p* (B) orbitals produces two molecular orbitals having different energies. The lowest energy orbital is indicated as a bonding orbital; the highest energy orbital is referred to as an anti-bonding orbital.

As shown in Figure 5.4, the two molecular orbitals formed by the linear combination of either *s* or *p* atomic orbitals occupy different energy levels. The molecular orbital with the lower energy is called the bonding orbital, while the one with higher energy is referred to as the anti-bonding orbital.

The term "anti-bonding orbital" arises from the fact that if electrons occupy or transition into this orbital, the molecule becomes unstable, breaking the chemical bond. To illustrate this concept, let us consider the simplest case: the hydrogen molecule (H_2). The two *s* orbitals of the hydrogen atoms resemble those shown in Figure 5.4A. Each hydrogen atomic orbital contains one electron. When two hydrogen atoms combine to form H_2, these two electrons must be distributed among the available molecular orbitals depicted in Figure 5.4A. Since electrons fill the lowest energy state first, both electrons occupy the bonding orbital, resulting in a more stable H_2 molecule compared to two separate hydrogen atoms. However, if one or both electrons were to occupy the antibonding orbital, the energetic stability of the molecule would be lost. As a result, the bond would break, and the two hydrogen atoms would separate.

5.2 THE STRUCTURE OF WATER: HYBRIDIZATION AND TETRAHEDRAL GEOMETRY

The previous paragraph provided a brief and simplified overview of orbital theory. We established that the linear combination of *i* atomic orbitals results in the same number of molecular orbitals, with half classified as bonding and the other half as anti-bonding. However, additional considerations must be made to explain the geometry of the water molecule

In the 1930s, Linus Pauling introduced the concept of hybridization. In simple terms, the valence orbitals of oxygen (i.e., the *s* and *p* orbitals in the

$n = 2$ shell) undergo a linear combination to form hybrid orbitals, which exhibit the following characteristics:

1. They remain atomic orbitals within the valence shell.
2. Their number equals the number of orbitals involved in the hybridization process.
3. They are isoenergetic (i.e., they have the same energy level).
4. They do not exist in isolated atoms but form only when an atom interacts chemically with other atomic systems[4].

Thus, in the water molecule, oxygen's valence orbitals undergo sp^3 hybridization, where the three p orbitals combine with the s orbital to form four equivalent, isoenergetic sp^3 hybrid orbitals, each possessing 25% (i.e., 1/4) s-character. The geometric shape of sp^3 hybrid orbitals is illustrated in Figure 5.5. Their spatial orientation is depicted in Figure 5.5A, while their overall representation is shown in Figure 5.5B. According to the Valence Shell Electron Pair Repulsion (VSEPR) theory, the four sp^3 hybrid orbitals adopt a tetrahedral arrangement (Figure 5.5B).

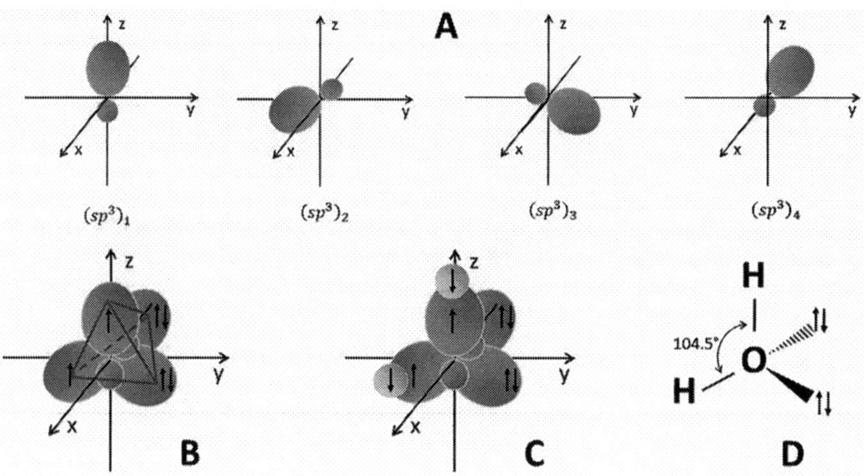

Figure 5.5 Geometric representation in polar coordinates of the four sp^3 orbitals (**A**) and the set of the same orbitals "seen" as a whole with the representation of the tetrahedral geometry in full and dashed (**B**). The arrows in (**B**) represent the six valence shell electrons of oxygen filling the sp^3 orbitals according to the Aufbau rules. The hydrogen atoms can bind to the oxygen only along the direction of the tetrahedron axes (**C**). The structure of water is, then, a distorted tetrahedron with a HÔH angle of 104.5° (**D**).

The six valence electrons in the hybridized oxygen atom occupy the four sp^3 orbitals following the Aufbau principle. As a result, two lone pairs (lp) and two unpaired electrons are present (Figure 5.5B). The two hydrogen atoms can approach the oxygen atom only along the directions of the two tetrahedral axes

[4] Pauling, L. 1931. The nature of the chemical bond. Application of results obtained from quantum mechanics and from a theory of paramagnetic susceptibility to the structure of molecules. Journal of the American Chemical Society. 53(4): 1367–1400.

where the unpaired electrons reside (Figure 5.5C). Once this occurs, covalent bonds form between hydrogen and oxygen, resulting in the tetrahedral structure of the water molecule (Figure 5.5D).

It is important to note that the geometry of the water molecule is a distorted tetrahedron, with an H–Ô–H bond angle of approximately 104.5° instead of the ideal 109.5° expected for a perfect tetrahedral arrangement. This deviation arises from the electrostatic repulsion between electron pairs.

Lone pairs exert stronger repulsion than bonding pairs because bonding electrons experience partial shielding from the positive nuclei of both bonded atoms (H and O). In contrast, lone pair electrons are influenced primarily by the oxygen nucleus, resulting in greater repulsion between lone pairs. As a consequence, lone pairs push bonding pairs closer together, reducing the H–Ô–H bond angle compared to the lp–Ô–lp angle, giving the water its characteristic bent molecular shape.

5.3 THE ELECTRONIC DISTRIBUTION IN THE WATER MOLECULE AND THE HYDROGEN-BOND

The tetrahedral structure of water has significant implications for environmental processes, particularly in soil chemistry and geochemical interactions. To fully understand its role in these systems, it is essential to introduce the hydrogen bond (H-bond) concept, which governs many of water's unique properties in natural environments.

A hydrogen bond forms between a molecular group containing a hydrogen atom covalently bonded to a more electronegative atom (hydrogen donor) and another group capable of donating electron density (hydrogen acceptor) (Figure 5.6).

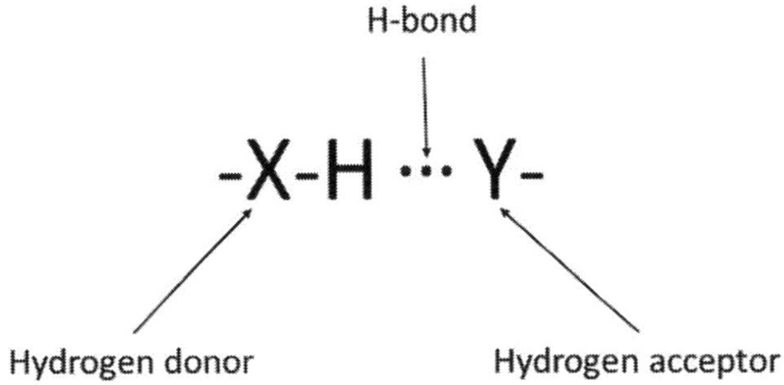

Figure 5.6 Representation of the hydrogen bond (the dotted line) is also indicated as H-bond. X, the so-called hydrogen donor, is an atom whose electronegativity is higher than the hydrogen. Y, the hydrogen acceptor, is an atom capable of being an electron donor.

Electronegativity measures an atom's tendency to attract a bonding pair of electrons. Consequently, when a covalent bond forms between two atoms with different electronegativities, the electron density is higher near the more electronegative atom.

Figure 5.7 illustrates the electrostatic potential surface mapped onto the electron density iso-surface of the water molecule[5]. The darker region represents the more electronegative oxygen atom, which carries the highest negative charge, while the lighter region corresponds to the less electronegative hydrogen atoms, which carry the highest positive charge. Since the centers of positive and negative charge do not coincide, the water molecule behaves as a dipole.

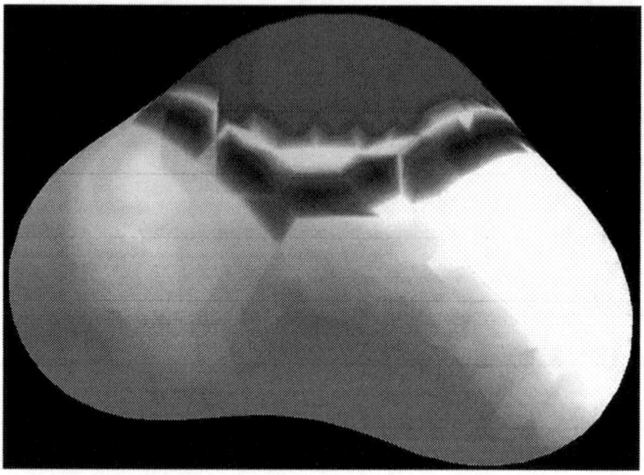

Figure 5.7 Representation of the electrostatic potential surface plotted on the iso-surface of the electron density of the water molecule. The color code shows the charge distributions in the molecule. The darker the color, the more negative the charge is. Therefore, oxygen carries the largest amount of negative charge, while the hydrogen atoms bear the largest amount of positive charge. This figure was drawn by using the ArgusLab software (http://www.arguslab.com/arguslab.com/ArgusLab.html).

When two water molecules approach each other, they align so that the more electronegative oxygen atom of one molecule is positioned near the partially positive hydrogen atom of the other. This alignment facilitates the formation of a hydrogen bond, also known as an H-bridge. In this interaction, the oxygen atom of one molecule acts as the hydrogen acceptor, while the oxygen of the other molecule, which is covalently bonded to hydrogen, serves as the hydrogen donor.

From a chemical-physical perspective, the hydrogen bond can be described as a combination of several intermolecular forces, including exchange repulsion, attractive dispersion, electrostatics, polarization, and charge transfer. Among these, charge transfer is particularly noteworthy. Although it contributes only partially to H-bonding in water, it is key in determining the directionality of hydrogen bonds between water molecules. This phenomenon involves the transfer of electrons from an occupied orbital of one molecule to the unoccupied orbitals of another, making it conceptually similar to a directional covalent bond. The

[5] The electrostatic potential surface represents the distribution of electric fields across a molecule, highlighting regions of partial positive and partial negative charge. It visually depicts how charge is distributed, helping to understand molecular interactions such as dipole-dipole forces and hydrogen bonding.

strength of an H-bond is influenced by its O$\hat{H}$O bond angle. The strongest H-bond occurs at 180°, while the weakest occurs at 130° (Figure 5.8)[6]. The scheme in Figure 5.8 also highlights the flexibility of H-bonds.

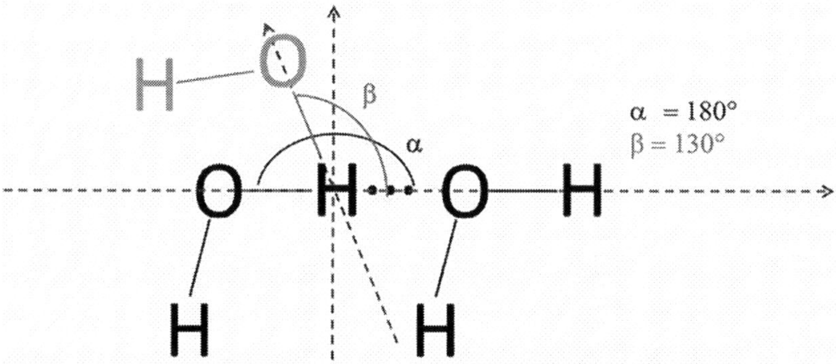

Figure 5.8 Representation of the geometry of the H-bond (dotted line). The strongest H-bond is achieved when the angle O$\hat{H}$O is 180°. When this angle is around 130° (see the grey molecule), the weakest H-bond realizes.

In liquid water, the O$\hat{H}$O bond angle can vary between 130° and 180°. Notably, a single water molecule contains two hydrogen atoms and two lone pairs, allowing it to donate two hydrogen atoms and accept two more. As a result, each water molecule can form up to four hydrogen bonds.

The simplest water cluster consists of five H_2O molecules arranged in a tetrahedral configuration (Figure 5.9).

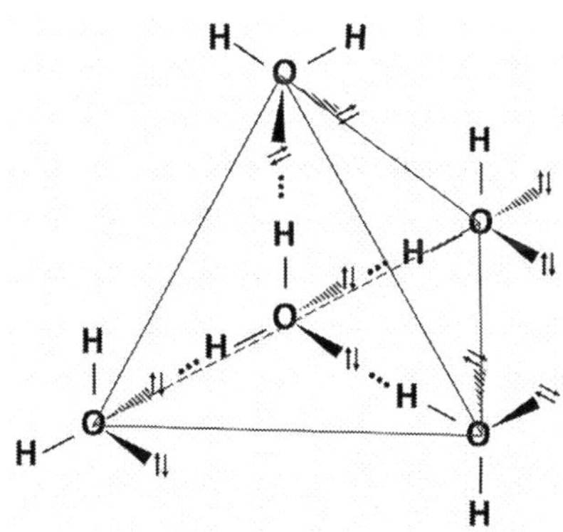

Figure 5.9 Water cluster consisting of five molecules, with four positioned at the vertices of a tetrahedron (represented by solid lines). The central water molecule forms hydrogen bonds (H-bridges) with all the surrounding molecules.

It is worth noting that the average H-bonds in a molecular cluster made by *n* different H_2O units is 2.5. This does not mean that each H_2O molecule forms

[6] Brown, I.D. 1976. On the geometry of O–H…H hydrogen bond. Acta Crystallographica A. 32: 24–31

two and a half H-bonds, but some molecules are engaged in four H-bonds, while others can form one, two, or three hydrogen bindings. The average of all the H-bonds is, hence, 2.5[7].

5.4 THE STRUCTURE OF THE WATER CLUSTERS

Figure 5.9 represents the basic water cluster where only five different H_2O units are related to each other via H-bonds. However, if we consider a number n of different water molecules (that is $(H_2O)_n$), the structure depicted in Figure 5.10A is obtained. Here, $n = 14$, corresponding to three interconnected basic tetrahedral clusters, has been accounted for. This ordered structure represents how solid water (i.e., the ice) should appear at the microscopic level. In particular, the oxygen atoms are positioned at the corners of "hexagonal" lattice structures with a "chair" conformation (Figure 5.10B). These lattices, sharing common sides, collectively delineate a "plane" of hexagonal structures, each of which is linked with those above and below via hydrogen bonds. Examining the ice structure between adjacent planes makes it clear that the coordination still involves hexagonal lattices, albeit with a "boat" conformation (Figure 5.10C).

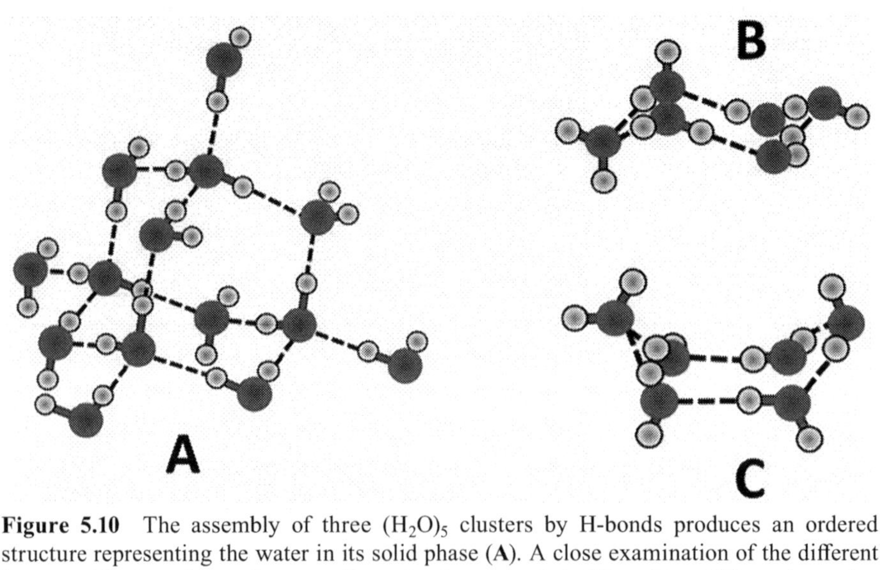

Figure 5.10 The assembly of three $(H_2O)_5$ clusters by H-bonds produces an ordered structure representing the water in its solid phase (**A**). A close examination of the different hexagonal lattices shows that a conformation between chair (**B**) and boat (**C**) conformation is possible. The dark balls are the oxygen atoms, while the light dark ones are the hydrogen atoms. The H-bonds are depicted as dashed lines.

It is worth noting that the molecular coordination of liquid water is similar to that of ice. However, unlike ice, the structure of liquid water is highly flexible and

[7] Muthachikavil, A.V., et al. 2021. Distinguishing weak and strong hydrogen bonds in liquid water—A potential of mean force-based approach. The Journal of Physical Chemistry B. 125: 7187–7198.

constantly evolving. In this context, the term "structure" in reference to liquids has only a statistical meaning—representing an average configuration that forms and dissipates billions of times per second. An example of a distorted liquid water structure is shown in Figure 5.11. As mentioned earlier, not all water molecules participate in forming four hydrogen bonds. Some molecules engage in all four, while others form only one, two, three, or none at all.

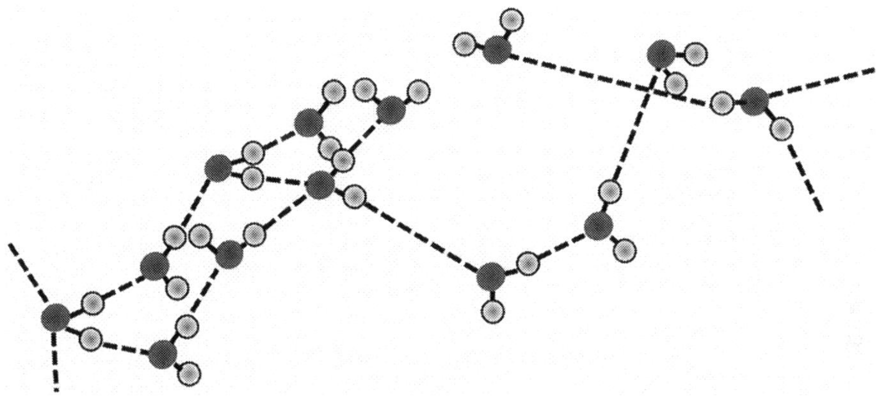

Figure 5.11 Representation of the distorted liquid water structure.

5.5 THE COOPERATIVE AND ANTI-COOPERATIVE NATURE OF THE HYDROGEN-BOND

Consider a water molecule that can act as both a hydrogen donor and a hydrogen acceptor (Figure 5.12). When this molecule donates one of its hydrogen atoms to a second water molecule acting as a hydrogen acceptor, a redistribution of electron density occurs. Specifically, the donated hydrogen becomes less electropositive. As a result, the lone-pair region of the oxygen atom—where this hydrogen is covalently bonded—becomes more electronegative, making it more likely to accept another hydrogen from a third water molecule, thereby forming a second hydrogen bond (Figure 5.12). Following the formation of the second H-bond, the electron density of the water molecule shifts again: the lone-pair region becomes less electronegative, while the second covalently bonded hydrogen becomes more electropositive. This change favors the formation of a third H-bond involving this hydrogen while simultaneously discouraging the formation of a fourth H-bond at the remaining lone pair (Figure 5.12).

The process by which the formation of one H-bond enhances the likelihood of another is known as H-bond cooperativity. Conversely, when an existing H-bond inhibits the formation of another, this effect is referred to as H-bond anti-cooperativity.

Quantum mechanical calculations have shown that H-bond cooperativity and anti-cooperativity are influenced by the spatial arrangement of water molecules. For example, in a homodromic configuration, where all H-bonds are aligned in the same direction, a water hexamer $[(H_2O)_6]$ exhibits the strongest cooperative effect (Figure 5.13A). Conversely, the strongest anti-cooperative effect is observed in a completely antidromic configuration, where H-bonds are oriented in different directions (Figure 5.13B).

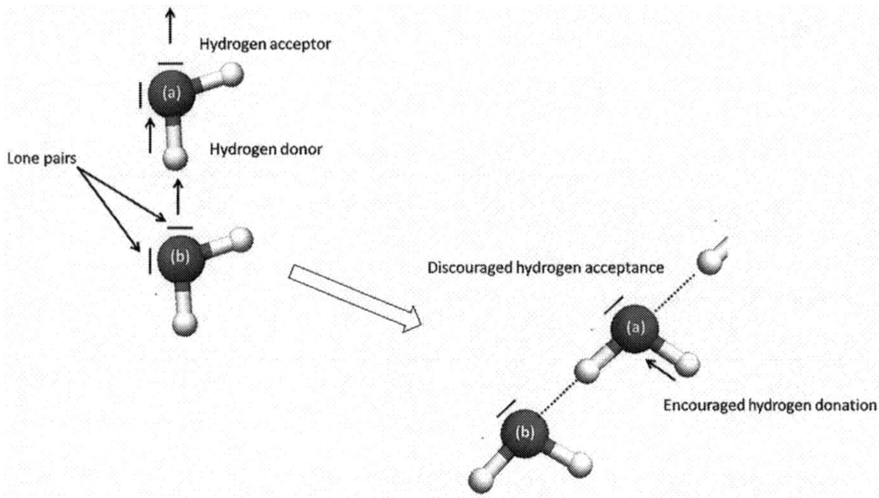

Figure 5.12 Scheme of H-bond cooperativity and anti-cooperativity effect. The water molecule (a) on the left-hand side of the figure can act as both a hydrogen donor and a hydrogen acceptor. The donation of the hydrogen atom allows the formation of a H-bond with the water molecule (b). This H-bond enables a redistribution of the electron density in (a) so that the lone pair zone of this molecule becomes more electron-rich. The arrows indicate the change in the electron density. As a consequence, the formation of the (b)-(a) H-bond encourages the hydrogen acceptor characteristic of the water (a), thus favoring the formation of the second H-bond reported on the right-hand side of this Figure. The second H-bond encourages the hydrogen donation property of a molecule (a) while it discourages the hydrogen acceptance characteristic of the same system. The condition where the formation of a H-bond favors that of a second H-bond is referred to as H-bond cooperation. When the formation of the H-bond discourages that of a second H-bond, H-bond anti-cooperativity occurs.

5.6 THE TRANSIENT NATURE OF THE HYDROGEN-BOND

The average lifetime of a hydrogen bond in liquid water is approximately 10 picoseconds (*ps*)[8]. This means that a hydrogen bond can form and break about ten trillion times per second. Notably, breaking a hydrogen bond is associated with forming a covalent bond and vice versa. The curved arrows on the left-hand side of Figure 5.14 illustrate how the disappearance of a covalent bond leads to

[8] Liu, J., X. He, J.Z.H. Zhang and L.-W. Qi. 2018. Hydrogen-bond structure dynamics in bulk water: insights from ab initio simulations with coupled cluster theory. Chemical Science. 9: 2065.

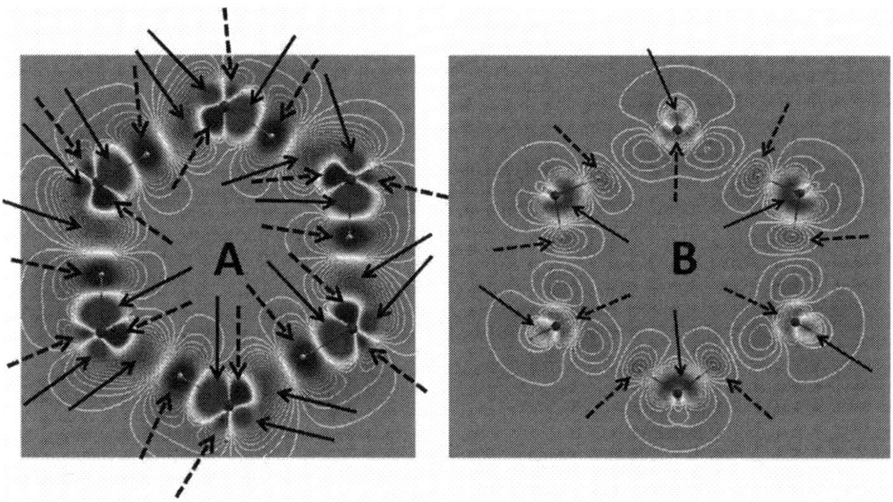

Figure 5.13 In (**A**), The differential electron charge density is depicted for the water hexamer in the cyclic homodromic configuration. Here, the dark and white balls represent oxygen (O) and hydrogen (H) atoms, respectively. The regions full arrows indicate an increase in electron density, while those dashed arrows indicate a decrease in electron density. In (**B**), the water hexamer has been reported in the total antidromic configuration. The arrow code is the same as in (**A**). The comparison between (**A**) and (**B**) reveals that the hexamer in (**A**) shows the strongest H-bond cooperativity, whereas the hexamer in (**B**) has the strongest anti-cooperativity. Re-elaborated from P.L. Silvestrelli (2017) Hydrogen bonding characterization in water and small molecules, The Journal of Chemical Physics 146: 244315, with the permission of AIP Publishing.

the formation of a new hydrogen bond. As a result, the bond distribution shown on the right-hand side of Figure 5.14 is achieved.

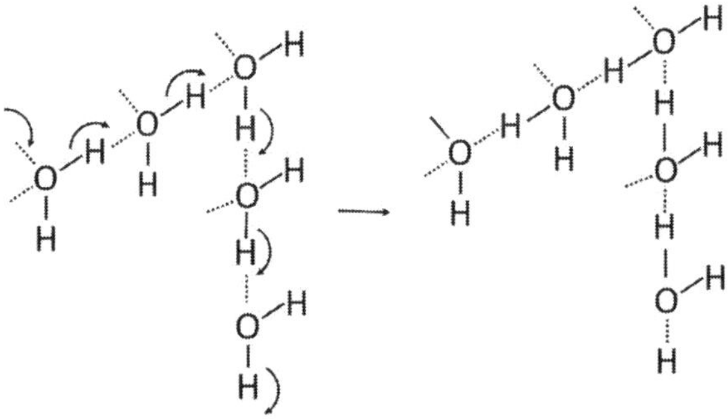

Figure 5.14 A cluster of water molecules is held together by H-bonds (dotted lines). The transient H-bond nature is evidenced by the curved arrows. The latter shows that a new covalent bond is formed when the H-bond breaks down and vice versa. When this occurs, a new bond distribution inside the cluster is achieved.

The transient nature of the hydrogen bond leads to important considerations regarding water's self-ionization equilibrium, described by equation (5.7):

$$2H_2O \rightleftarrows H_3O^+ + OH^- \quad (5.7)$$

In aqueous solutions, these ions do not exist in isolation but are instead solvated[9], meaning they are surrounded by water molecules that help reduce their charge density[10]. Quantum mechanical calculations indicate that each ion is stabilized by approximately three water molecules, which minimizes the charge density of both hydronium (H_3O^+) and hydroxide (OH^-) (Figure 5.15). Based on this, a more accurate representation of these species is $H_9O_4^+$ for hydronium and $H_7O_4^-$ for hydroxide rather than their simplified forms. Consequently, equation (5.7) can be rewritten as:

$$8H_2O \rightleftarrows H_9O_4^+ + H_7O_4^- \quad (5.8)$$

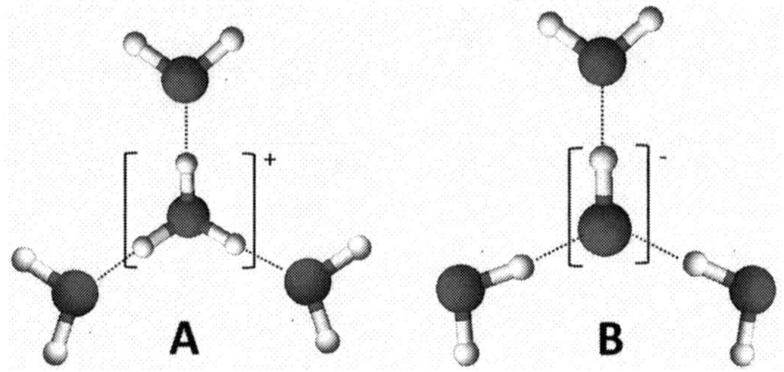

Figure 5.15 Solvated (or hydrated) forms of the ions hydronium (**A**) and hydroxide (**B**). According to calculations, three molecules of water are needed to minimize the charge density of both ions.

The solvated H^+ ion moves through the solution via proton hopping, exploiting the transient nature of hydrogen bonding. This mechanism enables the rapid transfer of protons from one position to another. Hydrogen bonds create a dynamic network where some regions are short and strong (corresponding to OH covalent bonds), while others are long and weak ($HO \cdots H$ hydrogen bonds). The continuous interconversion between hydrogen bonds and covalent linkages, as illustrated in Figure 5.16, facilitates the rapid migration of positive charge across different sites. This same process applies to the hydroxide ion (OH^-), allowing for its efficient transport in solution.

According to the mechanism described above, also known as the Grötthus mechanism, the positive charge carried by the proton (H^+) diffuses in solution not only due to concentration gradients[11] but also through the interconversion between

[9] Solvation is a general term indicating that a solute in solution is surrounded by the solvent molecules. The solute-solvent interaction is the basis for the mechanism of dissolution, which will be discussed below.

[10] The term "charge density" refers to the amount of charge per unit volume. Therefore, ions with the same charge may show different charge densities according to their size. As an example, the charge density of Na^+ is lower than that of K^+.

[11] The effect of the concentration gradient on solute diffusion in solution is the object of a paragraph below.

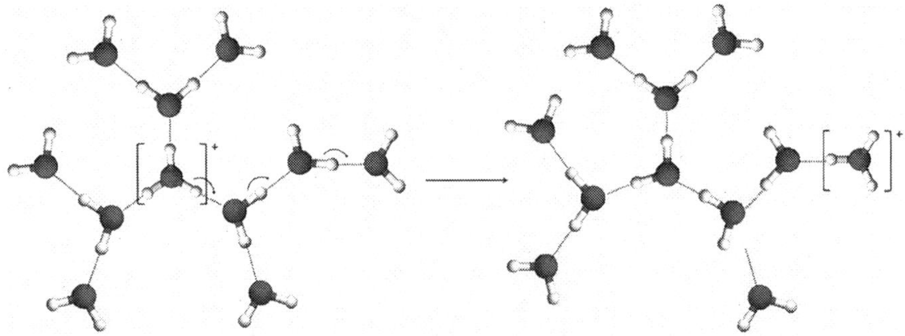

Figure 5.16 Several water molecules surround the hydronium. Due to the interconversion between the H-bonds and the covalent bindings, it moves among several positions. This process is also referred to as Grötthus mechanism.

hydrogen bonds and covalent bonds. Notably, charge migration (both positive and negative) does not occur in a continuous manner but follows alternating phases of rapid activity and inactivity[12]. During active phases, protons can "hop" distances of 0.4–0.8 nm on a sub-picosecond timescale, followed by long resting periods. Considering the approximate spherical shape of a water molecule with a diameter of ~0.3 nm, the 0.4–0.8 nm jump distance corresponds to the charge being transferred across approximately 1 to 3 water molecules.

5.7 THE HYDROGEN BOND AND THE ANOMALIES OF WATER

Water is one of the smallest molecules by molecular weight, at approximately 18 g mol^{-1}. Given its low mass, one might expect it to be highly volatile. However, comparing water to hydrogen sulfide (H_2S)—a structurally similar molecule where sulfur replaces oxygen—reveals an unexpected trend. Since sulfur is heavier than oxygen, H_2S has nearly twice the molecular weight of water. Based on this, one would expect H_2S to have lower volatility than water. Yet, at 1 atm pressure, H_2S boils at –60°C, while water boils at 100°C—a striking 160°C difference. This discrepancy arises from the presence of hydrogen bonding in water, which H_2S lacks[13]. The absence of H-bonds in H_2S results in much weaker intermolecular forces, leading to a significantly lower boiling point. In other words, hydrogen bonds anchor water molecules, making it more difficult to transition from the liquid to the gas phase.

Now, let us examine Figure 5.17, which compares the density changes of water and benzene over a temperature range extending from their freezing points

[12] Offei-Danso, A., et al. 2023. The collective burst mechanism of angular jumps in liquid water. Nature Communications. 14: 1345.

[13] Sulfur belongs to the same group of the periodic table as oxygen. However, their electronic configurations differ. Oxygen is [He] $2s^2 2p^4$, while sulfur is [Ne] $3s^2 3p^4$. As a result, the H–S bond is less polarized than the H–O bond. Due to this lower polarity, H_2S lacks the ability to form hydrogen bonds.

(5.5°C for benzene, 0°C for water) to their boiling points (80°C for benzene, 100°C for water).

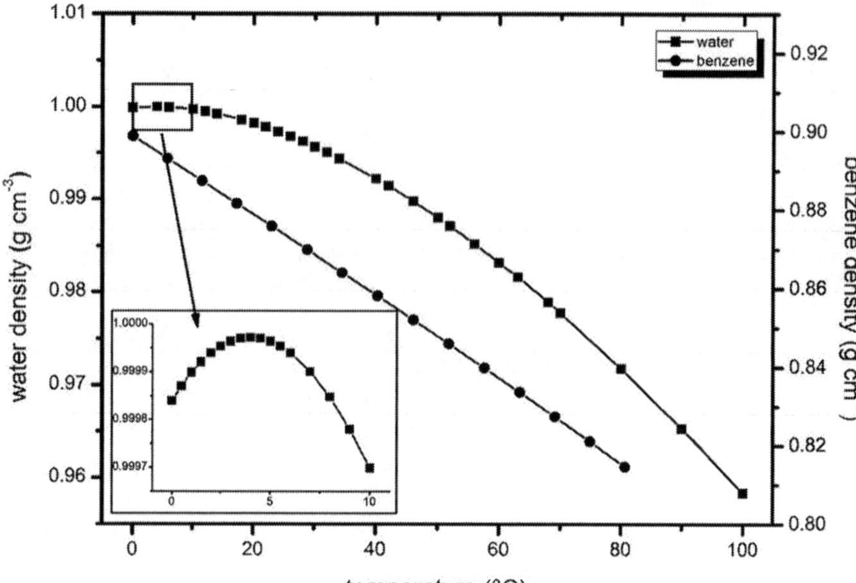

Figure 5.17 Density changes of benzene and water as temperature rises between the freezing and the boiling points (0/100°C for water and 5.5/80°C for benzene). The box highlights water behavior in the temperature range 0–10°C.

Benzene behaves like most typical liquids, with its density increasing as temperature decreases. This occurs because, during cooling, molecular motion slows down, allowing molecules to pack more closely together as they approach the solid state. At this point, molecules reach their minimum possible distance, maximizing density. Once benzene solidifies, further temperature reductions no longer affect its density.

Water exhibits an anomalous behavior in its density-temperature relationship. Like most liquids, water's density increases as it cools. However, below 3.98°C, its density decreases (as shown in the box in Figure 5.17). This anomaly arises from water's ability to form hydrogen bonds (H-bonds), which influence its molecular arrangement at lower temperatures.

As previously stated, directionality and flexibility are key characteristics of hydrogen bonds (Figure 5.8). In the liquid state, these properties allow water molecules to adopt the distorted conformations shown in Figure 5.11. However, when the temperature drops below 3.98°C, the directionality of hydrogen bonds becomes dominant over flexibility, leading water molecules to arrange into the more ordered structure depicted in Figure 5.10. This ordering creates voids, reducing the number of water molecules per unit volume compared to liquid water at slightly higher temperatures. In other words, between 3.98°C and 0°C, water expands as it cools, resulting in a lower density than its liquid state. This explains why ice floats on water. This phenomenon has significant implications for porous systems like rocks and soils. As discussed in Chapter 3, the expansion of water upon freezing plays a key role in rock decomposition and soil formation. Also, freezing-thawing cycles in soils can contribute to soil erosion and nutrient loss[14].

Another key consequence of ice floating on water is its crucial role in sustaining aquatic life. If ice were denser than water, it would sink, eventually causing entire water bodies to solidify, making life impossible. Instead, floating ice forms an insulating layer, helping to retain heat in the underlying liquid water, thus allowing life to persist even in freezing conditions.

Additionally, the density maximum at around 4°C, combined with ice's lower density, leads to other important hydrodynamic phenomena. In freshwater bodies, all layers must reach approximately 4°C before freezing can occur. This property also drives density-driven thermal convection, facilitating seasonal mixing in deeper temperate waters. This process transports oxygen to deeper regions, supporting aquatic ecosystems.

Finally, the high heat capacity[15] of oceans and seas allows them to function as thermal reservoirs, regulating Earth's climate by absorbing and storing heat. As a result, the temperature of large water bodies fluctuates only one-third as much as land temperatures. This thermal moderation plays a crucial role in regulating Earth's climate, as seen in phenomena like the Gulf Stream, which transports tropical warmth to northwestern Europe (see Chapter 4 for details on the water cycle).

Additionally, the compressibility of water reduces sea level by approximately 40 meters, effectively increasing the available land area by about 5%.

5.8 THE DIFFERENT FORMS OF ICE

In the previous paragraphs, ice was introduced, noting that it floats on liquid water due to its lower density. However, it is worth noting that ice can exist in different forms than that reported in Figure 5.10, depending on the relative positions of the water molecules involved in its formation during either temperature variation or the pressure to which water is subjected. In particular, the ice reported in Figure 5.10 is indicated as ice I_h, and it forms naturally under the standard temperature and pressure conditions (i.e., 0°C and 1 atm)[16]. The subscript h lays for hexagonal. This means that water molecules form hexagons like those depicted in Figure 5.10A and B. In the upper atmosphere, a metastable form of ice indicated

[14] Lan, H., et al. (2021) The cyclic expansion and contraction characteristics of a loess slope and implications for slope stability. Scientific Reports. 11: 2250.
– Zhang, L., et al. 2021. The influence mechanism of freeze-thaw on soil erosion: A review. Water. 13: 1010.

[15] The heat capacity (also indicated as C) represents the amount of thermal energy needed to produce a unit change in temperature for a given system.

[16] IUPAC Compendium of Chemical Terminology, 3rd ed. International Union of Pure and Applied Chemistry; 2006. Online version 3.0.1, 2019. https://doi.org/10.1351/goldbook.S05921.

as ice I_c, occurs[17]. The subscript *c* refers to the cubic symmetry. In other words, in the ice I_c crystals, the oxygen atoms of the different water molecules are placed as in a diamond[18] (Figure 5.18A).

Ice II is a rhombohedral form of ice obtained by compressing ice I_h at a pressure of around 2,000 atm and a temperature of ≈ –75°C. Alternatively, it can also be retrieved by quickly decompressing ice V. The dimensions of the ice II rhombohedron are $a = b = c = 778$ pm, and $\alpha = \beta = \gamma = 113.1°$ (Figure 5.18B).

Following heating to –23°C, at the pressure of ≈2,000 atm, ice II is transformed into ice III. The latter is a tetragonal structure with dimensions $a = b = 666.6$ pm, $c = 693.6$ pm, and $\alpha = \beta = \gamma = 90°$ (Figure 5.18B).

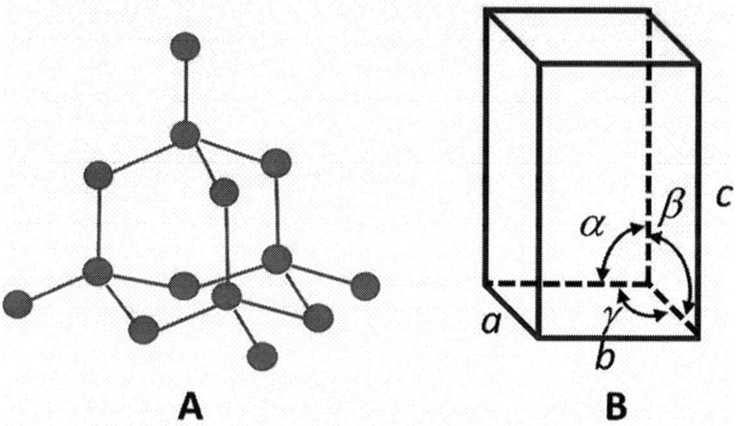

Figure 5.18 Structure of a diamond (**A**). The balls represent the oxygen atoms in water molecules. The lines are for the eyes only. They have been drawn to give an idea of the tridimensional diamond structure. Tetragonal lattice (**B**) with an indication of the sizes, i.e., the length of the edges (*a*, *b*, and *c*) and the angles (α, β, and γ) between the polyhedron faces. According to the different edges and angle values, different crystalline structures can be described. For example, when $a \neq c \neq b$ and $\alpha \neq \beta \neq \gamma$ a triclinic system is achieved, as $a = c = b$, and $\alpha = \beta = \gamma$ a cubic lattice is obtained. When $\alpha = \beta = \gamma = 90°$, a real cube is retrieved. The cubic geometry occurs also when $\alpha = \beta = \gamma \neq 90°$. In this case, a rhombohedral system can be described[19].

Ice IV was obtained in 1935 by compressing water at a pressure of ≈ 5,200 atm and a temperature of around –6°C[20]. The crystalline structure of this form of ice was unveiled only between the end of the seventies and the beginning of the

[17] Murray, B.J., et al. 2005. The formation of cubic ice under conditions relevant to Earth's atmosphere. Nature. 434: 202–205.

[18] Geiger, P., et al. 2014. Proton Ordering of Cubic Ice I_c: Spectroscopy and computer simulations. Journal of Physical Chemistry C. 118: 10989–10997.

[19] A complete discussion on the crystal structures can be found in Müller, U. 2013. *Symmetry Relationships between Crystal Structures: Applications of Crystallographic Group Theory in Crystal Chemistry*. Oxford University Press, 1st ed. ISBN: 9780199669950

[20] Bridgman, P.W. 1935. The pressure-volume-temperature relations of the liquid, and the phase diagram of heavy water. The Journal of Chemical Physics. 3: 598–605.

eighties of the 20th century[21, 22] when the combination of several spectroscopic techniques revealed that ice IV contains a rhombohedral unit cell of dimensions $a = b = c = 760$ pm, and $\alpha = \beta = \gamma = 70.1°$ (Figure 5.18B). In particular, the structure encompasses two distinct types of water molecules and four non-equivalent types of H-bonds. The structural aspect includes the H-bond traversing the center of a six-ring of water molecules, linking non-adjacent structural components[21].

Ice V shows a monoclinic crystal shape (a = 922 pm, b = 754 pm, c = 1,035 pm; $\alpha = \gamma = 90°$, $\beta = 109.2°$ in Figure 5.18B), obtained by cooling water to –20°C at ≈ 4,900 atm.

Ice VI, characterized by a density of 1.31 g cm^{-3}, possesses a tetragonal cell with dimensions $a = b = 627$ pm, $c = 579$ pm, and $\alpha = \beta = \gamma = 90°$ (Figure 5.18B). Each cell accommodates ten water molecules. The structural arrangement comprises hydrogen-bonded chains of water molecules, reminiscent of the tectosilicate chains found in fibrous zeolites. These chains are interconnected laterally to form an open, zeolite-like framework. This framework fills cavities with a second framework identical to the first. Although the two frameworks interpenetrate without interconnection, the complete structure can be conceptualized as a "self-clathrate." This structural feature naturally facilitates the attainment of high density in tetrahedrally linked framework structures[23].

Ice VII, obtained at -50°C under a pressure of around 22,000 atm, has a body-centered cubic structure where $a = b = c = 330$ pm, and $\alpha = \beta = \gamma = 90°$ (Figure 5.18B)[24]. The large density of Ice VII is due to the intersection between two different body-centered cubic structures, which enable contraction along the vertical direction and extension along the horizontal direction.

Ice VIII is obtained by cooling ice VII to temperatures below 5°C under pressures above approximately 9,800 atm. It comprises two intersecting cubic ice lattices, where each oxygen atom is surrounded by eight closest neighbors but is connected to only four neighbors with tetrahedrally coordinated hydrogen bonds. The sizes of the cell unit in ice VIII are $a = b = 444.9$ pm, $c = 641.3$ pm, and $\alpha = \beta = \gamma = 90°$ (Figure 5.18B)[25].

Ice IX, derived from the low-temperature variation of Ice III, exhibits a fully proton-ordered structure featuring two distinct types of water molecules. One type occupies a site lacking point symmetry, while the other type aligns along a twofold axis. Both types of water molecules contribute to a three-dimensional framework by forming four hydrogen bonds. The sizes of the ice IX crystal are $a = b = c = 673$ pm, and $\alpha = \beta = \gamma = 90°$ (Figure 5.18B)[26].

---

[21] Engelhardt, H. and B. Kamb. 1978. X-ray determination of the structure of ice IV. Journal of Glaciology. 21: 51–53.

[22] Engelhardt, H. and B. Kamb (1981). Structure of ice IV, a metastable high-pressure phase. The Journal of Chemical Physics. 75: 5887–5899.

[23] Kamb, B. 1965. Structure of Ice VI. Science. 150: 205–209.

[24] Kamb, B. and Davis, B.L. 1964. Ice VII, the densest form of ice. *In*: Proceedings of the National Academy of Sciences. 52: 1433–1439.

[25] Besson, J.M., et al. 1994. Variation of interatomic distances in Ice VIII to 10 GPa. Physical Review B. 49: 12540–12551.

[26] La Placa, S.J., et al. 1973. On a nearly proton-ordered structure for Ice IX. The Journal of Chemical Physics. 58: 567–580.

With increasing pressure and the contraction of the O···O distance, ice VII transforms into cubic Ice X, where the protons are evenly distributed between the oxygen atoms in a molecular crystal. Each oxygen atom is surrounded by four hydrogens equidistantly placed in a tetrahedral arrangement. The oxygens are organized in a body-centered cubic structure, each with eight nearest neighbor oxygen atoms forming a cube. Meanwhile, the hydrogen atoms are surrounded by twelve nearest neighbor hydrogen atoms arranged in a body-centered truncated cube. The crystal sizes of ice X are $a = b = c = 278$ pm, and $\alpha = \beta = \gamma = 90°$ (Figure 5.18B)[27].

Ice XI exhibits an orthorhombic structure, containing eight molecules per unit cell. Its lattice parameters are $a = 446.5$ pm, $b = 785.9$ pm, and $c = 729.2$ pm (while $\alpha = \beta = \gamma = 90°$) at –268°C (Figure 5.18B)[28].

Ice XII can be obtained by heating high-density amorphous ice at a constant pressure of ≈ 8,000 atm from –196°C to approximately –90°C, at a rate of at least 15°C min^{-1}, and then recovering it at atmospheric pressure and –196°C. This form of ice adopts a tetragonal crystal structure with sizes $a = b = 830.4$ pm, $c = 402.4$ pm, and $\alpha = \beta = \gamma = 90°$ (Figure 5.18B). Ice XII features a unique screw-type hydrogen-bonded arrangement, forming a right-handed double helix, unlike other crystalline forms of ice. The smallest ring size in this structure consists of seven molecules[29].

Ice XIII is the metastable proton-ordered form of ice V, retrieved by doping water with 10 mM HCl at a temperature below –143°C and at a pressure of around 5,000 atm. The crystal structure is monoclinic with dimensions of $a = 924$ pm, $b = 747$ pm, $c = 1{,}030$ pm, and $\alpha = \gamma = 90°$, $\beta = 109.7°$ (Figure 5.18B)[30].

Ice XIV represents the proton-ordered state of Ice XII and is induced by doping with 10 mM HCl at a temperature below –155°C and at a pressure of ≈ 12,000 atm. The crystals exhibit an orthorhombic structure with cell dimensions of $a = 835.0$ pm, $b = 813.9$ pm, and $c = 408.3$ pm, with $\alpha = \beta = \gamma = 90°$ (Figure 5.18B), and 12 water molecules per unit cell[30].

Ice XV represents the hydrogen-ordered form of Ice VI. It remains thermodynamically stable at temperatures below approximately –143°C, and a pressure range of 8,000 to 15,000 atm. The orthorhombic crystal structure exhibits cell dimensions of $a = 623$ pm, $b = 624$ pm, and $c = 579$ pm, with angles $\alpha = 90.06°$, $\beta = 89.99°$, and $\gamma = 89.92°$ (Figure 5.18B). During the hydrogen-bonding ordering process from Ice VI to Ice XV, the lattice constants a and b contract, while there is a notable expansion in the c lattice constant[31].

---

[27] Hemley, R.J., et al. 1987. Static compression of H_2O-ice to 128 GPa (1.28 Mbar). Nature. 330: 737–740.

[28] Hirsch, T.K. and L. Ojamäe. 2004. Quantum-chemical and force-field investigations of ice Ih: Computation of proton-ordered structures and prediction of their lattice energies. Physical Chemistry B. 108:15856-15864.

[29] Lobban, C., et al. 1998. The structure of a new phase of ice. Nature. 391: 268–270.

[30] Salzmann, C.G., et al. 2006. The preparation and structures of hydrogen ordered phases of ice. Science. 311: 1758–1761.

[31] Salzmann, C.G., et al. 2009. Ice XV: A new thermodynamically stable phase of ice. Physical Review Letters. 103: 105701.

Ice XVI is the empty hydrate of neon first theorized in 2001[32] and then obtained in 2014 at ≈ –127°C and at negative pressures between 4,000 and 10,000 atm[33], that is, the pressures needed to degas the neon-clathrates. This ice phase is a face-centered cube with dimensions $a = b = c =$ 1,751 pm, and $\alpha = \beta = \gamma = 90°$ (Figure 5.18B).

Ice XVII is the empty hydrate of the hydrogen molecule obtained by degassing at around –170°C H_2 enriched ice[34]. It consists of layers of pentamers of hydrogen-bonded water molecules, which are made from $(H_2O)_{12}$ partial dodecahedra. These layers are arranged to generate hexagonal helices along the z-axis[102]. The sizes of the crystal cell unit are $a = b =$ 633 pm, $c =$ 606 pm, $\alpha = \beta = 90°$, and $\gamma =$ 120° (Figure 5.18B).

The use of laser-driven shockwaves able to simultaneously compress and heat liquid water to a pressure ranging between 9.9×10^5 and 3.9×10^6 atm and a temperature of 1,700–2,800°C revealed that, under these conditions, water becomes solid in a time measured in few nanoseconds[35]. The solid water is made of nanometer-sized grains showing a superionic behavior[36]. This face-centered-cubic ice is referred to as Ice XVII and shows lattice sizes of $a = b = = c =$ 340 pm and $\alpha = \beta = \gamma = 90°$ (Figure 5.18B).

Ice XIX is obtained by cooling ice VI to approximately –153°C at a pressure of around 2,000 atm and by using HCl as a dopant[37]. Its crystal unit is a tetragonal disphenoid having sizes of $a = b =$ 861.9 pm, $c =$ 559.30 pm, and $\alpha = \beta = \gamma = 90°$.

Although a list of the different ice phases has been described, the paragraph is far from exhaustive. A detailed study of the different ice phases is left to the readers, who can refer to the bibliography present in the notes accompanying this paragraph. Here, some more details concerning ice densities and the number of water molecules inside the crystal cell units have been reported in Table 5.1. However, the readers must account that the ice densities reported therein are not comparable with each other because those values were obtained in the stability conditions for each ice form.

---

[32] Kosyakov, V.I. and V.A. Shestakov. 2001. On the Possibility of the Existence of a New Ice Phase under Negative Pressures. Doklady Physical Chemistry, 376: 49–51. – Translated from Doklady Akademii Nauk. 376: 782–784.

[33] Falenty, A., et al. 2014. Formation and properties of ice XVI obtained by emptying a type sII clathrate hydrate. Nature. 516: 231–237.

[34] del Rosso, L., et al. 2016. New porous water ice metastable at atmospheric pressure obtained by emptying a hydrogen-filled ice. Nature Communications. 7: 13394.

[35] Millot, M., et al. 2019. Nanosecond X-ray diffraction of shock-compressed superionic water ice. Nature. 569: 251–255.

[36] Simply speaking, the superionic behavior refers to the movement of the ionic hydrogen within a lattice made by oxygen ions that occupy fixed positions.

[37] Gasser, T.M., et al. 2021. Calorimetric investigation of hydrogen-atom sublattice transitions in the ice VI/XV/XIX Trio. Journal of Physical Chemistry B. 125: 11777–11783.

Table 5.1 Density values and number of water molecules in the crystal unit cell of the different ice phases. All the values were obtained under the stability conditions for each form[38]

Ice Form	Density (g cm^{-3})	Number of H_2O in the Crystal Unit Cell	Ice Form	Density (g cm^{-3})	Number of H_2O in the Crystal Unit Cell
I_h	0.92	12	X	2.51	9
I_c	0.93	8	XI	0.92	8
II	1.17	12	XII	1.29	12
III	1.14	12	XIII	1.23	28
IV	1.27	16	XIV	1.29	12
V	1.23	28	XV	1.30	10*
VI	1.31	10	XVI	0.81	136†
VII	1.70	9	XVII	0.85	12
VIII	1.46	8	XVIII	1.25	14
IX	1.16	8	XIX	1.32	

*The number of water molecules in the crystal cell unit of Ice XV have been deduced by reading the references in notes 23 and 24.

†The enormous number of water molecules in the crystal cell unit of ice XVI, when compared to the other ice phases, is due to the presence of sixteen pentagonal dodecahedron small cages and eight hexadecahedron large cages in the unit cell. All the cited voids allow ice XVI density to be the lowest in relation to the other ice phases[33].

5.9 THE PHASE DIAGRAM OF WATER

According to the content of the previous paragraphs, water may lie in different physical states. Together with the traditional three phases (i.e., vapor, liquid, and solid, that is, ice), water can provide up to twenty different solid phases, each with specific properties. Therefore, the water phase diagram can be depicted in Figure 5.19, where most, but not all, of the different forms of ice are reported.

It must be remembered that, apart from ice I_h, all the other forms of ice do not exist in nature, and they are obtained in the lab under the specific T-P conditions reported above. Noticeably, when dealing with the concept of "nature," one refers to the conditions found on the planet Earth. There are reasonable motifs to think that at the extreme pressures and temperatures relevant for the deep interior of some icy giant exoplanets as well as planets in our solar system (e.g., Uranus), ice form like Ice XVIII may be present[35]. This suggests that the presence of water on other planets does not necessarily indicate the presence of life as we know it. Consequently, the interplanetary journeys in search of life are likely to

[38] Details about the physical properties of the different forms of ice can be found at: https://water.lsbu.ac.uk/water/ice_phases.html.

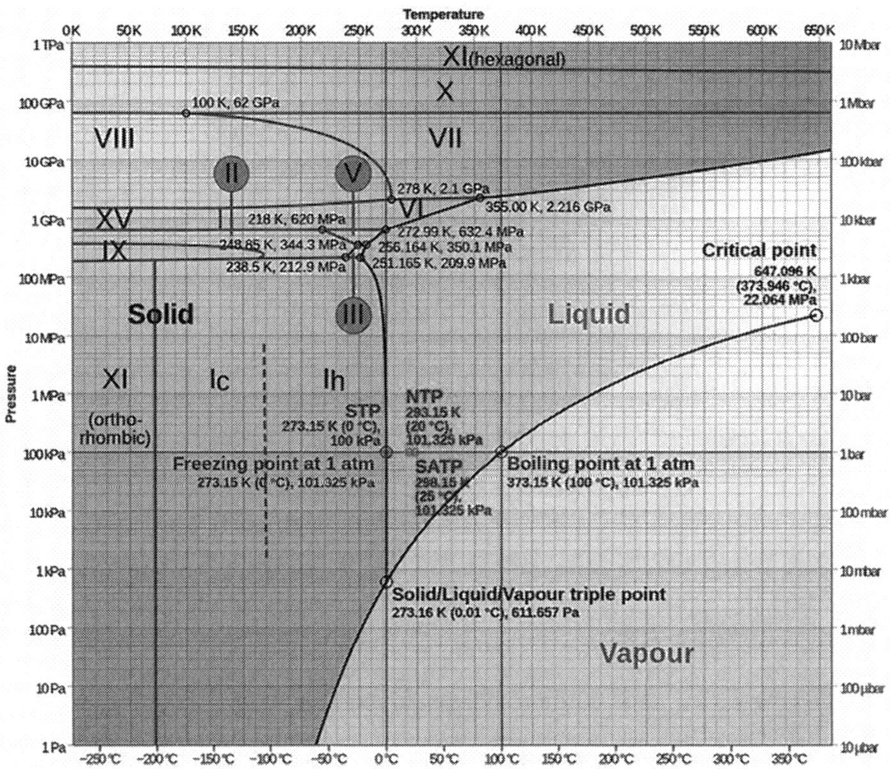

Figure 5.19 Lin-Log water phase diagram. The three typical water phases are reported: ice, liquid, and vapor. The ice phase, however, is also split into different ice states with all lab origins. STP stands for "standard temperature and pressure," that is 0°C and 1 atm; NTP stands for "normal temperature and pressure," which is 20°C and 1 atm; SATP means "standard ambient temperature and pressure," that is 25°C and 1 atm. This graph has been downloaded from Wikimedia Commons and it is licensed under the Creative Commons Attribution-Share Alike 3.0 Unported license.

be frustrating because we are seeking life forms similar to our own, while we should be looking for something different that is something we are currently unable to recognize and, therefore, detect.

5.10 A SIMPLISTIC THERMODYNAMIC APPROACH TO THE DISSOLUTION PROCESS IN WATER

The dissolution of a solute in a solvent can be understood as a process where existing molecular interactions are disrupted to form new ones. From a thermodynamic perspective, this process can be broken down into three fundamental steps, which can be expressed in the following reactions:

1. Disruption of Solute-Solute Interactions:

$$(\text{Solute})_n \leftrightarrows n\,\text{Solute} \tag{5.9}$$

In this step, the interactions between solute molecules (such as ionic bonds in salts or intermolecular forces in molecular compounds) are broken. This requires an input of thermal energy, making the process endothermic (i.e., $\Delta H_{(5.9)} > 0$)[39].

2. Disruption of Solvent-Solvent Interactions:

$$(\text{Solvent})_m \leftrightarrows m\,\text{Solvent} \tag{5.10}$$

Similarly, solvent molecules (in this case, water) must move apart to create space for the solute particles. This process also requires energy, as the hydrogen bonds or van der Waals forces between solvent molecules are disrupted, making it endothermic (i.e., $\Delta H_{(5.10)} > 0$).

3. Formation of Solute-Solvent Interactions:

$$n\,\text{Solute} + m\,\text{Solvent} \leftrightarrows (\text{Solute})_n - (\text{Solvent})_m \tag{5.11}$$

Once the solute and solvent particles are free, new interactions form between them, such as ion-dipole forces (in the case of ionic compounds dissolving in water). This process is exothermic, as energy is released when solute-solvent interactions form (i.e., $\Delta H_{(5.11)} < 0$).

The overall process can be summarized by adding these reactions together:

$$(\text{Solute})_n + (\text{Solvent})_m \rightarrow (\text{Solute})_n - (\text{Solvent})_m \tag{5.12}$$

This equation represents the dissolution of the solute in the solvent, forming a solution. The net energy involved in this process is given by the enthalpy of solvation (ΔH_{solv}), which accounts for the energy required to break solute-solute and solvent-solvent interactions and the energy released from solute-solvent interactions:

$$\Delta H_{\text{solv}} = \Delta H_{(5.9)} + \Delta H_{(5.10)} + \Delta H_{(5.11)} \tag{5.13}$$

[39] In thermodynamics, the term enthalpy (H) describes the amount of thermal energy that a system can exchange with the environment. H is given by the sum of the system's internal energy (U) and the product of its pressure (p) and volume (V): $H = U + pV$. The internal energy of a system with well-defined boundaries comprises the total kinetic energy from molecular motions and the potential energy associated with the vibrational motions and electrical energy of the atoms within the molecules. Enthalpy is a state function, meaning it describes the condition of a system irrespective of the process by which it arrived at that condition. The absolute value of the enthalpy for a system's state cannot be provided. Only a relative enthalpy value (ΔH) can be supplied, which is obtained by comparing the system's state before and after the transformation. For spontaneous transformations, $\Delta H < 0$, such as, for example, in the case of a ball rolling down the side of a hill. Conversely, when $\Delta H > 0$, the transformation does not spontaneously occur, as in the case of a ball which is unable to go up a hill.

Here, ΔH_{solv} can be either positive, negative, or zero, depending on the balance between the energy required to break the initial interactions and the energy released when new ones are formed:

— If $\Delta H_{solv} < 0$, the dissolution is exothermic, meaning that more energy is released when solute-solvent interactions form than is required to break solute-solute and solvent-solvent bonds. This often leads to a spontaneous dissolution process.

— If $\Delta H_{solv} > 0$, the dissolution is endothermic, meaning that an external energy input (e.g., heating) is needed for the dissolution to occur.

— If $\Delta H_{solv} = 0$, the dissolution process is thermoneutral, meaning the energy gained and lost during dissolution balances out, leading to an ideal solution[40].

While the enthalpy variations play a significant role in determining whether dissolution is exothermic or endothermic, the entropy[41] variations (ΔS) of the system also play a key role in determining spontaneity. Entropy measures disorder or the number of possible configurations in the system. The disruption of solute-solute and solvent-solvent interactions (Reactions 5.9 and 5.10) generally increases entropy because the separated molecules or ions have more freedom of movement. This increase in disorder means that the entropy change for these reactions is typically positive: $\Delta S_{(5.9)} > 0$, and $\Delta S_{(5.10)} > 0$.

When the solute-solvent interactions form (Reaction 5.11), the system experiences a more complex entropic change. The formation of a solute-solvent complex may reduce the freedom of some particles, but the overall disorder in the system often increases due to the mixing of solute and solvent. Therefore: $\Delta S_{(5.11)} > 0$.

The total entropy change during dissolution is known as the entropy of solvation (ΔS_{solv}), is given by:

$$\Delta S_{solv} = \Delta S_{(5.9)} + \Delta S_{(5.10)} + \Delta S_{(5.11)} \tag{5.14}$$

For a spontaneous dissolution, the entropy increase from solute-solvent interactions must outweigh any decrease caused by the disruption of solute and solvent interactions, meaning: $\Delta S_{(5.11)} > \Delta S_{(5.9)} + \Delta S_{(5.10)}$. In many cases, the mixing of solute and solvent increases the number of configurations available to both molecules, driving the entropy change to be positive.

---

[40] A solution can be considered ideal when the following conditions are fulfilled: 1. The solute can be considered point-like; 2. The interactions among the solute molecules can be neglected; 3. The interactions between solute and solvent are homogeneous throughout the solution; that is, they are the same in all directions. All the aforementioned conditions can be realized only when extreme dilution is considered.

[41] According to statistical thermodynamics, entropy is a state function (see note 39 for the definition of state function) that quantifies the degree of molecular disorder or randomness within a system. As per the second law of thermodynamics, a transformation occurs spontaneously only if there is an increase in entropy ($\Delta S > 0$) from the initial state to the final state of the system.

The Gibbs free energy[42] (ΔG_{solv}) combines the effects of enthalpy and entropy to determine whether the dissolution process is spontaneous. The Gibbs free energy change is given by:

$$\Delta G_{solv} = \Delta H_{solv} - T\Delta S_{solv} \tag{5.15}$$

where T is the temperature in Kelvin. For a dissolution to be spontaneous, the Gibbs free energy must decrease, meaning $\Delta G_{solv} < 0$.

As a general remark, it is possible to report that:

- If $\Delta H_{solv} < 0$, and $\Delta S_{solv} > 0$, the dissolution is always spontaneous at all temperatures.
- If $\Delta H_{solv} > 0$, but $T\Delta S_{solv}$ is large enough, the dissolution can still be spontaneous. In this case, the entropy term compensates for the positive enthalpy, making $\Delta G_{solv} < 0$.
- If $\Delta H_{solv} > 0$, and $\Delta S_{solv} < 0$, the dissolution is non-spontaneous, as both factors oppose the process.

5.11 GIBBS FREE ENERGY AND ACTIVITY

The variation in Gibbs free energy during the dissolution process is related to the amount of solute in the solution. Under ideal conditions (see note 40 for the meaning of "ideality"), the relationship is given by Equation (5.16):

$$\Delta G_{solv} = RT \ln x_i \tag{5.16}$$

where R is the universal gas constant (8.314 $J \cdot K^{-1} \cdot mol^{-1}$), T is the temperature in Kelvin, and x_i is the mole fraction[43] of the i-th component of the solution.

However, in non-ideal conditions, solute-solvent and solute-solute interactions deviate from ideality. In such cases, Equation (5.16) becomes Equation (5.17), which incorporates the activity coefficient γ_i, a factor that accounts for these interactions:

$$\Delta G_{solv} = RT \ln \gamma_i x_i = RT \ln a_i \tag{5.17}$$

[42] The Gibbs free energy (also referred to as the letter G) is a state function (see note 39 for the definition of state function) measuring the maximum amount of work that must be done to transform a system from the initial to the final state. In this context, the term "work" refers to the energy exchanged through mechanical interactions. For instance, in a chemical reaction occurring in a container with a piston, the formation of gaseous products exerts pressure on the piston, pushing it outward and thus performing work.

[43] Given a number of different solutes in solution, the mole fraction of the i-th solute is defined as the ratio between the number of moles of the i-th solute (n_i) and the sum of the number of moles of all the m chemical species present in solution ($\Sigma_1^m n_i$), including the solvent: $x_i = \frac{n_i}{(\Sigma_1^m n_i)}$. Due to this definition, the mole fraction is dimensionless, and it shows values in the range $0 \le x_i \le 1$, with 0 when the i-th species is absent and 1 when it is the only component of the solution. In equation (16), the mole fraction can also be replaced by the molar concentration. The meaning of the equation does not change.

Here, γ_i measures the deviation from ideality while a_i represents the activity[44] of the i-th component in solution, which reflects its "effective concentration"[44] under non-ideal conditions.

As a general rule, $0 \leq \gamma_i \leq 1$, where $\gamma = 1$ signifies ideal conditions (i.e., no significant interactions between the solute and solvent molecules occur).

Activity can also be described in terms of "fugacity"[45], a thermodynamic property related to the escaping tendency of a species from the solution. For a component i, the activity can be expressed as the ratio of the fugacity of the species in solution (f_i) to the fugacity in its standard state (f_i^0). When water is the solvent, the activity of water (a_w) can be related to vapor pressure:

$$a_w = \frac{f_w}{f_w^0} = \frac{P_w}{P_w^0} \tag{5.18}$$

Here, P_w is the vapor tension[46] of water in solution, and P_w^0 is the vapor tension of pure water.

The activity of water a_w ranges between 0 and 1, with $a_w = 1$ indicating pure water (where all water molecules are available for solute interactions). As a_w decreases, fewer water molecules are available to hydrate the solute, which limits solubility. When the solute concentration becomes sufficiently high, and water molecules are no longer available to hydrate the solute effectively, saturation is reached. At this point, any additional solute will precipitate out of the solution.

5.12 THE IONIC STRENGTH

For electrolyte-containing[47], the ionic strength is a measure of the total concentration of ions (both positive and negative) in an aqueous solution, which is weighted by the square of their charges. It is defined by the relationship (5.19):

---

[44] Activity is a dimensionless parameter that reflects the reactivity tendency of a given chemical species. The higher the activity, the greater is its attitude to transform. Activity depends not only on the concentration of the species under observation, but also on the presence and concentration of other species, each influencing the reactivity of the former. Activity is also referred to as the "effective concentration" in solution, representing the reduced reactivity of a given chemical due to its interactions with other systems. In other words, the higher the a_i value, the greater the tendency of the i-th species to react. Therefore, the tendency of the i-th species in solution to react is the greatest in ideal conditions. In Chapter 7, the concept of effective concentration is explored in greater depth.

[45] Fugacity measures the tendency of a molecule to flee from the surface of a liquid/solid phase. The stronger the molecule is anchored to the surface, the lower the fugacity. Conversely, as the molecule is weakly anchored, its ability to flee increases.

[46] The term "vapor tension" refers to the pressure of the water vapor that is in equilibrium with the liquid phase.

[47] Electrolytes are chemical systems that provide cations and anions when dissolved in water. Electrolytes can be classified as either strong or weak. Strong electrolytes, such as NaCl, are completely ionized in water, whereas weak electrolytes, like acetic acid and phosphoric acid, are not fully ionized.

$$I = \frac{1}{2}\sum_{i=1}^{n} c_i (z_i)^2 \tag{5.19}$$

Here c_i is the concentration (in mol L^{-1}) of the i-th ion (either cation or anion), and z_i is the valence (i.e., the charge number) of the ion. The squared charges ensure that all contributions to the ionic strength are positive, regardless of whether the ions are cations or anions. Figure 5.20 shows the typical ion concentrations found in various natural water systems. It illustrates that hydrothermal waters, oceans, and brine have the highest saline concentrations, while rainwater and groundwater exhibit the lowest salt content. The varying ionic strengths observed in different water systems significantly influence the types of living organisms that can thrive within each specific environment. However, this topic is beyond the scope of the present book.

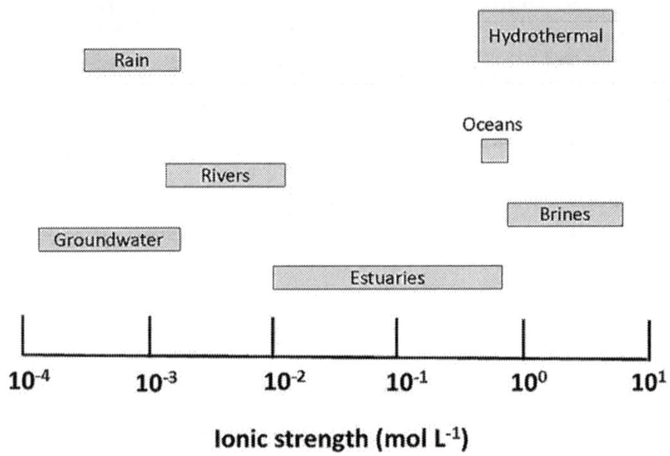

Figure 5.20 Ionic strengths of various naturally occurring water systems. The figure was obtained by evaluating the data reported here: https://www.epa.gov/caddis/ionic-strength.

5.13 THE DEBYE-HÜCKEL THEORY: THE RELATIONSHIP BETWEEN IONIC STRENGTH AND ACTIVITY COEFFICIENT

The relationships between the activity coefficient and the ionic strength were discovered by the Dutch chemist Peter Debye and the German physical chemist Erich Hückel in the first half of the 20th century. Their theory is based on the following assumptions:

1. **The electrolytes are strong:** They are completely dissociated into ions in solution.
2. **Ions are spherical and carry specific charges:** Both cations and anions are considered spheres having charges z_+e, and z_-e, respectively, where $z_\pm$ is the valence (charge number) of the ion (positive for cations and negative for anions), and e is the elementary charge ($e = 1.602 \times 10^{-19}$ C).

3. **The solvent is a continuous medium:** The solvent is viewed as a continuous medium characterized by its dielectric constant (ε_r)[48] and does not interfere with the electric fields produced by the dissolved ions.
4. **An ionic atmosphere surrounds each ion:** Each ion is surrounded by an ionic atmosphere that contains both cations and anions, with a net charge opposite to that of the central ion. Specifically, the ionic atmosphere around a cation predominantly contains anions and vice versa.
5. **Interactions are governed by Coulomb's law:** The force (F) of the interaction between cations and anions in solution is governed by Coulomb's law, which can be expressed as:

$$F = \frac{z_+ z_- e^+ e^-}{4\pi\epsilon_0 \varepsilon_r r^2} \tag{5.20}$$

where ε_r is the permittivity of the solvent[48], ε_0 is the vacuum permittivity, which is a constant with a value of 8.854×10^{-12} $C^2 \cdot kg^{-1} \cdot m^{-3} \cdot s^2$, and r is the distance between the ions.

The results of the Debye-Hückel theory can be expressed as in equation (5.21):

$$\log \gamma_i = \frac{-Az_i^2\sqrt{I}}{1 + Ba_i\sqrt{I}} \tag{5.21}$$

Here, γ_i is the activity coefficient of the i-th ionic species; A is a constant that depends on the solvent's dielectric constant and temperature (for water at 25°C, $A = 0.5085$ $mol^{-1/2}$ $L^{1/2}$); z_i is the valence (charge number) of the ionic species; I is the ionic strength; B is a corrective factor that is influenced by the temperature; a_i is the effective ionic radius of the i-th ionic species. Table 5.2 reports the values for the parameters A, and B at different temperatures.

Table 5.2 Values of the A and B parameters as affected by the temperature

Temperature			
°C	**K**	A **($mol^{-1/2}$ $L^{1/2}$)**	B **($\times 10^{-10}$ $L^{-1/2}$ $mol^{-1/2}$ m^{-1})**
0	273	0.4883	0.3241
5	278	0.4921	0.3249
10	283	0.4960	0.3258
15	288	0.5000	0.3262
20	293	0.5042	0.3273
25	298	0.5085	0.3281
30	303	0.5130	0.3290
35	308	0.5175	0.3297
40	313	0.5221	0.3306

[48] The dielectric constant also referred to as electric permittivity is a physical quantity that measures the tendency of a material to counteract the intensity of an electric field generated inside it. Therefore, in the case of water, the dielectric constant measures water ability to resist the electric field generated by the electric charges (positive and negative) dissolved within it.

Equation (5.21) also reveals that in extremely dilute conditions—specifically, when the concentration of the ionic species approaches zero (concentration → 0) — the ionic strength I likewise approaches zero ($I \to 0$). Under these circumstances, the term $Ba\sqrt{I}$ becomes negligibly small compared to 1, such that $1 \gg Ba_i\sqrt{I}$. Consequently, Equation (5.21) simplifies and can be approximated as shown in Equation (5.22):

$$\log \gamma_i = -Az_i^2\sqrt{I} = 0 \tag{5.22}$$

Therefore, it follows that under extremely diluted conditions, the activity coefficient γ_i approaches unity (i.e., γ_i = 1).

The extended Debye-Hückel equation is effective for calculating activity coefficients of ions in dilute electrolyte solutions, typically up to ionic strengths of about 0.1 mol L^{-1} when using molarity. However, as the ionic strength increases, ion-ion interactions become more significant, and the assumptions underlying the extended Debye-Hückel equation become less valid. Under these conditions, the Davies equation in (5.23) provides an empirical modification that extends the applicability to higher ionic strengths up to approximately 0.5 mol L^{-1}.

$$\log \gamma_i = -Az_i^2\left(\frac{\sqrt{I}}{1+\sqrt{I}} - 0.3I\right) \tag{5.23}$$

The Davies equation omits the ion size parameter, simplifying calculations when precise a_i values are unavailable. Moreover, it introduces an empirical term (i.e., $-0.3I$) to account for deviations at higher ionic strengths due to short-range ion-ion interactions not considered in the Debye–Hückel theory.

The extended Debye-Hückel and Davies equations are suitable for calculating activity coefficients in dilute to moderately concentrated electrolyte solutions, typically up to ionic strengths of about 0.5 mol L^{-1}. However, when dealing with solutions of higher ionic strength—where ion-ion interactions, ion pairing, and short-range forces become significant—the assumptions of these simpler models are no longer valid. Under such conditions, the Pitzer equations must be employed. They provide a more accurate and comprehensive framework for modeling the thermodynamic behavior of concentrated electrolyte solutions by incorporating both long-range electrostatic interactions and short-range specific ion effects.

One of the forms of the Pitzer model is reported in Equation (5.24):

$$\ln \gamma_\pm = z_+ z_- f_\gamma + \sum_i m_i\left(2\beta_{AX}^{(0)} + 2\beta_{AX}^{(1)} \exp\left(-\alpha\sqrt{I}\right)\right) + M_i M_j C_\psi \tag{5.24}$$

where:

— f_γ is the Debye-Hückel term given by: $f_\gamma = \frac{A\sqrt{I}}{1+b\sqrt{I}} + \frac{2}{b} ln\left(1+b\sqrt{I}\right)$, with A having the meaning previously described, and b is an adjustable parameter, often set to 1.2 $L^{1/2}$ $mol^{-1/2}$.

— $\beta_{MX}^{(0)}$ and $\beta_{MX}^{(1)}$ are empirical interaction parameters specific to the ion pair A^{z+} and X^{z-}. These parameters account for the short-range binary interactions between ions.

- α is a parameter typically set at 2.0 $L^{1/2}$ $mol^{-1/2}$, influencing the exponential term.
- C_ϕ is a ternary interaction parameter accounting for interactions involving three ions. This term is often small and sometimes neglected when dealing with dilute solutions.
- M_i, and M_j are the molar concentrations (mol L^{-1}) of the ions in solution.

5.14 SOLUBILITY AND SATURATION

The phenomena of solubility and saturation were introduced earlier, emphasizing that solubility is often linked to the interaction between solute particles and solvent molecules, such as water. When no more solvent is available to interact with the solute, the solute's dissolution ceases, forming a saturated solution and possible precipitation. However, explaining solubility solely in terms of the availability of solvent molecules is an oversimplification. To fully understand why different materials exhibit varying solubility and saturation points, we must consider additional parameters, such as temperature, pressure, and the nature of both the solute and the solvent.

Among these, temperature is one of the most straightforward factors influencing solubility. In general, increasing the temperature tends to increase solubility, particularly for endothermic dissolution processes. This is because higher temperatures provide additional kinetic energy, enhancing the movement of solute and solvent molecules, thus increasing the likelihood of interactions that promote dissolution. Furthermore, the increased thermal energy weakens the intermolecular forces within both the solute and the solvent, facilitating the breakdown of the solute's lattice structure and promoting solvation, the process in which solute particles are surrounded by solvent molecules. However, this trend is not universal. For some compounds, such as cesium sulfate ($Cs_2(SO_4)_3$)), solubility decreases with increasing temperature because the dissolution process is exothermic. In such cases, as the temperature rises, the equilibrium shifts toward the undissolved state, following Le Chatelier's principle, which explains how systems adjust to minimize changes in conditions.

To better illustrate this, consider the dissolution of cesium sulfate more precisely as in Equation (5.25)[49]:

$$Cs_2(SO_4)_3 \leftrightarrows 2Cs_{aq}^{2+} + 3(SO_4^{2-})_{aq} + \text{heat} \tag{5.25}$$

In this reaction, heat is a product of dissolution, meaning that raising the temperature drives the equilibrium toward the solid form of the salt, thus reducing its solubility.

While temperature is a significant factor, it is important to note that the relationship between temperature and solubility is not linear, and this is clearly observed in the solubility behavior of various salts (as shown in Figure 5.21). Sodium chloride (NaCl), for example, shows minimal variation in solubility

---

[49] The subscript *aq* in (5.25) means that the chemical species is hydrated.

across a range of temperatures, whereas salts like potassium chloride (KCl) and ammonium chloride (NH_4Cl) exhibit lower solubility than NaCl at lower temperatures but greater solubility as the temperature increases.

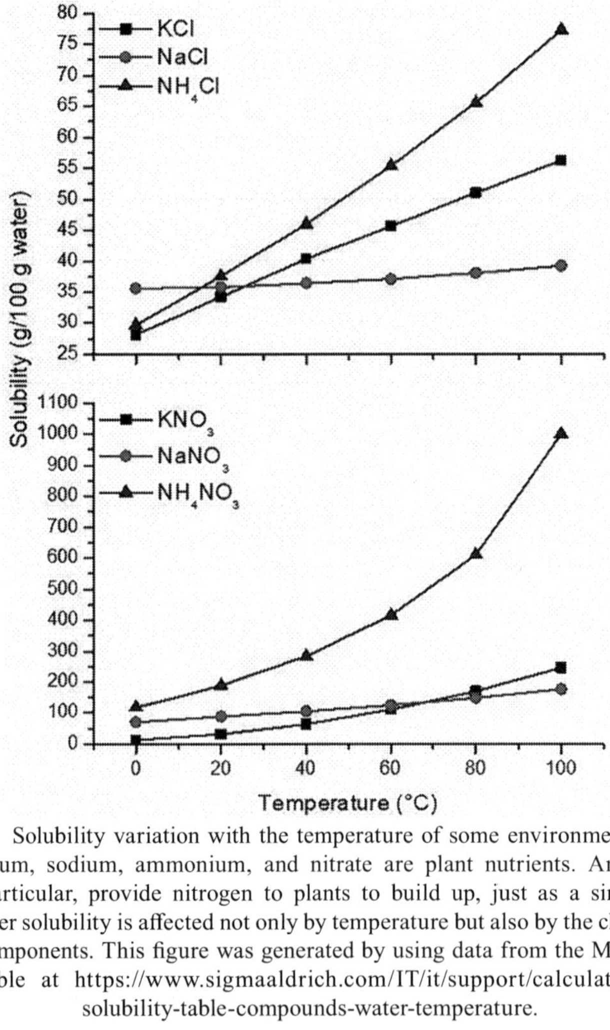

Figure 5.21 Solubility variation with the temperature of some environmentally relevant salts. Potassium, sodium, ammonium, and nitrate are plant nutrients. Ammonium and nitrate, in particular, provide nitrogen to plants to build up, just as a simple example, proteins. Water solubility is affected not only by temperature but also by the chemical nature of the salt components. This figure was generated by using data from the Merck solubility table available at https://www.sigmaaldrich.com/IT/it/support/calculators-and-apps/solubility-table-compounds-water-temperature.

These differences can be partially explained by examining the crystal structures of the salts. Sodium chloride and potassium chloride both crystallize in a face-centered cubic lattice, with ionic bond strengths inversely proportional to the distance between their ions. According to Coulomb's law, the stronger the ionic interactions, the lower the solubility, which explains why NaCl, with its shorter ionic distances, should theoretically be less soluble than KCl. However, at lower temperatures, sodium chloride is actually more soluble than both potassium and ammonium chloride. This is due to the partial covalent character present in the bonds of these salts, which becomes more significant when the ions have similar sizes. At lower temperatures, the covalent character predominates, reducing the solubility of potassium and ammonium chloride compared to sodium chloride. As the temperature increases, the ionic character begins to dominate, resulting in higher solubility for KCl and NH_4Cl relative to NaCl.

Similar trends can be observed in nitrate salts, where differences in crystal lattice structures and ionic interactions affect their solubility at different temperatures. For example, sodium nitrate ($NaNO_3$) forms a body-centered cubic lattice, while potassium nitrate (KNO_3) and ammonium nitrate (NH_4NO_3) adopt an orthorhombic lattice, each with distinct bond lengths and coordination geometries. These structural factors, along with the balance between lattice energy[50] and solvation energy[51], explain the observed solubility trends for these salts.

In summary, solubility is governed by a complex interplay of factors, including temperature, the nature of ionic and covalent interactions, lattice energy, and solvation energy. While temperature often enhances solubility, especially for endothermic dissolutions, exceptions arise when the dissolution process is exothermic, as seen with cesium sulfate. Additionally, structural characteristics such as crystal lattice geometry and bond strengths play critical roles in determining the solubility behavior of various salts.

5.15 THE PARAMETERS FOR A QUALITATIVE/ QUANTITATIVE EVALUATION OF SOLUBILITY

All the discussion above deals with the solubility of inorganic electrolytes in water. However, soil systems also contain organic matter, which is critical in defining soil quality, as will be detailed in Chapter 11. The nature of organic matter is vast, ranging from water-soluble electrolyte systems to materials that hardly solubilize in water. Nevertheless, all forms of organic matter can significantly impact soil quality. Therefore, understanding the mechanisms behind the solubility of natural organic matter (NOM)—the organic substances present in environmental compartments—is crucial to comprehending soil characteristics.

---

[50] Lattice energy refers to the amount of energy required to separate one mole of an ionic solid into its individual gaseous ions. It is a measure of the strength of the forces holding the ions together in a crystal lattice. The greater the lattice energy, the stronger the ionic bonds within the solid, and typically, the harder it is for the solid to dissolve in a solvent like water. In other words, a salt with a high lattice energy requires more energy to break apart its ionic bonds, making it less likely to dissolve. Conversely, a salt with lower lattice energy dissolves more easily because the bonds between its ions are weaker.

[51] Solvation energy (or hydration energy, when water is the solvent) refers to the energy released when solute particles, such as ions or molecules, are surrounded and stabilized by solvent molecules. In the case of an ionic compound dissolving in water, solvation energy is the energy released as water molecules surround and interact with the individual ions. Solvation energy is important because it competes with lattice energy. For a salt to dissolve, the solvation energy must be strong enough to overcome the lattice energy that holds the ions together in the solid form. If the solvation energy exceeds the lattice energy, the salt dissolves more easily.

In this context, the Hildebrand solubility parameter (δ) offers a valuable tool for predicting the solubility behavior of complex systems such as NOM in soil solutions[52]. While its original use was focused on synthetic polymers, its application in soil science may help elucidate how organic matter dissolves or interacts with other components in the soil environment. By considering the solubility of NOM in relation to soil solution, researchers can gain insights into processes such as nutrient cycling, pollutant transport, and overall soil fertility. This is essential because organic matter can range from highly soluble fulvic acids, which are integral to nutrient availability, to more complex humic substances that are less soluble but crucial for soil structure and long-term carbon storage. In addition, since organic matter's solubility affects its sorption-desorption behavior in soil, the solubility parameter can be used to model how NOM may influence the mobility of nutrients and contaminants. Understanding the interactions between NOM and the aqueous phase or mineral surfaces can offer predictions about whether organic compounds will remain mobile or become adsorbed, thus altering their bioavailability and environmental impact.

From a chemical-physical point of view, the Hildebrand parameter is linked to the cohesive energy density (CED)[53] of a material, which describes the energy required to separate molecules from each other. Specifically, the Hildebrand solubility parameter is defined in Equation (5.26):

$$\delta = \sqrt{\frac{\Delta H_v - RT}{V_m}} \tag{5.26}$$

Here, ΔH_v is the energy of vaporization (the energy required to vaporize one mole of the material under investigation), RT accounts for the temperature correction, where R is the universal gas constant (8.314 J mol^{-1} K), T is the absolute temperature, and V_m is the molar volume (that is, the volume occupied by one mole of the studied material).

As stated above, this parameter provides insight into the intermolecular forces within a substance, which are the main contributors to the cohesive energy in any system. In simpler terms, the Hildebrand parameter quantifies the energy needed to overcome the molecular attractions within a material, thereby offering an estimate of how easily a substance will dissolve in a given solvent.

---

[52] Zalamena, G., et al. 2023. The Hildebrand solubility parameter and its importance in the scientific and technological scenario of flow assurance operations. Petroleum Chemistry. 63(8): 886–89.
– Velasco, P.Q., et al. 2019. The application of the surface energy-based solubility parameter theory for the rational design of polymer-functionalized MWCNTs. Physical Chemistry Chemical Physics. 21: 5331.

[53] Cohesive energy density (CED) is a measure of the amount of energy required to separate the molecules of a substance from each other, per unit volume. It quantifies the strength of the intermolecular forces holding the substance together, such as van der Waals forces, hydrogen bonds, and dipole-dipole interactions. The higher the CED, the stronger the attractions between molecules, and thus, more energy is needed to break them apart. This concept is key in understanding solubility, as substances with similar CED values tend to dissolve in each other ("like dissolves like").

One of the fundamental principles in solubility theory is the concept of "like dissolves like," meaning that substances with similar solubility parameters are more likely to dissolve in each other. For example, non-polar solvents with low Hildebrand parameters are more effective at dissolving non-polar solutes, while polar solvents with higher δ values are better suited for dissolving polar solutes.

When the solubility parameter of a solvent closely matches that of a solute, the energy required to disrupt the solute-solute interactions is nearly the same as that needed to introduce the solute into the solvent. In such cases, the substance will dissolve easily. Conversely, when there is a large disparity between the solubility parameters, the energy cost to dissolve the substance becomes high, leading to poor solubility.

Table 5.3 Values of the Hildebrand parameter for some systems

Sample	Hildebrand parameter (δ, MPa$^{1/2}$)
n-pentane	14.4
Ethyl acetate	18.2
Acetone	19.7
Ethanol	26.2
Methanol	29.7
Glycerol	36.2
Water	48.0

Table 5.3 reports the Hildebrand parameter for some organic and inorganic systems. While it can explain the solubility of non-polar substances like n-pentane in ethyl acetate because of the similarity between their δ values, it cannot account for the high solubility of polar compounds like acetone and ethyl alcohol in water, as seen experimentally. In fact, the Hildebrand parameters of these two systems are quite different from each other (19.7 and 48.0 MPa$^{1/2}$, respectively). This is because the Hildebrand parameter primarily applies to non-polar and weakly polar systems, where van der Waals forces dominate. It does not consider specific interactions like hydrogen bonding or ionic interactions, which are crucial for the solubility of polar compounds and typical among the polar components of natural organic matter. To address these limitations, more sophisticated models, such as the Hansen Solubility Parameters (HSP), were developed.

The Hansen Solubility Parameters, developed by Charles Hansen in the 1960s, breaks solubility into three distinct components, providing a more nuanced and accurate solubility prediction across a wide range of materials, including solvents, polymers, and natural organic matter. The three HSP components are:

1. Dispersion forces (δ_D). These forces, also known as London dispersion forces, arise from the interaction of temporary dipoles induced in molecules. They dominate non-polar systems and represent the basic van der Waals forces responsible for cohesion in materials like hydrocarbons.
2. Polar forces (δ_P). They refer to the interactions between molecules that possess permanent dipoles. These forces are much stronger than dispersion forces and are characteristic of molecules with asymmetrical charge distributions, such as acetone or ethyl acetate.

3. Hydrogen bonding (δ_H). A deep analysis of hydrogen bonding has already been done in the previous paragraphs. However, very briefly, H-bond refers to the specific attractive force between a hydrogen atom bonded to a highly electronegative atom (such as oxygen or nitrogen) and another electronegative atom in a nearby molecule. This interaction is particularly strong and plays a crucial role in the solubility of compounds like water, alcohols, and amines.

Together, these three components create a more detailed "solubility space," allowing a better understanding of how different materials will interact. The total solubility parameter can be represented as:

$$\delta^2 = \delta_D^2 + \delta_P^2 + \delta_H^2 \tag{5.27}$$

By treating solubility as a balance between these forces, the HSP model allows for more precise predictions, particularly in cases where hydrogen bonding or polarity significantly affects solubility, such as in biological systems or with complex natural organic matter.

The HSP model introduces the concept of solubility distance (Ra), which measures the "distance" between two substances (typically a solute and a solvent). A smaller R_a value indicates that the two substances are more likely to dissolve in each other, as their combined dispersion, polar, and hydrogen bonding parameters are similar. This "distance" is calculated as with *Sol* and *Solv* standing for solute and solvent, respectively:

$$R_a = \sqrt{4\left(\delta_D^{Sol} - \delta_D^{Solv}\right)^2 + \left(\delta_P^{Sol} - \delta_P^{Solv}\right)^2 + \left(\delta_H^{Sol} - \delta_H^{Solv}\right)^2}$$

The factor of 4 applied to the dispersion term reflects its relatively lower importance compared to polar and hydrogen bonding interactions, which play a larger role in many solubility phenomena.

While HSP is a powerful model, it is not without limitations. The model assumes that solubility is a linear combination of the three components, which may not always hold in complex systems. It does not directly account for ionic interactions, which are important in systems involving salts or strong electrolytes.

5.16 HYDROPHILICITY AND HYDROPHOBICITY

Hydrophilicity refers to the ability of a substance to attract and interact with water. Hydrophilic materials tend to dissolve or absorb water because their molecular structures allow them to bond with water easily. These materials are usually either polar or charged, meaning they have regions with partial positive or negative charges. Since water itself is polar, it interacts well with other polar substances. For instance, molecules with hydroxyl (–OH) or carboxyl (–COOH) groups can form hydrogen bonds with water, making them hydrophilic.

Charged molecules like salts and acids are also highly hydrophilic. When they come into contact with water, their ions separate and interact with the water molecules, forming a kind of shell around them called a hydration shell (see above). This interaction helps stabilize the system by reducing the overall electrostatic energy.

Hydrophilicity can also be defined for solid systems when their surfaces interact with water molecules. Let us consider a drop of water made by a hundred thousand molecules. The H-bonds among the different water molecules define the so-called cohesion forces. As the water molecules are placed on a solid surface, they tend to interact with the functional groups of the surface through interactions (e.g., H-bonds and Van der Waals interactions), which define the so-called adhesive forces. When adhesive forces predominate over the cohesive ones, the solid surface is hydrophilic in nature. One way to measure how hydrophilic a surface is involves looking at the contact angle formed when a water droplet lands on it. A small contact angle, typically less than 90°, means that the water spreads out over the surface, indicating strong hydrophilicity (Figure 5.22A). Glass is a great example of a hydrophilic material since water droplets tend to spread out completely when they come into contact with it.

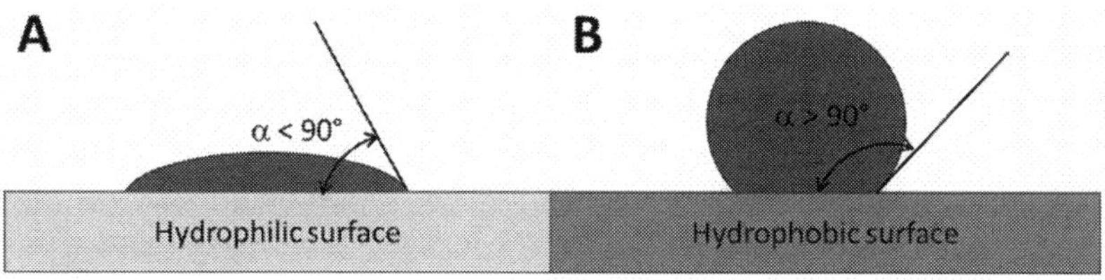

Figure 5.22 The contact angle (α) between a drop of water and a solid surface indicates the wettability of the surface. If the adhesive forces (i.e., the interactions between water molecules and the solid surface) dominate over the cohesive forces (i.e., the interactions among water molecules, such as hydrogen bonds), the surface is hydrophilic (**A**). Conversely, when the cohesive forces dominate over the adhesive ones, the surface is hydrophobic (**B**).

Hydrophobicity, in contrast, describes a material's tendency to repel water. Hydrophobic substances do not dissolve in water and resist forming a hydration shell. Instead, they often clump together in water, forming structures like micelles, where the water-averse parts are hidden from the surrounding liquid.

Hydrophobic substances are typically non-polar, meaning they do not have the kind of charge separation seen in polar molecules. Because of this, water molecules prefer to interact with each other rather than with hydrophobic material. This leads to the minimization of contact between water and the non-polar molecules, a phenomenon known as the hydrophobic effect.

The hydrophobic effect is driven by thermodynamics. When non-polar substances are introduced into water, the water molecules form a cage-like structure around them, which is energetically unfavorable. To reduce this energy and increase entropy, the system forces the non-polar molecules to group together, minimizing the area exposed to water. This effect drives the formation of biological structures like lipid bilayers in cell membranes and the self-assembly of micelles in detergents.

In the case of interactions between water and solid surfaces, hydrophobicity occurs when the cohesion forces predominate over the adhesive ones. Hydrophobicity is measured using the same contact angle method as hydrophilicity, but this time, a large contact angle (greater than 90°) indicates hydrophobicity (Figure 5.22B). Materials like Teflon®, for example, have contact angles close to 120°, showing that water beads up and rolls off rather than spreading out. Some surfaces, known as superhydrophobic, can have contact angles greater than 150°, making them extremely water-repellent and useful for applications like self-cleaning surfaces.

The contact angle described above also indicates the wettability (W) of a solid surface. A hydrophilic surface is more "wettable" than a hydrophobic one. Therefore, a water droplet spreads more on a hydrophilic surface than on a hydrophobic surface (Figure 5.22).

EXERCISES

1. Write the electronic configuration of oxygen.
2. What is meant by "hybrid orbital"?
3. Draw the Lewis structure of water.
4. What are hydrogen bonds, and why are they important for the properties of water?
5. Explain why ice floats on water.
6. What is the dielectric constant, and how does it affect the solubility of salts in water?
7. Describe the three fundamental steps of the dissolution process of a solute in a solvent.
8. What is the activity of a chemical species in solution?
9. Define ionic strength and discuss its importance in natural systems.
10. Explain the Debye-Hückel theory and its use in calculating ion activity coefficients in solution.

FURTHER READING

Ball, P. 1999. H_2O, A Biography of Water, 1st Ed. Weidenfeld and Nicolson. ISBN: 9780753810927.

Chandler, D. 2005. Interfaces and the driving force of hydrophobic assembly. Nature 437(7059): 640–647.

Desiraju, G.R. and T. Steiner. 1999. The Weak Hydrogen Bond in Structural Chemistry and Biology, 1st Ed. Oxford University Press. ISBN: 9780198509707.

Essington, M.E. 2003. Soil and Water Chemistry. An Integrative Approach, 1st Ed. CRC Press. ISBN: 9780429209697.

Franks, F. 2000. Water, A Matrix of Life, 2nd Ed. RSC. ISBN: 9780854045839.

Hobbs, P.V. 1974. Ice Physics, 1st Ed. Oxford University Press. ISBN: 9780199587711.

Israelachvili, J.N. 2011. Intermolecular and Surface Forces, 3rd Ed. Academic Press. ISBN: 9780123919274.

Levine, I.N. 2014. Quantum Chemistry, 7th Ed. Pearson Education, Inc. ISBN: 978-0321803450

Maréchal, Y. 2007. The Hydrogen Bond and the Water Molecule, 1st Ed. Elsevier. ISBN: 978-0444559302.

Mittal, K.L. (ed.). 2009. Contact Angle, Wettability and Adhesion. VSP. ISBN: 978-0429088384.

McQuarrie, D.A. 2008. Quantum Chemistry, 2nd Ed. University Science Books, Sausalito CA (USA). ISBN: 9781891389504.

Pauling, L. and E.B. Wilson. 1985. Introduction to Quantum Mechanics with Applications to Chemistry, 1st Ed. Dover Publications Inc., NY (USA). ISBN: 9780486648712.

Tanford, C. 1980. The Hydrophobic Effect: Formation of Micelles and Biological Membranes, 2nd Ed. John Wiley and Sons. ISBN: 9780471048930.

Zhang, P. and G. Li. 2019. Superhydrophobic surfaces: A review on fundamentals, applications, and challenges. Journal of Applied Physics. 125(1): 011301.

Chapter **6**

The Chemical Equilibria in Ideal Conditions

6.1 INTRODUCTION

Chemical equilibria in aqueous solutions form the foundation of many processes, among which natural systems play the major role. Understanding these equilibria is essential for comprehending the complexities of environmental science, biological interactions, and numerous industrial applications, as well. At its core, the concept of chemical equilibrium refers to a state where the rates of the forward and reverse reactions in a system are equal, leading to constant concentrations of the reactants and products over time. However, this seemingly static state belies the dynamic nature of molecular interactions occurring continuously at the microscopic level. The principles governing these equilibria in aqueous solutions are deeply influenced by the unique properties of water, as discussed in Chapter 5, where the nature and chemistry of water were extensively explored.

The remarkable properties of water, such as its high dielectric constant, hydrogen bonding capabilities, and dual role as both a solvent and a reactant profoundly impact on chemical equilibria in aqueous environments. Water's role is not limited to merely dissolving substances; it also actively participates in chemical reactions through processes like hydrolysis and acid-base interactions, significantly affecting the distribution of ions in solution. The dissociation of water itself lays the groundwork for understanding *pH*, a critical parameter that dictates the acidic or basic nature of aqueous systems. This understanding is fundamental when analyzing the behavior of acids, bases, and their conjugate pairs under ideal conditions, which forms the primary focus of this chapter.

In aqueous solutions, the equilibrium position can shift in response to changes in concentration, pressure, temperature, and the presence of other solutes. These shifts are guided by Le Chatelier's principle, which predicts how a system at equilibrium responds to external disturbances to re-establish a new state of

DOI: 10.1201/9781003780212-6

balance. For instance, adding acids or bases to an aqueous solution directly influences the hydrogen ion concentration, thereby altering the equilibrium states of weak acids and bases. Understanding these principles is critical for ideal systems, where simplifying assumptions about solute-solvent interactions allow us to predict and manipulate these shifts with mathematical precision.

The study of chemical equilibria in ideal conditions also extends to redox reactions, where electron transfer processes govern the oxidation state of chemical species in solution. In this context, ideal conditions assume negligible interactions between ions, allowing for straightforward calculations of redox potentials using the Nernst equation. These concepts are crucial in many laboratory and industrial processes where controlled environments are maintained to ensure predictable outcomes.

While this chapter will focus on the principles of chemical equilibria under ideal conditions, it is essential to note that real-world aqueous systems often deviate from these assumptions. Factors such as ionic strength, complexation, and the presence of competing ions lead to non-ideal behavior, which significantly influences the chemical dynamics of solutions in natural and practical settings. These deviations from ideality are particularly relevant in soil and environmental chemistry, where interactions between ions and soil particles create complex equilibria.

However, the detailed exploration of these non-ideal conditions and their impact on soil solutions will be reserved for Chapter 7, where the intricate behavior of chemical species in soils will be examined. There will be a discussion on how factors like ion exchange, adsorption, and variations in soil composition influence the chemical equilibria in these non-ideal systems. This distinction allows Chapter 6 to provide a clear and focused analysis of ideal conditions, laying a solid foundation before delving into the more complex real-world scenarios presented in the subsequent chapter.

This chapter builds upon the concepts introduced in Chapter 5 by exploring the theoretical framework of chemical equilibria in aqueous solutions under ideal conditions. The focus will be on examining the mathematical models that describe these equilibria, such as equilibrium constants, and investigating their practical applications in controlled environments. The objective is to gain a comprehensive understanding of the behavior of acids, bases, and their conjugate pairs in ideal systems, thereby laying the groundwork for a deeper exploration of non-ideal conditions in the subsequent chapter.

6.2 A GENERAL OVERVIEW OF THE CHEMICAL EQUILIBRIA IN IDEAL CONDITIONS

According to Note 40 in Chapter 5, a solution can be considered ideal when the following conditions are met:

1. The solute can be treated as point-like, meaning the size and volume of the solute molecules are negligible compared to those of the solvent

molecules. This assumption simplifies calculations by ignoring the physical dimensions of the solute.

2. Solute-solute interactions are negligible, implying that interactions among solute molecules do not significantly affect the solution's behavior. This assumption is valid in cases of extreme dilution, where the solute molecules are sufficiently separated.
3. Solute-solvent interactions are homogeneous, meaning these interactions are uniform in all directions throughout the solution. This ensures an even distribution of solute molecules in the solvent and justifies using simple models to describe the system.

Under these conditions, which approximate extreme dilution, a generic equilibrium, such as the one described in equation (6.1), is governed by the equilibrium expression outlined in relationship (6.2):

$$aA + bB \leftrightarrows cC + dD \tag{6.1}$$

$$K_{eq}^{T} = \frac{[C]^c [D]^d}{[A]^a [B]^b} \tag{6.2}$$

The relationship in (6.2) represents the equilibrium constant expression, which reflects the equilibrium condition of the reaction. While it is related to Le Chatelier's principle, it is not itself the mathematical expression of the principle. Le Chatelier's principle states that if a system at equilibrium experiences a change in concentration, pressure, or temperature, the system will adjust to counteract the change and re-establish equilibrium. For instance, if the concentration of reactants (A or B) increases, the system will shift towards producing more products (C and D) to restore equilibrium. Conversely, an increase in product concentrations will drive the system toward forming more reactants. The symbol K_{eq}^{T} refers to the equilibrium constant, where the subscript *eq* stands for "equilibrium," and the superscript *T* indicates that the value of K_{eq}^{T} is temperature dependent. Since the equilibrium constant is sensitive to temperature changes, altering the temperature will shift the balance between reactants and products, influencing both the forward and reverse reaction rates.

To illustrate how temperature affects equilibrium constants, consider the dissociation of water:

$$2H_2O \leftrightarrows H_3O^+ + OH^- \tag{6.3}$$

The equilibrium constant for this reaction, known as the ionic product of water, is given by:

$$K_w = [H_3O^+][OH^-] \tag{6.4}$$

Figure 6.1 illustrates how the K_w values change as a function of temperature (squares and solid line). The K_w value at 0°C is 1.14×10^{-15} M^2. As the temperature increases, the dissociation constant also increases, indicating that the equilibrium progressively shifts toward the formation of more H_3O^+, and OH^-.

Under conditions of electroneutrality, where $[H_3O^+] = [OH^-]$, relationship (6.5) follows:

$$K_w = [H_3O^+]^2 \tag{6.5}$$

From (6.5), it is clear that the *pH* decreases as the temperature increases, as shown by the circles and the dotted line in Figure 6.1 (for a better understanding of the *pH* meaning, the reader is addressed to the paragraphs below). However, this does not mean that water becomes more acidic at higher temperatures. Rather, the shift in equilibrium due to increased temperature reduces the *pH* while maintaining electroneutrality. Therefore, even though the *pH* drops, the solution remains neutral because the concentrations of H_3O^+, and OH^- are equal at all temperatures.

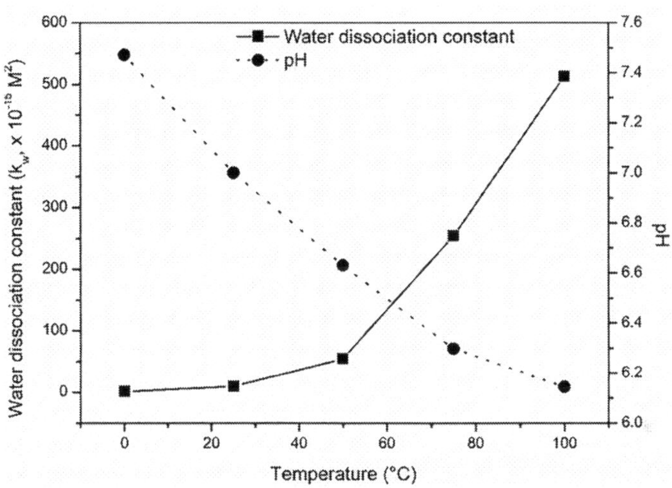

Figure 6.1 Variation of the water dissociation constant and pH at different temperatures[1].

Qualitatively, the formation of more H_3O^+, and OH^- with rising temperature can be explained by the fact that higher temperatures generally provide more kinetic energy to the molecules involved in a chemical system. In the case of water dissociation, increasing temperature affects the equilibrium by supplying additional energy, which favors the endothermic direction of the reaction. Water dissociation is an endothermic process, meaning it absorbs heat. According to Le Chatelier's principle, when the temperature of the system is increased, the equilibrium will shift in the direction that absorbs the added heat to counteract the change. In this case, the dissociation of water into H_3O^+, and OH^- absorbs heat, so the equilibrium shifts to the right (toward more dissociation), producing higher concentrations of both ions. This increase in dissociation means that more water molecules break apart to form hydronium and hydroxide ions as the temperature rises, hence explaining why their concentrations increase with temperature. This

[1]The K_W, and pH values used for this Figure have been taken from Silvestroni, p. 1999. Fondamenti di Chimica. CEA, 10th Ed. (in Italian) ISBN: 9788808084019

shift results in a lower pH, even though the solution remains neutral because the concentrations of both ions remain equal.

From all that has been written above, it comes that for endothermic reactions in aqueous solutions, increasing temperature will generally increase the value of the equilibrium constant, shifting the equilibrium toward products. Increasing temperature usually decreases the equilibrium constant for exothermic reactions, favoring the reactants.

6.3 THE *pH*

pH is defined as the decimal logarithm of the inverse of the hydrogen ion concentration. In other words:

$$pH = \log_{10}\frac{1}{[H_3O^+]} = -\log_{10}[H_3O^+] \tag{6.6}$$

Similarly, the *pOH*, which represents the concentration of hydroxide ions in solution, is given by:

$$pOH = \log_{10}\frac{1}{[OH^-]} = -\log_{10}[OH^-] \tag{6.7}$$

Being the relationship between *pH* and *pOH* governed by the dissociation of water, the corresponding relationship is expressed as:

$$pK_w = pH + pOH \tag{6.8}$$

At 25°C, where the ionic product of water, K_w, equals 1×10^{-14} M^2, the following holds true:

$$pK_w = 14 = pH + pOH \tag{6.9}$$

Under standard ambient conditions[2], a neutral solution is characterized by equal concentrations of hydrogen and hydroxide ions, resulting in $pH = pOH = 7$. Hence, at 25°C, $pH < 7$, and $pOH > 7$ indicate acidic conditions, whereas $pH > 7$, and $pOH < 7$ reflect basic or alkaline conditions.

It is essential to note that the *pH* and *pOH* values corresponding to neutrality are highly dependent on temperature. As previously discussed, the neutral *pH* at higher temperatures deviates from 7. For instance, a *pH* of 6.1 at 100°C does not imply an acidic solution. Instead, it reflects an increase in the concentration of hydrogen ions due to the temperature-dependent shift in the dissociation of water. The neutrality condition remains intact as long as the concentrations of H_3O^+, and OH^- are equal despite the lower *pH* at elevated temperatures.

[2] Standard ambient conditions correspond to a temperature of 25°C, and an atmospheric pressure of 1 bar = 0.986 atm = 100 kPa.

6.4 THE ACID-BASE EQUILIBRIA IN IDEAL CONDITIONS

Acid-base equilibria describe the balance between acids and bases and their corresponding dissociated forms in aqueous solutions. These equilibria are pivotal to understanding many processes, including those occurring in environmental systems, biological functions, and chemical manufacturing. In the context of ideal solutions, acid-base equilibria are governed by simplified thermodynamic principles that assume minimal interactions between solute molecules and a uniform distribution of these molecules throughout the solvent. These assumptions help make calculations more straightforward, especially for highly diluted solutions where complex intermolecular forces are negligible.

As discussed in the previous paragraph, acid-base equilibria are governed by equilibrium constants, just like all chemical equilibria. These constants also provide information about the strength of acids and bases, specifically indicating the extent of dissociation in aqueous solutions.

For acids, the equilibrium constant is known as the acid dissociation constant (K_a), and for bases, it is the base dissociation constant (K_b).

Consider a generic dissociation reaction for monoprotic acid:

$$HA + H_2O \leftrightarrows A^- + H_3O^+ \tag{6.10}$$

The equilibrium expression for this reaction is:

$$K_a = \frac{[A^-][H_3O^+]}{[HA]} \tag{6.11}$$

The larger the K_a, the stronger the acid, as it indicates a greater tendency for the acid to dissociate. Conversely, a smaller K_a represents a weaker acid that dissociates to a lesser extent.

For bases, the equilibrium is described similarly, where a base B reacts with water to produce hydroxide ions:

$$B + H_2O \leftrightarrows HB + OH^- \tag{6.12}$$

The base dissociation constant (K_b) follows:

$$K_b = \frac{[HB][OH^-]}{[B]} \tag{6.13}$$

Like K_a, the magnitude of K_b indicates the strength of the base. Strong bases have large K_b values, indicating that they readily accept protons and form hydroxide ions in solution.

One of the most important relationships in acid-base chemistry is between K_a, K_b, and the ionic product of water, which is constant at a given temperature. For any conjugate acid base pair, the product of their dissociation constants equals:

$$K_w = K_a \times K_b \tag{6.14}$$

The relationship (6.14) is quite useful because it allows the calculation of K_b once K_a is known, and vice versa, it permits the retrieval of K_a if K_b is known.

All have been written above for a monoprotic acid (HA), and a monoprotic base (B) is valid for polyprotic systems as well.

Polyprotic acids undergo stepwise dissociation in aqueous solutions, where each molecule can release more than one proton in successive stages. Unlike monoprotic acids, which undergo a single dissociation step, polyprotic acids dissociate in multiple steps, each with its equilibrium constant. Each dissociation occurs sequentially, and the acid becomes weaker with each successive proton release as the remaining conjugate base becomes less likely to give up another proton. Common examples of polyprotic acids include sulfuric acid (H_2SO_4), phosphoric acid (H_3PO_4), and carbonic acid (H_2CO_3).

The general dissociation process of a polyprotic acid, H_nA, can be represented by the following steps:

1. The first dissociation, where the acid loses its first proton, is characterized by an acid dissociation constant K_{a1}:

$$H_nA + \mathrm{H_2O} \leftrightharpoons H_{n-1}A + \mathrm{H_3O^+}$$

2. The second dissociation, losing the second proton, is governed by K_{a2}:

$$H_{n-1}A + \mathrm{H_2O} \leftrightharpoons H_{n-2}A + \mathrm{H_3O^+}$$

3. For subsequent dissociations, similar expressions apply, with decreasing dissociation constants, K_{a3}, K_{a4}, ..., K_{an}

Each dissociation constant reflects the strength of the acid at that specific dissociation step. Typically, the first dissociation is the strongest, and each successive proton is removed with increasing difficulty, leading to smaller equilibrium constants for the second and subsequent dissociations ($K_{a1} > K_{a2} > \cdots > K_{an}$). This trend is due to the increased negative charge on the conjugate base formed after each dissociation, which makes it harder to lose additional protons. The same considerations apply also to polyprotic bases.

Understanding the stepwise dissociation of polyprotic acids (or bases) is crucial for calculating pH and predicting the concentrations of different protonated species in solution at equilibrium. Multiple dissociation steps also affect buffer capacity, as polyprotic acids (or bases) can resist changes in pH over a broader range of pH values, particularly near each dissociation point. However, the effect of temperature on each dissociation step must also be considered, as described previously in relation to how dissociation constants shift with temperature.

6.5 THE BUFFER CAPACITY

Buffer capacity is a fundamental concept in acid-base chemistry, particularly in systems that require stability in pH despite the addition of small amounts of acid or base. A buffer is a solution that contains an equal amount of a weak acid and its conjugate base (or an equal amount of a weak base and its conjugate acid) and resists changes in *pH* when small amounts of hydronium ions (H_3O^+) or hydroxide ions (OH^-) are added. Buffer capacity refers to the amount of acid or base that a buffer solution can neutralize before a significant change in *pH* occurs. In ideal conditions, where interactions are minimized, and the solution behaves as a perfect mixture, buffer capacity is governed by the equilibrium constants of the acid-base pair and the concentrations of the buffering components.

A buffer solution works because of the equilibrium between a weak acid and its conjugate base or a weak base and its conjugate acid. An example is given in the chemical equations (6.10) and (6.12), the first referring to a weak acid and the second to a weak base. By transforming the relationship in (6.11) by applying the base 10 logarithm, it comes:

$$\log_{10} K_a = \log_{10}\left(\frac{[A^-][H_3O^+]}{[HA]}\right)$$

$$\log_{10} K_a = \log_{10}\left(\frac{[A^-]}{[HA]}\right) + \log_{10}([H_3O^+])$$

This yields:

$$-\log_{10}\left([H_3O^+]\right) = \log_{10}\left(\frac{[A^-]}{[HA]}\right) - \log_{10} K_a$$

From which the Henderson-Hasselbalch equation is derived:

$$pH = pK_a + \log_{10}\left(\frac{[A^-]}{[HA]}\right) \tag{6.15}$$

The Henderson-Hasselbalch equation measures the effectiveness of a buffer solution. When $[A^-] = [HA]$, the logarithmic term becomes zero, and the *pH* equals the pK_a:

$$pH = pK_a \tag{6.16}$$

In ideal conditions, where interactions between solute molecules are minimal and the solution behaves uniformly, the Henderson-Hasselbalch equation accurately predicts the *pH* of a buffer solution based on the ratio of the conjugate base to the weak acid. The closer the *pH* is to the pK_a of the weak acid, the buffer will be more effective at resisting changes in *pH*. This is because at a *pH* near the pK_a, the concentrations of the weak acid and its conjugate base are nearly equal, providing maximum capacity for neutralizing added acid or base.

The buffer capacity is highest when the *pH* equals the pK_a of the weak acid because the system can absorb both H_3O^+ and OH^- equally well. As the *pH* moves away from the pK_a, either too acidic or too basic, the buffer capacity decreases. This is because one component of the buffer (either the weak acid or the conjugate base) becomes too depleted to effectively neutralize either additional H_3O^+ or OH^-.

For buffer solutions made by equal amounts of a weak base and its conjugate acid, the Henderson-Hasselbalch equation becomes:

$$pOH = pK_b + \log_{10}\left(\frac{[\text{HA}]}{[\text{A}^-]}\right) \tag{6.17}$$

Thus, the maximum buffer capacity is achieved when:

$$pOH = pK_b \tag{6.18}$$

The primary factor affecting buffer capacity is the concentration of the buffer components. In ideal conditions, where the solution is dilute and solute-solvent interactions are negligible, the concentration of the weak acid and its conjugate base (or weak base and its conjugate acid) directly determines the buffer's capacity. A buffer with higher concentrations of the weak acid and its conjugate base (and vice versa) will have a greater ability to neutralize added acids and bases than a buffer with lower concentrations.

Another critical factor is the proximity of the *pH* to the pK_a (or the *pOH* to the pK_b). Buffers are most effective when the *pH* of the solution is close to the pK_a of the acid (or pK_b of the base). At this point, the concentrations of the weak acid and conjugate base are roughly equal, allowing the buffer to neutralize both acids and bases effectively. If the *pH* drifts too far from the pK_a, the buffer loses its effectiveness because the balance between the acid and base components shifts significantly in one direction, leaving the solution unable to neutralize either additional acid or base efficiently.

Buffer capacity (β) can be quantified by the amount of strong acid or base required to change the *pH* of 1 liter of buffer solution by 1 *pH* unit. Mathematically, buffer capacity is defined as:

$$\beta = \frac{\text{moles of acid or base added}}{\Delta pH} \tag{6.19}$$

In ideal conditions, where the solution behaves predictably according to the laws of chemical equilibria, buffer capacity is proportional to the concentrations of the buffering components and inversely related to the change in *pH*.

The effect of temperature on buffer capacity is also worth considering. Temperature affects the dissociation constants (K_a or K_b) of the weak acid or base, shifting the equilibrium and, consequently, the *pH* of the buffer solution. Higher temperatures typically increase the dissociation of weak acids and bases, thereby altering the balance between the acid and its conjugate base. As discussed earlier, the dissociation constants are temperature-dependent, influencing the buffer's *pH* and capacity to resist *pH* changes. In practical terms, while ideal conditions often assume temperature stability, real-world applications must account for the temperature sensitivity of buffers, particularly in natural systems.

6.6 THE SOLUBILITY EQUILIBRIA

Solubility equilibria, describing the balance between the dissolved ions of a solute and the undissolved solid phase in a saturated solution, play a crucial role in naturally occurring systems, particularly in the interactions between water and soil minerals. These equilibria govern the dissolution and precipitation of minerals, as well as the availability of essential nutrients and contaminants in soils and aquatic environments. In natural systems, factors such as pH, temperature, and the presence of other ions significantly influence solubility equilibria, and understanding these processes is essential for predicting the chemical behavior of soils and the surrounding environment. The solubility of minerals in soil and water impacts everything from plant nutrient availability to the mobility of potentially harmful elements, making it a key aspect of environmental chemistry and soil science.

In soil systems, many minerals dissolve in water to form ions, establishing a balance between the solid mineral phase and the dissolved ions in solution. This balance is known as solubility equilibrium. For example, when calcium carbonate ($CaCO_3$) dissolves in soil water, it dissociates into calcium and carbonate ions:

$$CaCO_{3(s)} \rightleftarrows Ca^{2+}_{(aq)} + CO^{2-}_{3(aq)} \tag{6.20}$$

At equilibrium, the rate at which the solid $CaCO_3$ dissolves equals the rate at Ca^{2+} and CO_3^{2-} recombine to form the solid mineral. This dynamic equilibrium is governed by the solubility product (*sp*) constant (K_{sp}) for calcium carbonate, which is expressed as:

$$K_{sp} = [Ca^{2+}][CO_3^{2-}] \tag{6.21}$$

K_{sp} value determines how much of the mineral can dissolve in the soil water under given conditions, influencing the concentration of calcium and carbonate ions available in the solution. In natural systems, it is affected by environmental factors such as pH, temperature, and the presence of other dissolved species.

In soil environments, the concepts of saturation and supersaturation are crucial for understanding the behavior of minerals. A saturated solution in soil water contains the maximum concentration of dissolved ions that the system can hold at equilibrium. For example, a saturated solution of calcium carbonate in soil water contains as much Ca^{2+} and CO_3^{2-} as can dissolve, and any additional solid $CaCO_3$ will remain undissolved.

A supersaturated solution contains more dissolved ions than expected at equilibrium. Supersaturation can occur when changes in environmental conditions, such as temperature or evaporation, temporarily increase the concentration of ions in the solution. In a supersaturated state, excess dissolved ions may precipitate out of the solution, forming solid minerals until equilibrium is restored. For example, in arid regions, evaporation of water from the soil can lead to the supersaturation of salts like calcium sulfate ($CaSO_4$), or sodium chloride (NaCl). As water evaporates, the concentration of dissolved ions increases, potentially exceeding the solubility limits and resulting in the precipitation of these salts, a process often observed in the formation of salt crusts on the soil surface.

The common ion effect plays a significant role in the solubility equilibria of minerals in soils, particularly when multiple dissolved species are present. The presence of a common ion can reduce the solubility of minerals due to the shifting of equilibrium, as described by Le Chatelier's Principle. For example, in soils where calcium sulfate is present, the addition of calcium ions from another source, such as the dissolution of calcium carbonate, can reduce the solubility of $CaSO_4$.

The common ion effect is especially important in regulating nutrient availability in soils. For instance, phosphate ions (PO_4^{3-}) are often limited in availability because of the presence of calcium or magnesium ions, which combine with phosphate to form low-solubility minerals such as calcium phosphate ($Ca_3(PO_4)_2$). In calcareous soils, where calcium is abundant, the common ion effect can significantly reduce the availability of phosphate, a critical nutrient for plant growth, by precipitating it as an insoluble salt.

The solubility of minerals in soil is often highly dependent on the *pH* of the surrounding environment. For many minerals, changes in *pH* can alter the equilibrium between the dissolved ions and the solid phase, significantly affecting the availability of nutrients or the mobility of contaminants. For example, carbonate-containing minerals, like calcium carbonate, are more soluble in acidic conditions. In these conditions, in fact, the following reaction steps occur:

$$CaCO_{3(s)} + H_3O^+_{(aq)} \rightleftharpoons HCO^-_{3(aq)} + Ca^{2+}_{(aq)} + H_2O$$

$$HCO^-_{3(aq)} + H_3O^+_{(aq)} \rightleftharpoons H_2CO_{3(aq)} + H_2O$$

$$H_2CO_{3(aq)} + H_3O^+_{(aq)} \rightleftharpoons CO_{2(g)} + H_3O^+_{(aq)} + H_2O$$

In other words, in acidic conditions, calcium carbonate dissolves, producing carbon dioxide.

In alkaline soils, minerals like iron and aluminum hydroxides become less soluble. In basic conditions, hydroxide ions react with metal ions to form insoluble hydroxides, such as iron hydroxide ($Fe(OH)_3$). The formation of these insoluble hydroxides reduces the availability of essential micronutrients like iron, making them less accessible to plants in alkaline soils. Thus, *pH* adjustments in soil management are critical for ensuring that nutrients remain bioavailable.

Temperature also profoundly affects the solubility of minerals in soil water. In general, the solubility of many salts and minerals increases with temperature, as higher temperatures provide the energy needed to break the bonds in the solid phase, allowing more ions to dissolve. For instance, the solubility of gypsum ($CaSO_4 \cdot 2H_2O$) increases with temperature, meaning warmer soils can hold more dissolved calcium and sulfate.

However, not all minerals follow this trend. Some salts, such as sodium chloride (NaCl), exhibit little change in solubility with increasing temperature, while others, like calcium sulfate, may become less soluble at higher temperatures in certain conditions. In naturally occurring systems, these solubility changes are significant in controlling the precipitation and dissolution of salts in different climates. In cold regions, for example, reduced temperatures can lead to the precipitation of salts that are otherwise soluble, contributing to seasonal variations in soil chemistry.

To illustrate how to apply the solubility product constant in calculating the solubility of a solid compound, let's consider the solubility equilibrium of calcium carbonate given in equation (6.20). For calcium carbonate, the solubility product is given by $K_{sp} = 5.9 \times 10^{-9}\ M^2$. At equilibrium, the concentrations of the dissociated ions are equal, meaning $[Ca^{2+}] = [CO_3^{2-}]$. Let the solubility of calcium carbonate be represented by the letter s, so that $s = [Ca^{2+}] = [CO_3^{2-}]$. This allows us to express the solubility product relationship as:

$$K_{sp} = [Ca^{2+}][CO_3^{2-}] = s^2$$

Substituting the given value of the solubility product into this equation:

$$s^2 = 5.9 \times 10^{-9}\ M^2$$

Solving for s, we find:

$$s = \sqrt{K_{ps}} = 7.7 \times 10^{-5}\ M$$

This result indicates that 7.7×10^{-5} moles of calcium and carbonate ions can dissolve in 1 liter of water under these conditions.

To consider the effect of the common ion on the solubility equilibrium of calcium carbonate, let's imagine that the solution already contains 0.1 moles of calcium ions. In this scenario, the dissociation of $CaCO_{3(s)}$ must account for the additional calcium ions already present in the solution. The concentration of calcium ions can now be expressed as $0.1 + s$, where s is the solubility of the salt. The concentration of carbonate ions remains s. The solubility product expression in the presence of the common ion now becomes:

$$K_{sp} = [Ca^{2+}][CO_3^{2-}] = (0.1 + s)s$$

Expanding this expression leads to the following:

$$K_{sp} = 0.1s + s^2$$

By solving the second-order equation, the solubility of calcium carbonate in a solution containing 0.1 moles of Ca^{2+} is: $s = 1.5 \times 10^{-8}$ M. This result shows that the solubility of calcium carbonate in a solution containing 0.1 moles of calcium ions is significantly reduced to 1.5×10^{-8} M, which is three orders of magnitude lower than the solubility in pure water without the presence of the additional calcium ions. This example highlights the common ion effect, where the presence of a common ion in the solution decreases the solubility of a slightly soluble salt. Understanding this principle is essential in predicting and controlling the behavior of solubility equilibria in various chemical and environmental contexts, including soil chemistry and natural water systems.

6.7 THE REDOX EQUILIBRIA IN IDEAL CONDITIONS

Redox reactions, or reduction-oxidation reactions, are fundamental chemical processes involving the transfer of electrons between chemical species. These reactions are central to many natural processes, and environmental chemistry. In ideal conditions, where simplifying assumptions are made about solute behavior and interactions, redox equilibria can be described using theoretical models that facilitate the prediction of electron transfer and the overall balance of the system. The basic principles of redox equilibria in these conditions rely on concepts such as oxidation states, electrode potentials, and the Nernst equation, which together provide a framework for understanding how redox reactions proceed.

At the heart of redox chemistry are two simultaneous processes: oxidation, where a species loses electrons, and reduction, where a species gains electrons. The species that loses electrons is called the reducing agent (or reductant), while the one that gains electrons is called the oxidizing agent (or oxidant). In a redox reaction, the oxidizing agent gains the electrons lost by the reducing agent, ensuring that the total number of electrons remains balanced.

An example of a simple redox reaction is the oxidation of zinc by copper ions in solution:

$$\mathrm{Zn}_{(s)} + \mathrm{Cu}^{2+}_{(aq)} \rightarrow \mathrm{Zn}^{2+}_{(aq)} + \mathrm{Cu}_{(s)}$$

In this reaction, zinc undergoes oxidation as it loses two electrons to form $\mathrm{Zn}^{2+}_{(aq)}$:

$$Zn_{(s)} \rightarrow \mathrm{Zn}^{2+}_{(aq)} + 2e^{-}$$

Simultaneously, copper ions (Cu^{2+}) in the solution undergo reduction as they gain two electrons to form solid copper:

$$\mathrm{Cu}^{2+}_{(aq)} + 2e^{-} \rightarrow \mathrm{Cu}_{(s)}$$

These semi-reactions illustrate the fundamental concept that oxidation and reduction always occur together in a coupled manner. The complete reaction is a balance between the transfer of electrons from the reducing agent to the oxidizing agent.

The driving force for redox reactions is the difference in redox potential, often represented as E_h, which measures the tendency of chemical species to accept or donate electrons. A high E_h value indicates an environment that favors oxidation, while a low E_h value indicates reducing conditions.

Given a general reduction reaction in the form:

$$Ox + ne^{-} \rightleftharpoons Red$$

Being *Ox* the oxidated species, *Red* the reduced one, and ne^{-} the *n* electrons (e^{-}) involved in the reaction, the redox potential is obtained according to the Nernst equation reported in (6.22):

$$E_h = E_0 - \frac{RT}{nF} \ln(Q) \tag{6.22}$$

where

- E_h is the redox potential under real conditions.
- E_0 is the standard redox potential, representing the potential of a reduction semi-reaction relative to a standard reference, usually the hydrogen: $2H^{+} + 2e^{-} \rightarrow H_2$ ($E_0 = 0$ V).
- R is the gas constant ($8.314\,\mathrm{J\,mol^{-1}\,K^{-1}}$).
- T is the temperature (in Kelvin).

- n is the number of electrons transferred.
- F is the Faraday constant, 9.6×10^4 C mol^{-1}.
- Q is the reaction quotient, defined as the ratio of the concentrations of products to reactants, each raised to the power of their stoichiometric coefficients.

The standard reduction potential quantitatively measures the driving force behind a redox reaction. Species with a higher E_0 value have a greater tendency to gain electrons and undergo reduction, while those with lower E_0 values are more likely to lose electrons and undergo oxidation. The relative difference between the reduction potentials of two half-reactions determines the direction in which the redox reaction will proceed. For example, if the reduction potentials for the copper and zinc semi-reactions are as follows:

$$\mathrm{Cu}^{2+}_{(aq)} + 2e^- \rightarrow \mathrm{Cu}_{(s)},\ E_0 = +0.34\ \mathrm{V}$$

$$Zn_{(s)} \rightarrow \mathrm{Zn}^{2+}_{(aq)} + 2e^-,\ E_0 = -0.76\ \mathrm{V}$$

The more positive potential of copper indicates that it is a stronger oxidizing agent compared to zinc. Thus, copper ions will readily accept electrons from zinc in a redox reaction under ideal conditions.

The use of the standard redox potential E_0 allows to achieve the electrochemical series. This is a list of elements, and their ions arranged according to their standard reduction potentials. This series provides a convenient way to predict the outcome of redox reactions under ideal conditions by comparing the relative positions of the species involved. As stated above, the more positive the reduction potential, the higher its tendency to accept electrons and undergo reduction. In other words, if the following semi-reactions are accounted for:

$$\mathrm{Ag}^{+}_{(aq)} + e^- \rightarrow \mathrm{Ag}_{(s)},\ E_0 = +0.80\ \mathrm{V}$$

$$\mathrm{Fe}^{3+}_{(aq)} + e^- \rightarrow \mathrm{Fe}^{2+}_{(aq)},\ E_0 = +0.77\ \mathrm{V}$$

Silver ions have a higher reduction potential than iron ions, indicating that silver is more readily reduced. Therefore, in an ideal redox reaction, silver will be reduced preferentially over iron when both species are in solution. The electrochemical series also allows for the determination of spontaneity in redox reactions. A reaction is considered spontaneous if the reduction potential of the overall redox reaction is positive, meaning that the transfer of electrons from the reducing agent to the oxidizing agent is energetically favorable.

Table 6.1 reports the standard redox potentials of some chemical species that also have environmental importance.

In order to understand how to handle redox potentials to establish the spontaneity of a redox reaction, let us consider the oxidation of nitrogen gas ($N_{2(g)}$) to nitrogen dioxide ($NO_{2(g)}$), in the presence of oxygen, which is reduced to water. This is a process that occurs in various natural settings. The balanced semi-reactions for this redox process are:

Oxidation (occurring at the anode of the imaginary cell):

$$N_{2(g)} + 2H_2O_{(l)} \rightarrow 2NO_{2(g)} + 4H^+ + 4e^-$$

Reduction (occurring at the cathode of the imaginary cell):

$$O_{2(g)} + +4H^+ + 4e^- \rightarrow 2H_2O_{(l)}$$

where $E_0^{\text{anode}} = +1.15$ V, and $E_0^{\text{cathode}} = +1.23$ V.

Table 6.1 Standard redox potentials for some species present in the environment[3]

Oxidized Species	$+e^-$	Reduced Species	E_0 (V)
$Li^+_{(aq)}$	$+1e^-$	$Li_{(s)}$	–3.04
$Na^+_{(aq)}$	$+1e^-$	$Na_{(s)}$	–2.71
$O_{2(g)} + 4H^+$	$+4e^-$	$2H_2O_{(l)}$	+1.23
$SO^{2-}_{4(aq)} + 4H^+$	$+2e^-$	$SO_{2(g)} + 2H_2O$	+0.20
$NO^-_{3(aq)} + 3H^+$	$+2e^-$	$HNO_{2(g)} + H_2O$	+0.94
$NO^-_{3(aq)} + H_2O$	$+2e^-$	$NO^-_{2(aq)} + 2OH^-$	+0.01
$ClO^-_{3(aq)} + 3H_2O$	$+6e^-$	$Cl^- + 6OH^-$	+0.62
$MnO^-_{4(aq)} + 2H_2O$	$+3e^-$	$MnO_{2(s)} + 4OH^-_{(aq)}$	+0.59
$2NO_{2(g)} + 4H^+$	$+4e^-$	$N_{2(g)} + 2H_2O$	+0.27

The standard potential for the overall reaction is calculated as:

$$E_0^{\text{tot}} = E_0^{\text{cathode}} - E_0^{\text{anode}} = 1.23 - 1.15\text{ V} = 0.08\text{ V}$$

Now, suppose the partial pressures of nitrogen, oxygen, and nitrogen dioxide gases deviate from their standard conditions with the following values:

$$[N_{2(g)}] = P_{N_2} = 0.80\text{ atm}$$

$$[O_{2(g)}] = P_{O_2} = 0.21\text{ atm}$$

$$[NO_{2(g)}] = P_{NO_2} = 0.0010\text{ atm}$$

The quotient Q in the Nernst equation is:

$$Q = \frac{P_{NO_2}^2}{P_{N_2} P_{O_2}^2} = \frac{(0.0010)^2\text{ atm}^2}{(0.80)(0.21)^2\text{ atm}^3} = 2.84 \times 10^{-5}\text{ atm}$$

By replacing this Q value in the Nernst equation at the temperature of 25°C (i.e., 298 K), it comes:

[3] The data have been taken from the tables included in Küster-Thiel (1993) Rechentafeln für die Chemische Analytik, Ammoniaklaboratorium BASF AG, Ludwigshafen/Rhein (Germany), 104th Ed.

$$E_h = 0.08\ \text{V} - \frac{0.0591}{4} \ln(2.84 \times 10^{-5}) = 0.15\ \text{V}$$

The cell potential under these non-standard conditions is, therefore, approximately 0.15 V, indicating that the reaction is more favorable under these conditions compared to the standard state.

6.8 THE GRAPHICAL REPRESENTATION OF THE CHEMICAL EQUILIBRIA

Chemical equilibria, often described using mathematical models and equations, can also be effectively represented through graphical methods. While equations are fundamental to predicting and quantifying the behavior of chemical species under different conditions, they often require extensive calculations to interpret the data fully. Conversely, graphs provide a powerful visual representation of these equilibria, allowing us to understand complex chemical relationships at a glance. The well-known adage, "a picture is worth a thousand words," is particularly true in chemistry, where graphs can convey intricate information quickly and intuitively. By translating mathematical data into visual formats, graphs not only communicate the same information as equations but also make it more accessible, enabling chemists to analyze trends, identify patterns, and draw conclusions with greater ease. In this way, graphical representation is an essential tool that simplifies the interpretation of chemical equilibria and enhances the decision-making process in both research and practical applications.

6.9 REPRESENTING ACID-BASE EQUILIBRIA

Let us consider the equilibrium of a monoprotic acid in the form:

$$HA + H_2O \rightleftharpoons A^- + H_3O^+$$

Two conditions are valid for the given equilibrium:

$$K_a = \frac{[A^-][H_3O^+]}{[HA]}$$

and

$$[HA] + [A^-] = C$$

where, C is the total concentration of the acid/base pair at the equilibrium. From the conditions above, it comes:

$$[HA] = C - [A^-],$$

and

$$[A^-] = \frac{[HA]K_a}{[H_3O^+]}$$

The combination of the last two relationships gives:

$$[\mathrm{A}^-] = \frac{(C - [\mathrm{A}^-])K_a}{[\mathrm{H_3O^+}]}$$

That, after rearrangement, becomes:

$$[\mathrm{A}^-] = \frac{CK_a}{K_a + [\mathrm{H_3O^+}]}$$

By applying the same manipulation to obtain the trend for [HA], it comes that

$$[\mathrm{HA}] = \frac{C[\mathrm{H_3O^+}]}{K_a + [\mathrm{H_3O^+}]}$$

When Ka ≪ $[H_3O^+]$, it is possible to ignore the K_a value, and thus:

1. $[\mathrm{HA}] = \dfrac{C[\mathrm{H_3O^+}]}{K_a + [\mathrm{H_3O^+}]} = \dfrac{C[\mathrm{H_3O^+}]}{[\mathrm{H_3O^+}]} = \mathrm{C}$. After application of the decimal logarithm, $\log_{10}([\mathrm{HA}]) = \log_{10}(\mathrm{C})$
2. $[\mathrm{A}^-] = \dfrac{CK_a}{K_a + [\mathrm{H_3O^+}]} = \dfrac{CK_a}{[\mathrm{H_3O^+}]}$, which becomes: $\log_{10}(\mathrm{C}) = \log_{10}([\mathrm{A}^-]) + pK_a - pH$

   When $K_a \gg [\mathrm{H_3O^+}]$, the concentration of the hydrogen ion can be neglected, and:
3. $\log_{10}(\mathrm{C}) = \log_{10}([\mathrm{HA}]) + pK_a - pH$
4. $\log_{10}([\mathrm{A}^-]) = \log_{10}(\mathrm{C})$.

By considering a generic monoprotic acid with a $K_a = 1.8 \times 10^{-5}$ M and a $C = 0.1$ M, it is easy to draw the graph reported in Figure 6.2.

In the graph reported in Figure 6.2, the crossing point corresponds to the condition $[\mathrm{HA}] = [\mathrm{A}^-]$. This is the region where the solution can resist changes in *pH* upon the addition of small amounts of acid or base and reveals the maximum buffer capacity occurs when $pH = pK_a$. In addition, one of the most useful features of the graph in Figure 6.2 is its ability to estimate the relative concentrations of species at a given *pH*. For example, the y-axis represents the fraction of the total concentration *C* that exists as either HA or A^-. Identifying the *pH* on the x-axis and reading the corresponding points on the curves makes it possible to determine which species is dominant at any specific *pH*. In particular, the graph in Figure 6.2 shows that HA is the predominant species in solution up to $pH \approx 4.7$, while A^- predominates when $pH > 4.7$.

In the case of polyprotic acids, as represented in Figure 6.3, the graph becomes even more informative. Each dissociation step corresponds to its own

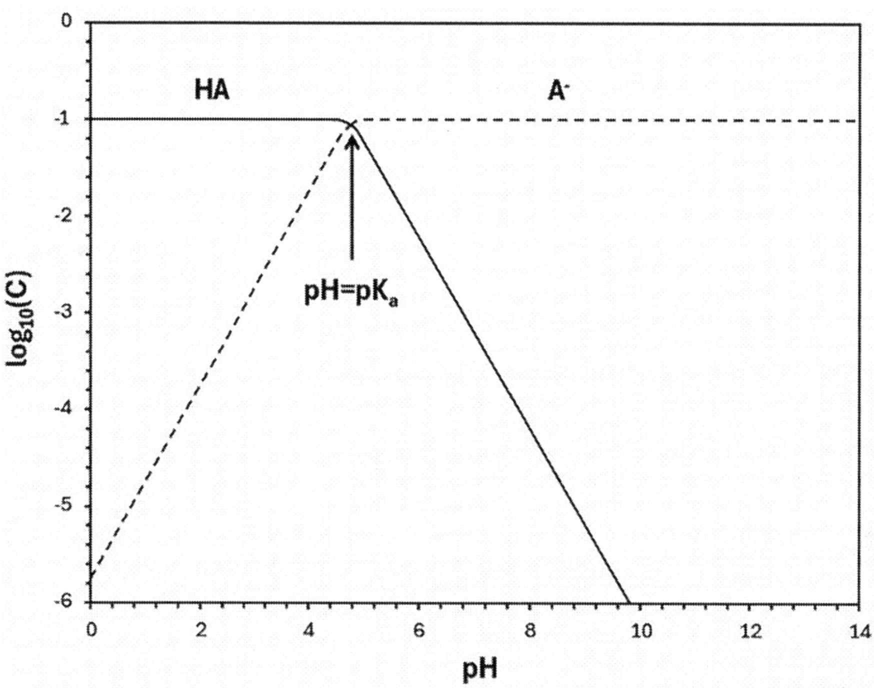

Figure 6.2 Graphical representation of the HA/A$^-$ equilibrium when $K_a = 1.8 \times 10^{-5}$ M, and $C = 0.1$ M. The crossing point when $pH = pK_a$ corresponds to the condition $[HA] = [A^-]$.

pK_a and the graph will show multiple crossing points where the concentrations of different species are equal. For example, with carbonic (H_2CO_3) and phosphoric (H_3PO_4) acids, the graph shows the relative concentration of each chemical species dominating in different *pH* ranges. Therefore, H_2CO_3 predominates when $pH < 6$, HCO_3^- is the predominant species when $\sim 6 < pH < 10$, while the carbonate ion (CO_3^{2-}) predominates when $pH > 10$. For what concerns phosphoric acid, the triprotic species (H_3PO_4) is predominant when $pH < 2$, $H_2PO_4^-$ is the most abundant species when $2 < pH < 7$, HPO_4^{2-} is predominant when $7 < pH < 12$, and the phosphate ion (PO_4^{3-}) is the most abundant as *pH* rises above 12. Moreover, the different crossing points also allow us to identify the various buffer capacities of the different acid/base pairs present in each solution. In particular, Figure 6.3A shows that carbonic acid reveals two buffer points. The first one corresponds to the condition $[H_2CO_3] = [HCO_3^-]$, and $pH = pKa_1$, while the second buffer point is due to the condition $[HCO_3^-] = [CO_3^{2-}]$, and $pH = pKa_2$. For phosphoric acid, which is a triprotic acid, three buffer points can be identified, each corresponding to one of the following conditions, respectively (Figure 6.3B):

— $[H_3PO_4] = [H_2PO_4^-]$, and $pH = pKa_1$

— $[H_2PO_4^-] = [HPO_4^{2-}]$, and $pH = pKa_2$

— $[HPO_4^{2-}] = [PO_4^{3-}]$, and $pH = pKa_3$.

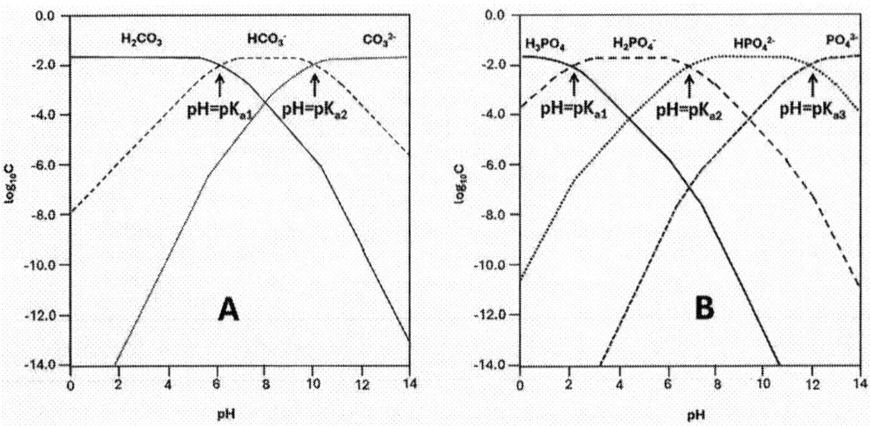

Figure 6.3 Graphical representation of the multiple equilibria given by carbonic (H_2CO_3) (**A**), and phosphoric (H_3PO_4) (**B**) acids.

Based on the considerations above, it can be hypothesized that under highly basic conditions, when carbonate and phosphate ions predominate, the simultaneous presence of calcium ions may lead to the precipitation of calcium carbonate and calcium phosphate salts. This process can remove plant-available forms of calcium and phosphorus, reducing their availability for plant nutrition.

6.10 DISTRIBUTION CURVES FOR ACID-BASE SPECIES

Distribution curves (also known as speciation graphs) for an acid (or a base) are graphs that show the molar fractions or relative percentage concentrations of the different species present in a solution as a function of *pH*. These curves are highly effective for visualizing acid-base equilibria, especially in the case of weak acids and bases.

To understand how to build up these curves, let us consider the simple case of a generic weak monoprotic acid (HA) and its conjugate base (A^-).

According to what has been previously reported, the acid constant for the dissociation of monoprotic acid is:

$$K_a = \frac{[A^-][H_3O^+]}{[HA]}$$

To calculate the concentrations of the three species, namely, [HA], $[A^-]$, and $[H_3O^+]$, when the total concentration of the acid (C) and the total concentration of the protons ($[H_3O^+]_{tot}$) are known, additional equations are needed beyond the expression for the acid dissociation constant. The first additional equation is the conservation of the total amount of the acid:

$$C = [HA] + [A^-]$$

The second equation involves conserving the total amount of protons in the solution, accounting for all sources of hydrogen ions. In a simple system, this can be expressed as:

$$[H_3O^+]_{tot} = [HA] + [H_3O^+]_{water}$$

Where $[H_3O^+]_{water}$ represents the contribution of water's autoionization. In practical applications, especially when dealing with weak acids and bases, the contribution from the dissociation of water may be negligible compared to the proton concentration resulting from the acid dissociation itself.

Combining these equations allows for the calculation of the relative concentrations of [HA], and $[A^-]$ at any given *pH*. Once the concentrations are known, the molar fraction of each species can be calculated as:

$$\text{Molar fraction of HA} = \frac{[HA]}{C}$$

$$\text{Molar fraction of } A^- = \frac{[A^-]}{C}$$

By plotting these molar fractions against the *pH*, a distribution curve can be generated (Figure 6.4). The resulting graph shows how the predominant species shifts from the protonated form (HA) at low *pH* to the deprotonated form (A^-) at high *pH*. At the point where H = pK_a, [HA] = $[A^-]$, and each species represents 50% of the total concentration.

For polyprotic acids, similar principles apply, but the system requires additional equilibrium expressions to account for each dissociation step, leading to more complex distribution curves that reflect the sequential loss of protons as *pH* increases.

To understand how to build a distribution curve for a generic diprotic acid (H_2A) the following equilibria must be considered:

$$H_2A \rightleftharpoons HA^- + H^+ \tag{6.23}$$

$$HA^- \rightleftharpoons A^{2-} + H^+ \tag{6.24}$$

$$H_2A \rightleftharpoons A^{2-} + 2H^+ \tag{6.25}$$

The three equilibria (6.23) to (6.25) are described by the following equilibrium constants, respectively:

$$K_{a_1} = \frac{[H^+][HA^-]}{[H_2A]} \tag{6.26}$$

$$K_{a_2} = \frac{[H^+][A^{2-}]}{[HA^-]} \tag{6.27}$$

$$K_{a_1}K_{a_2} = \frac{[H^+]^2[A^{2-}]}{[H_2A]} \tag{6.28}$$

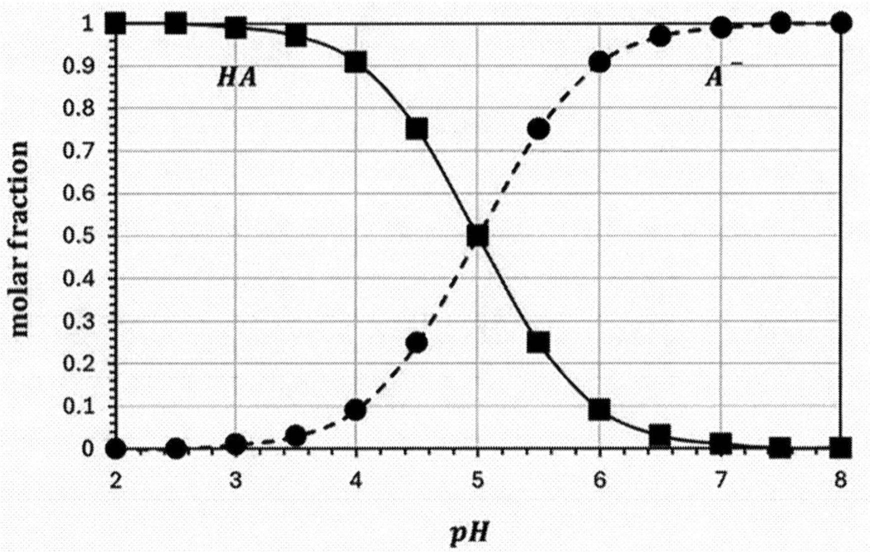

Figure 6.4 Distribution curves for generic monoprotic acid (HA) with a concentration of 0.1 M and acid constant of 1×10^{-5}. The crossing point describes the buffer condition where *pH* = *pKa*.

From which the following relationships can be obtained:

$$\frac{[\mathrm{HA}^-]}{[\mathrm{H_2A}]} = \frac{K_{a_1}}{[\mathrm{H}^+]} \tag{6.29}$$

$$\frac{[\mathrm{A}^{2-}]}{[\mathrm{H_2A}]} = \frac{K_{a_1} K_{a_2}}{[\mathrm{H}^+]^2} \tag{6.30}$$

The mass conservation condition is:

$$C_{tot} = [\mathrm{H_2A}] + [\mathrm{HA}^-] + [\mathrm{A}^{2-}] \tag{6.31}$$

The molar fractions of each chemical species in solution are:

$$\alpha_{\mathrm{H_2A}} = \frac{[\mathrm{H_2A}]}{C_{tot}} = \frac{[\mathrm{H_2A}]}{[\mathrm{H_2A}] + [\mathrm{HA}^-] + [\mathrm{A}^{2-}]}$$

$$\alpha_{\mathrm{HA}^-} = \frac{[\mathrm{HA}^-]}{C_{tot}} = \frac{[\mathrm{HA}^-]}{[\mathrm{H_2A}] + [\mathrm{HA}^-] + [\mathrm{A}^{2-}]}$$

$$\alpha_{A^{2-}} = \frac{[\mathrm{A}^{2-}]}{C_{tot}} = \frac{[\mathrm{A}^{2-}]}{[\mathrm{H_2A}] + [\mathrm{HA}^-] + [\mathrm{A}^{2-}]}$$

From which:

$$\alpha_{\mathrm{H_2A}} = \frac{1}{1 + \dfrac{[\mathrm{HA}^-]}{[\mathrm{H_2A}]} + \dfrac{[\mathrm{A}^{2-}]}{[\mathrm{H_2A}]}} \tag{6.32}$$

$$\alpha_{HA^-} = \frac{1}{1+\frac{[H_2A]}{[HA^-]}+\frac{[A^{2-}]}{[HA^-]}} \tag{6.33}$$

$$\alpha_{A^{2-}} = \frac{1}{1+\frac{[H_2A]}{[A^{2-}]}+\frac{[HA^-]}{[A^{2-}]}} \tag{6.34}$$

By replacing relationships (6.29) to (6.31) in the relationships (6.32), (6.33), and (6.34), the following final expressions can be obtained:

$$\alpha_{H_2A} = \frac{[H^+]^2}{[H^+]^2+K_{a_1}[H^+]+K_{a_1}K_{a_2}} \tag{6.35}$$

$$\alpha_{HA^-} = \frac{K_{a_1}[H^+]}{[H^+]^2+K_{a_1}[H^+]+K_{a_1}K_{a_2}} \tag{6.36}$$

$$\alpha_{A^{2-}} = \frac{K_{a_1}K_{a_2}}{[H^+]^2+K_{a_1}[H^+]+K_{a_1}K_{a_2}} \tag{6.37}$$

The application of the same principles to a triprotic acid (H_3A) let us obtain the final expressions (6.38) to (6.41):

$$\alpha_{H_3A} = \frac{[H^+]^3}{[H^+]^3+K_{a_1}[H^+]^2+K_{a_1}K_{a_2}[H^+]+K_{a_1}K_{a_2}K_{a_3}} \tag{6.38}$$

$$\alpha_{H_2A^-} = \frac{K_{a_1}[H^+]^2}{[H^+]^3+K_{a_1}[H^+]^2+K_{a_1}K_{a_2}[H^+]+K_{a_1}K_{a_2}K_{a_3}} \tag{6.39}$$

$$\alpha_{HA^{2-}} = \frac{K_{a_1}K_{a_2}[H^+]}{[H^+]^3+K_{a_1}[H^+]^2+K_{a_1}K_{a_2}[H^+]+K_{a_1}K_{a_2}K_{a_3}} \tag{6.40}$$

$$\alpha_{A^{3-}} = \frac{K_{a_1}K_{a_2}K_{a_3}}{[H^+]^3+K_{a_1}[H^+]^2+K_{a_1}K_{a_2}[H^+]+K_{a_1}K_{a_2}K_{a_3}} \tag{6.41}$$

The applications of the expressions (6.35) to (6.37), and (6.38) to (6.41) to carbonic (H_2CO_3; $K_{a_1} = 4.3 \times 10^{-7}$; $K_{a_2} = 5.6 \times 10^{-11}$) and phosphoric ($H_3PO_4$; $K_{a_1} = 7.2 \times 10^{-3}$; $K_{a_2} = 6.3 \times 10^{-8}$; $K_{a_3} = 4.0 \times 10^{-13}$) acids, respectively, leads to the distribution graphs depicted in Figures 6.5A and 6.5B.

The visual inspection of the distribution curves of carbonic (Figure 6.5A) and phosphoric (Figure 6.5B) acids allows us to conclude that the main forms of the two acids in the typical *pH* condition of the Mediterranean soils (i.e., 5.5–7.5) are HCO_3^-, and $H_2PO_4^-$.

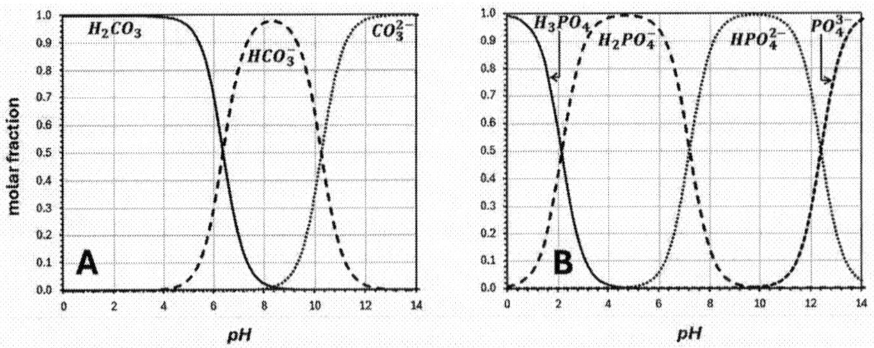

Figure 6.5 Distribution curves for carbonic (**A**) and phosphoric acid (**B**).

6.11 THE TITRATION CURVES

Titration is an analytical technique used to determine the concentration of an unknown analyte in a solution (also called the titrand) by gradually adding a solution of known concentration (the titrant) until the reaction between the two solutions is complete. The point at which the reaction is complete is known as the equivalence point. During titration, the volumes of both solutions are carefully measured, and an indicator is often used to signal the equivalence point, typically via a color change in acid-base titrations. The fundamental concept behind titration is that, at the equivalence point, the moles of the titrand (n_1) and the titrant (n_2) are stoichiometrically equivalent (i.e., $n_1 = n_2$). Therefore, given a precise volume of titrate (denoted as V_1) and a measured volume of titrant (denoted as V_2), the relationship between the concentrations and volumes of the titrand and titrant is expressed as:

$$C_1V_1 = C_2V_2 \tag{6.42}$$

Where C_1 is the unknown concentration of titrand, while C_2 is the known concentration of titrant. From this relationship, the unknown concentration of the titrand can be calculated using:

$$C_1 = \frac{C_2V_2}{V_1} \tag{6.43}$$

The graphical representation of a titration is called a titration curve, which plots pH as a function of the volume of titrant added during an acid-base titration. The shape of the curve provides important information about the dissociation characteristics of the acid or base being titrated, the equivalence point (where stoichiometrically equivalent amounts of acid and base have reacted), and the buffering capacity of the system. Sharp changes in the slope of the titration curve around the equivalence point indicate significant shifts in pH, while flatter regions reflect buffering zones where the pH changes more gradually despite the addition of titrant.

To understand how to draw a titration curve, let us consider 50 mL of a 0.1 M HCl solution which is titrated with increasing amounts of a 0.1 M NaoH solution at the temperature of 25°C. Four conditions must be accounted for.

1. Initial *pH* before NaOH addition.

   As HCl is a strong acid, it fully dissociates in solution:

   $$HCl \rightarrow H^+ + Cl^-.$$

   Therefore, the initial *pH* can be calculated as:

   $$[H^+] = 0.1\ M;\ pH = 1.$$

   The initial number of H^+ moles (which will be used in the following steps) are:

   $$n_{H^+} = 0.1\ M \times 0.05\ L = 0.005\ mol$$

2. *pH* after the addition of a small amount of NaOH before the equivalence point.

   Let us add 1 mL of 0.1 M NaOH. The number of NaOH moles is:

   $$n_{NaOH} = n_{OH^-} = 0.1\ M \times 0.001\ L = 1 \times 10^{-4}\ mol.$$

   These moles neutralize the same number of H^+ moles. Therefore, the remaining H^+ moles can be calculated as follows:

   $$n_{H^+} = 0.005\ mol - 0.0001\ mol = 0.0049\ mol.$$

   The concentration of H^+ after the addition of 1 mL of NaOH is:

   $$[H^+] = \frac{0.0049\ mol}{(0.050 + 0.001)L} = 0.0961\ M$$

   Provides $pH = 1.02$.

   If 10 mL of NaOH are added, using the same method, the *pH* would be 1.18. After adding 20 mL, *pH* increases to 1.35, and so on.

3. *pH* at the equivalence point.

   When the equivalence point is reached, the number of $[H^+]$ equals those of $[OH^-]$. Hence, $n_{H^+} = n_{OH^-}$. The volume of NaOH needed to completely neutralize HCl is 50 mL. For this reason, the *pH* at the equivalence point is 7.

4. *pH* after the equivalence point.

   Once the equivalence point is surpassed, there are no more H^+ from HCl remains in the solution. Only the excess OH^- from NaOH must be considered. For example, after adding 60 mL of NaOH, the excess OH^- is:

   $$[OH^-] = \frac{0.01\ mol}{0.110\ L} = 0.0909\ M$$

   corresponding to $pOH = -\log_{10}[OH^-] = 1.04$, and $pH = 14 - pOH = 12.96$.

Table 6.2 shows the *pH* concentration changes in the HCl solution as increasing amounts of NaOH are added.

Table 6.2 *pH* changes during the titration of 50 mL of a 0.1 M HCl solution with a 0.1 M NaOH solution

V_{NaOH} **(mL)**	***pH***
0	1.00
10	1.18
20	1.35
40	1.95
45	2.28
48	2.69
49	3.00
50	7.00
51	11.00
55	11.68
60	11.96
80	12.36
100	12.52

Figure 6.6A shows the graphical elaboration of the data from Table 6.2, with the equivalence point located at the center of the curve at $pH = 7$.

In addition, Figure 6.6 also displays titration curves for acetic acid (CH_3COOH, 50 mL, 0.1 M, $K_a = 1.8 \times 10^{-5}$), and phosphoric acid (H_3PO_4, 50 mL, 0.1 M, $K_{a_1} = 7.2 \times 10^{-3}$; $K_{a_2} = 6.3 \times 10^{-8}$; $K_{a_3} = 4.2 \times 10^{-13}$) titrated with 0.1 M NaOH (Figure 6.6B, and Figure 6.6C, respectively).

The titration curve of acetic acid reveals a buffering region and a smooth rise until the equivalence point near $pH = 8.72$ (corresponding to $pH = pK_a = 8.72$).

Phosphoric acid, being triprotic, exhibits multiple equivalence points, with distinct buffering regions and equivalence points for each dissociation step: $pH = pK_{a_1} = 4.65$, $pH = pK_{a_2} = 9.30$, and $\mathrm{pH} = pK_{a_3} = 12.3$.

6.12 BUFFER CAPACITY CURVES

A buffer capacity curve is a graphical representation that illustrates how the buffer capacity of a solution varies with changes in *pH.* The buffer capacity measures a solution's ability to resist changes in *pH* upon adding small amounts of acid or base. In a buffer capacity curve, the buffer capacity (often denoted as β) is plotted on the vertical axis, while the *pH* is plotted on the horizontal axis.

To build a buffer capacity curve, the following experimental steps must be considered:

1. **Choose a Buffer System:** Select a buffer system consisting of a weak acid and its conjugate base or a weak base and its conjugate acid.

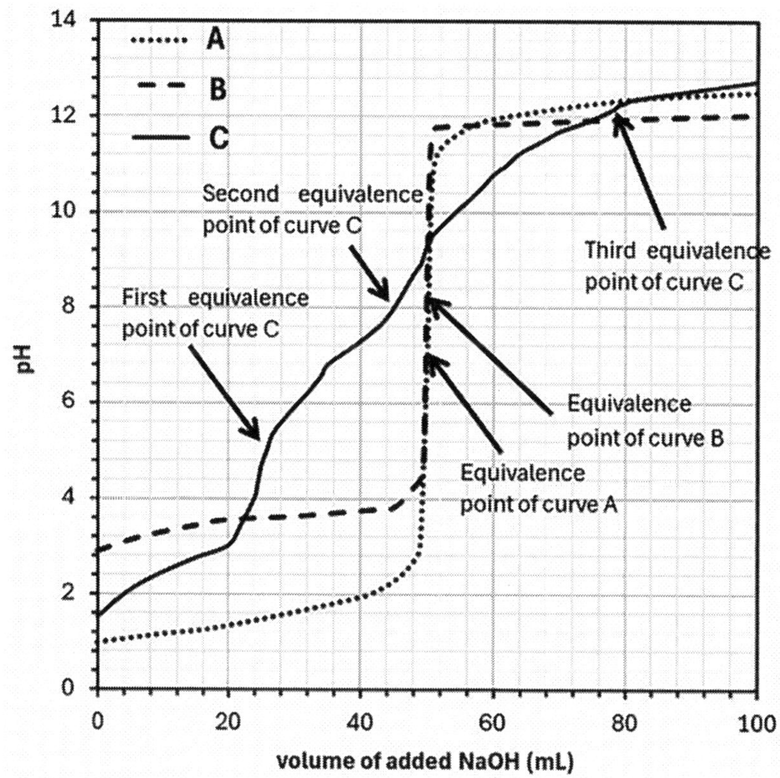

Figure 6.6 Titration curves were obtained by titrating 50 mL of a 0.1 M HCl solution (A), 50 mL of a 0.1 M acetic acid (CH_3COOH) solution (B), and 50 mL of a 0.1 M phosphoric acid (H_3PO_4) solution (C) with a 0.1 M NaOH solution.

For example, acetic acid (CH_3COOH) and its conjugate base acetate (CH_3COO^-).

2. **Prepare the Buffer Solution:** Prepare a buffer solution with known concentrations of the weak acid and its conjugate base. The ratio of these concentrations determines the initial pH of the solution, which can be calculated using the Henderson-Hasselbalch equation (6.11).
3. **Measure Buffer Capacity at Different pH Values:** Gradually add small, incremental amounts of strong acid (such as HCl) or strong base (such as NaOH) to the buffer solution. After each addition, measure the *pH* of the solution and calculate the buffer capacity using the formula:

$$\beta = \frac{\Delta n}{\Delta pH}$$

Where Δn is the number of moles of acid or base added per unit volume, and ΔpH is the resulting change in *pH*.

4. **Plot the Buffer Capacity Curve:** Plot the calculated buffer capacity (β) against the corresponding pH values. The curve typically shows a peak near the pK_a of the buffering species, indicating the region where the buffer is most effective. For a diprotic or polyprotic buffer, multiple peaks can appear, corresponding to the different pK_a values of the species.

The buffer capacity can also be mathematically derived. While the complete derivation is beyond the scope of this book, the final equations relating buffer capacity with pH changes are provided below.

For a monoprotic acid (HA) the buffer capacity is:

$$\beta = \frac{2.303CK_a[\mathrm{H}^+]}{(K_a + [\mathrm{H}^+])^2} \tag{6.44}$$

For a diprotic acid (H_2A):

$$\beta = 2.303\left(\frac{CK_{a_1}\left[H^+\right]}{\left(K_{a_1}+\left[H^+\right]\right)^2}+\frac{CK_{a_2}\left[H^+\right]}{\left(K_{a_2}+\left[H^+\right]\right)^2}\right) \tag{6.45}$$

For a triprotic acid (H_3A):

$$\beta = 2.303\left(\frac{CK_{a_1}[\mathrm{H}^+]}{(K_{a_1}+[\mathrm{H}^+])^2}+\frac{CK_{a_2}[\mathrm{H}^+]}{(K_{a_2}+[\mathrm{H}^+])^2}+\frac{CK_{a_3}[\mathrm{H}^+]}{(K_{a_3}+[\mathrm{H}^+])^2}\right) \tag{6.46}$$

In general, the equation for a generic polyprotic acid is[4]:

$$\beta = 2.303\sum_{i=1}^{n}\frac{CK_{a_i}[\mathrm{H}^+]}{\left(K_{a_i}+[\mathrm{H}^+]\right)^2} \tag{6.47}$$

In these equations, C is the total concentration of all species in solution, and K_{a_i} is the i-th acid constant for the i-th acid-base equilibrium.

Figure 6.7 shows the graphical representation of the buffer capacity for acetic, carbonic, and phosphoric acids, respectively. The peaks are centered at $pH = pK_{a_i}$, representing the points of maximum buffer capacity for each acid. These curves clearly visualize the pH values at which the buffering capacity is the greatest.

For soil systems, understanding buffer capacity is crucial for predicting how soils will respond to acid rain, fertilization, and other environmental inputs. Soils with high buffer capacity, such as those rich in calcium carbonate, can resist pH changes more effectively, protecting plants and microorganisms from the harmful effects of acidification. Conversely, soils with low buffer capacity are more vulnerable to pH shifts, which can lead to nutrient depletion and reduced fertility.

---

[4] King, D.W. and D.R. Kester. 1990. A general approach for calculating polyprotic acid speciation and buffer capacity. Journal of Chemical Education. 67: 932–933.
– Michałowska-Kaczmarczyk, A.M. and T. Michałowski. 2015. Dynamic buffer capacity in acid–base systems. Journal of Solution Chemistry. 44: 1256–1266.

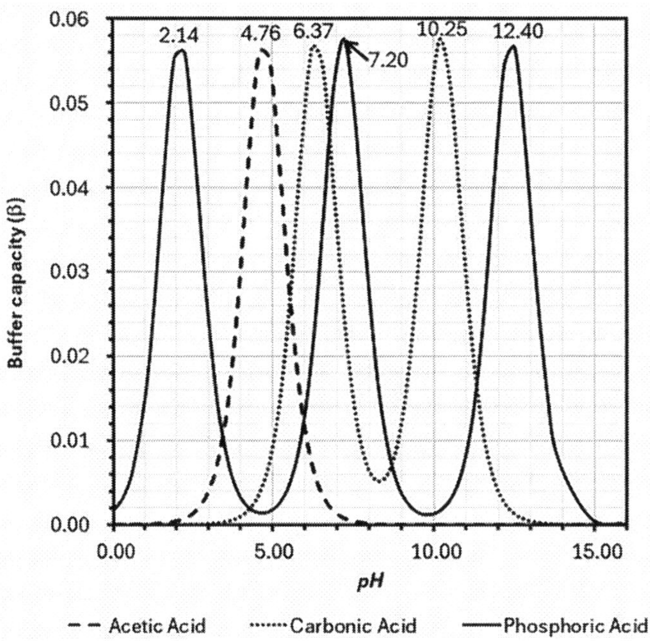

Figure 6.7 Buffer capacity curves for acetic, carbonic, and phosphoric acids. The maxima represents the condition at which the buffering capacity is maximum. All the curves were obtained using equations (6.40) to (6.42).

6.13 SOLUBILITY EQUILIBRIA AND GRAPHICAL REPRESENTATION: SOLUBILITY VS *pH*

Solubility equilibria describe the balance between a dissolved solute and its undissolved solid in a solution. The behavior of slightly soluble salts, particularly how their solubility changes in response to *pH*, can be visualized using graphs where the logarithm of the ion concentration is plotted against *pH*. These graphs are especially useful for salts whose solubility is *pH*-dependent, such as those containing carbonate, phosphate, or hydroxide ions.

The solubility of many compounds, particularly those containing ions derived from weak acids or bases, depends heavily on the *pH* of the solution. For example, the solubility of salts like calcium carbonate ($CaCO_3$) or aluminum hydroxide ($Al(OH)_3$) varies significantly with *pH* due to the common ion effect and the formation of new species at different *pH* levels.

For any sparingly soluble salt (*AB*), the solubility product constant (K_{sp}) governs the equilibrium between the solid phase and the dissolved ions:

$$AB_{(s)} \rightleftharpoons A^+_{(aq)} + B^-_{(aq)}$$

where $K_{sp} = [A^+_{(aq)}][B^-_{(aq)}]$.

At equilibrium, it holds that:

$$[A^+_{(aq)}] = [B^-_{(aq)}]$$

If one of the two species (e.g., $A^+_{(aq)}$, or $B^-_{(aq)}$) is *pH*-dependent, then a graph can be plotted where the concentration of one of the ions is plotted against *pH*. Such graphs provide insights into how *pH* affects solubility, allowing us to predict the behavior of the salt in various *pH* conditions.

Let us consider the solubility equilibrium of calcium carbonate which dissolves in water according to the following equilibrium

$$CaCO_{3(s)} \rightleftharpoons Ca^{2+}_{(aq)} + CO^{2-}_{3(aq)}$$

The solubility product constant is given by:

$$K_{sp} = [Ca^{2+}_{(aq)}][CO^{2-}_{3(aq)}] \tag{6.48}$$

At 25°C, its value is: 4.8×10^{-9}.

However, the concentration of carbonate ions in solution is *pH*-dependent because of the equilibria involving bicarbonate and carbonic acid:

$$H_2CO_{3(aq)} \rightleftharpoons HCO^-_{3(aq)} + H^+_{(aq)}$$

$$HCO^-_{3(aq)} \rightleftharpoons CO^{2-}_{3(aq)} + H^+_{(aq)}$$

These are governed by K_{a1} and K_{a2}, respectively:

$$K_{a1} = \frac{[HCO^-_{3(aq)}][H^+_{(aq)}]}{[H_2CO_{3(aq)}]} = 4.3 \times 10^{-7} \tag{6.49}$$

$$K_{a2} = \frac{[CO^{2-}_{3(aq)}][H^+_{(aq)}]}{[HCO^-_{3(aq)}]} = 5.6 \times 10^{-11} \tag{6.50}$$

The K_{a1} and K_{a2} values reported above are valid at 25°C. These equilibria affect the amount of carbonate ions available in solution, and thus the solubility of calcium carbonate is a function of *pH*.

Due to the *pH* dependence of the amount of carbonate ions, the solubility (*s*) of calcium carbonate is measured only by the calcium concentration in the solution at the equilibrium. Therefore,

$$K_{sp} = s \times [CO^{2-}_{3(aq)}] \tag{6.51}$$

and

$$s = \frac{K_{sp}}{CO^{2-}_{3(aq)}} \tag{6.52}$$

The carbonic acid equilibria can also be joined as:

$$H_2CO_{3(aq)} \rightleftharpoons CO^{2-}_{3(aq)} + 2H^+_{(aq)}$$

which is governed by:

$$K = K_{a1} \times K_{a2} = \frac{\left[CO^{2-}_{3(aq)}\right]\left[H^+_{(aq)}\right]^2}{\left[H_2CO_{3(aq)}\right]} = 4.3 \times 10^{-7} \tag{6.53}$$

The K value is only valid at 25°C.

The mass conservation balance can be written as:

$$C_{tot} = [H_2CO_{3(aq)}] + [HCO^{-}_{3(aq)}] + [CO^{2-}_{3(aq)}] \tag{6.54}$$

Where C_{tot} is the concentration of all the carbonate species at the equilibrium in solution.

The fraction of carbonate in solution can be calculated as:

$$\frac{[CO^{2-}_{3(aq)}]}{C_{tot}}$$

Which can be rewritten as:

$$\frac{[CO^{2-}_{3(aq)}]}{C_{tot}} = \frac{\left[CO^{2-}_{3(aq)}\right]}{[H_2CO_{3(aq)}] + [HCO^{-}_{3(aq)}] + [CO^{2-}_{3(aq)}]}$$

By dividing the right-hand side ratio by $[CO^{2-}_{3(aq)}]$:

$$\frac{\left[CO^{2-}_{3(aq)}\right]}{C_{tot}} = \frac{1}{\frac{\left[H_2CO_{3(aq)}\right]}{\left[CO^{2-}_{3(aq)}\right]} + \frac{\left[HCO^{-}_{3(aq)}\right]}{\left[CO^{2-}_{3(aq)}\right]} + 1} \tag{6.55}$$

The combination of equations (46) and (49) with (51) provides:

$$\frac{[CO^{2-}_{3(aq)}]}{C_{tot}} = \frac{1}{\frac{[H^{+}_{(aq)}]^2}{K_{a1} \times K_{a2}} + \frac{[H^{+}_{(aq)}]}{K_{a2}} + 1}$$

After rearrangement:

$$[CO^{2-}_{3(aq)}] = \frac{C_{tot} \times K_{a1} \times K_{a2}}{[H^{+}_{(aq)}]^2 + [H^{+}_{(aq)}] \times K_{a1} + K_{a1} \times K_{a2}} \tag{6.56}$$

Combining (6.56) in (6.52) and rearranging, the *pH*-dependence equation of the calcium carbonate solubility is obtained:

$$s = \frac{K_{sp}([H^{+}_{(aq)}]^2 + [H^{+}_{(aq)}] \times K_{a1} + K_{a1} \times K_{a2})}{C_{tot} \times K_{a1} \times K_{a2}} \tag{6.57}$$

The graphical representation of equation (6.57) is given in Figure 6.8A where it is possible to read that as *pH* increases, the amount of calcium ion in the solution drastically decreases because it precipitates in the form of calcium carbonate. The same applies to the solubility of calcium phosphate ($(Ca)_3(PO_4)_2$, Figure 6.8B), for which the solubility equation is given in (6.58)[5].

[5] For the calcium phosphate, $K_{sp} = 2.14 \times 10^{-6}$; $K_{a1} = 7.24 \times 10^{-3}$; $K_{a2} = 6.31 \times 10^{-8}$; $K_{a1} = 3.98 \times 10^{-13}$ at 25°C.

$$s = \sqrt[3]{\frac{K_{sp}\left([H^+_{(aq)}]^3 + [H^+_{(aq)}]^2 \times K_{a1} + [H^+_{(aq)}] \times K_{a1} \times K_{a2} + K_{a1} \times K_{a2} \times K_{a3}\right)}{C_{tot} \times K_{a1} \times K_{a2} \times K_{a3}}} \quad (6.58)$$

The graphical representation of solubility equilibria, such as those of calcium carbonate, calcium phosphate demonstrates how *pH* affects the solubility of salts with *pH*-dependent ions. In acidic conditions, calcium carbonate becomes more soluble due to the formation of bicarbonate and carbonic acid. However, as *pH* increases, the system shifts toward the precipitation of calcium carbonate. The same applies also to calcium phosphate.

For soil systems, this behavior has significant implications. In acidic soils, calcium may be more available to plants, but this can also contribute to the leaching of calcium ions, leading to nutrient depletion over time. In basic soils, calcium carbonate (and/or phosphate) precipitates, reducing the availability of calcium and potentially leading to calcium deficiencies in plants. Understanding these solubility dynamics is essential for predicting nutrient availability and managing soil health, particularly in regions with fluctuating *pH* levels, such as those impacted by acid rain or agricultural activities.

6.14 SOLUBILITY EQUILIBRIA AND GRAPHICAL REPRESENTATION: SOLUBILITY VS TEMPERATURE

As discussed earlier in this book, equilibrium constants are influenced by temperature changes, which significantly affect solubility equilibria. Table 6.3 presents the solubility product (K_{sp}) of calcium carbonate, along with the first (K_{a1}), and second (K_{a1}) dissociation constants for carbonic acid at two different temperatures: 25°C and 90°C.

Table 6.3 Solubility constant (K_{sp}) of calcium carbonate, and acid dissociation constants (K_{a1}, and K_{a2}) of carbonic acid at 25°C and 90°C

Constants	Temperature (°C)	
	25	90
K_{sp}	4.95×10^{-9}	$6.40 \times 10–^{8}$
K_{a1}	4.30×10^{-7}	2.60×10^{-7}
K_{a2}	5.60×10^{-11}	$7.50 \times 10–^{11}$

The data in Table 6.3 illustrate that the solubility product of calcium carbonate (K_{sp}) and the dissociation constants (K_{a1} and K_{a2}) of carbonic acid increase with temperature. This trend suggests that calcium carbonate becomes more soluble as temperature rises, and carbonic acid dissociates more readily at higher temperatures, albeit to a smaller extent for K_{a1} and K_{a2}.

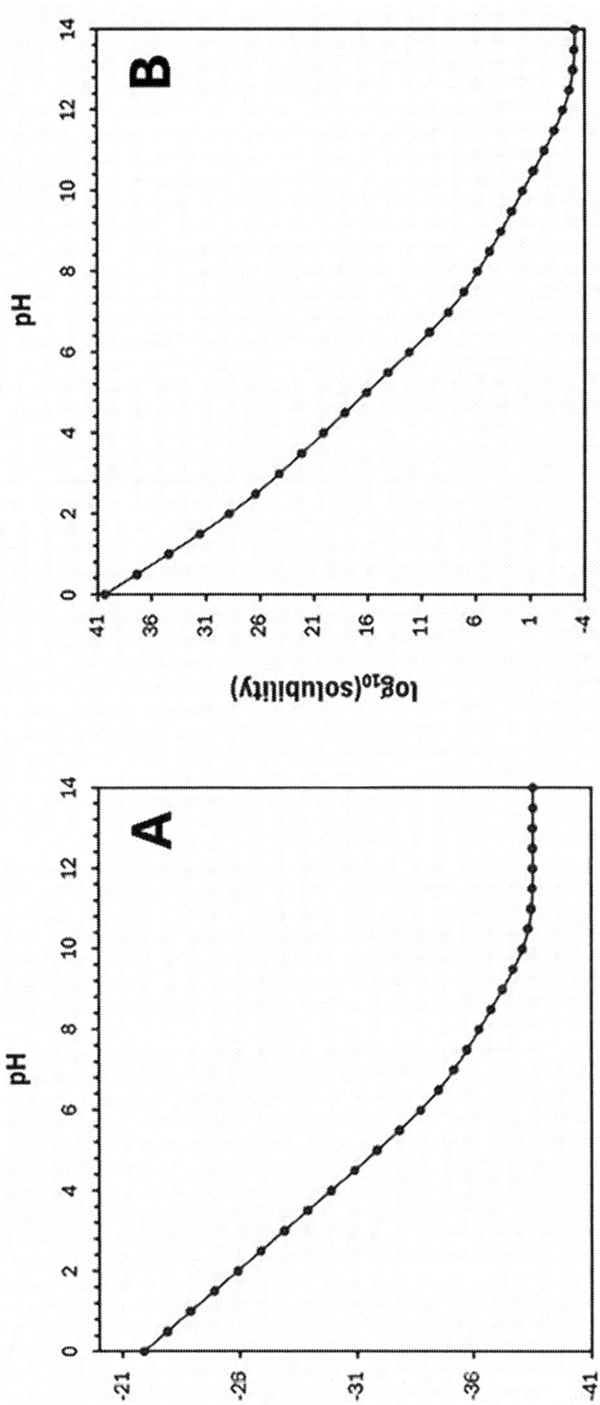

Figure 6.8 Solubility diagrams for calcium carbonate (**A**), and calcium phosphate (**B**). For the elaboration of this figure, a total concentration (C_{tot}) of 0.001 M was assumed.

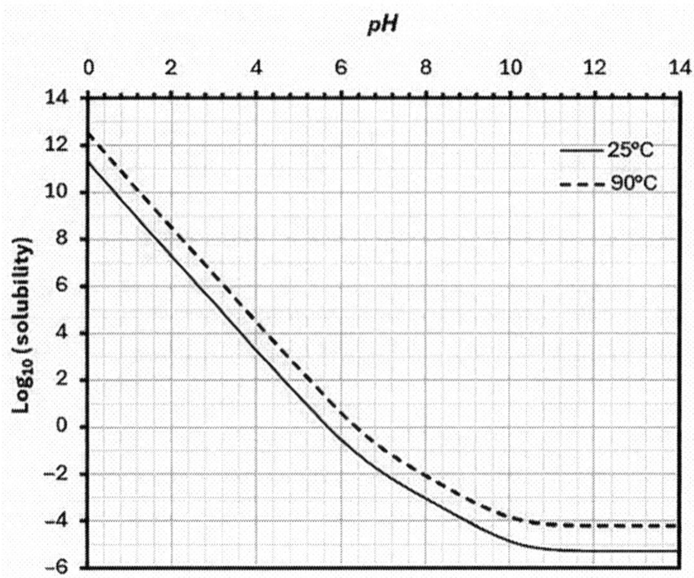

Figure 6.9 Solubility diagram for calcium carbonate at 25°C and 90°C. Data from Table 6.3 and equation (52) were used, with the total concentration of carbonic species in the solution set at 0.001 M.

Using the constants in Table 6.3 and equation (6.57), solubility diagrams were generated (Figure 6.9) to visualize the effect of temperature on calcium carbonate solubility. The diagrams in Figure 6.9 clearly show a substantial increase in calcium ion concentration in solution when the temperature increases from 25° to 90°C. This trend can be attributed to the greater K_{sp} value at higher temperatures, which allows more calcium carbonate to dissolve. Additionally, higher dissociation constants for carbonic acid at elevated temperatures contribute to an increased availability of carbonate ions, further supporting calcium carbonate solubility in solution.

6.15 SOLUBILITY EQUILIBRIA AND GRAPHICAL REPRESENTATION: THE EFFECT OF THE COMMON ION

To obtain a solubility diagram that shows the effect of the common ion (for example, for calcium carbonate with an added calcium or carbonate ion), the following steps must be considered:

1. Use the solubility constant (K_{sp}) as given in equation (6.48).
2. Consider the addition of a salt providing a common ion with calcium carbonate, such as calcium chloride (for Ca^{2+}) or sodium carbonate (for Ca_3^{2-}). According to Le Chatelier's principle, this added common ion will shift the equilibrium, favoring the formation of solid calcium carbonate

and thereby reducing its solubility. For example, if additional calcium ions are added, the concentration of carbonate ions must decrease to maintain, at the selected temperature, a constant K_{sp} value, and vice versa.

3. After adding, for instance, Ca^{2+} ions, the K_{sp} becomes:

$$K_{sp} = \left([Ca^{2+}]_{added} + [Ca^{2+}]_{equilibrium}\right) \times [CO_3^{2-}]$$

Assuming that $[Ca^{2+}]_{added} \gg [Ca^{2+}]_{equilibrium}$, this simplifies to:

$$K_{sp} \approx [Ca^{2+}]_{added} \times [CO_3^{2-}]$$

Therefore, the solubility (s) of calcium carbonate, represented by the concentration of CO_3^{2-}, can be determined as:

$$s = [CO_3^{2-}] = \frac{K_{sp}}{[Ca^{2+}]_{added}}$$

4. Using the solubility constant from Table 6.3, the graph in Figure 6.10 can be obtained.

Figure 6.10 shows that the solubility of calcium carbonate (i.e., $[CO_3^{2-}]$) decreases as the concentration of added calcium ions increases. However, solubility increases with temperature, with the highest solubility observed at 90°C, as previously described in the section on the temperature dependence of solubility.

6.16 THE REDOX DIAGRAMS

Understanding how to graph a redox reaction begins with the Nernst equation, as given in equation (6.22). For a generic redox reaction:

$$Ox + ne^- \rightarrow Red$$

The Nernst equation is:

$$E_h = E_0 - \frac{RT}{nF}\log_{10}\left(\frac{[Red]}{[Ox]}\right) \tag{6.59}$$

where $[Red]$, and $[Ox]$ represent the concentrations of the reduced and oxidized species, respectively. At room temperature (25°C), equation (6.59) simplifies to:

$$E_h = E_0 + \frac{0.0597}{n}\log_{10}\left(\frac{[Ox]}{[Red]}\right)$$

This can be further expressed in terms of pE, where $pE = \frac{E}{0.0597}$, resulting in:

$$pE_h = pE_0 + \frac{1}{n}\log_{10}\left(\frac{[Ox]}{[Red]}\right) \tag{6.60}$$

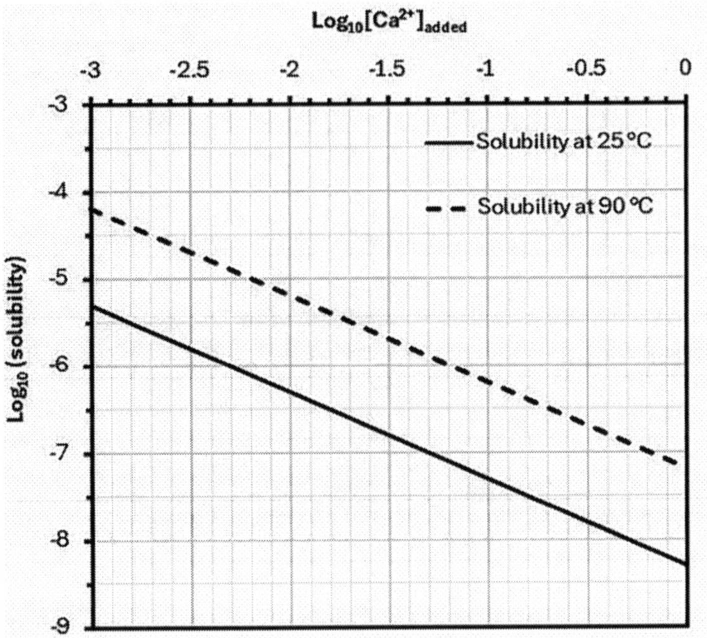

Figure 6.10 Variation in the solubility of calcium carbonate is affected by the common ion effect. Solubility changes are shown for varying concentrations of calcium ions at two different temperatures. As discussed in the temperature dependence section, solubility increases with temperature, with the highest solubility observed at the higher temperature.

Equation (6.59) allows for plotting the electrochemical potential as a function of the concentration of the reduced or oxidized species.
When protons are involved in the redox semi-reaction, as in:

$$Ox + ne^- + mH^+ \rightarrow Red$$

The Nernst equation becomes:

$$E_h = E_0 + \frac{0.0597}{n}\log_{10}\left(\frac{[Ox][\mathrm{H}^+]^m}{[Red]}\right) \tag{6.61}$$

Expanding this, yields:

$$E_h = E_0 + \frac{0.0597}{n}\log_{10}\left(\frac{[Ox]}{[Red]}\right) + \frac{0.0597 \times m}{n}\log_{10}\left([\mathrm{H}^+]\right)$$

Which can be rewritten as:

$$pE_h = pE_0 - \frac{1}{n}\log_{10}\left(\frac{[Red]}{[Ox]}\right) + \frac{m}{n} \times pH \tag{6.62}$$

Equation (6.62) enables the plotting of electrochemical potential versus pH, which is useful for understanding redox behavior in reactions involving protons.

6.17 THE REDOX DIAGRAMS: HOW TO GENERATE AN E_h-CONCENTRATION GRAPH

To generate a redox diagram for a system independent of *pH*, we can consider the iron redox species using the following two semi-reactions:

$$Fe^{3+} + e^- \rightarrow Fe^{2+},\ E_0^{Fe^{3+}/Fe^{2+}} = 0.767\ \text{V} \tag{6.63}$$

$$Fe^{2+} + e^- \rightarrow Fe,\ E_0 = -0.440\ \text{V} \tag{6.64}$$

For the Fe^{3+}/Fe^{2+} couple, the Nernst equation is:

$$E_h = E_0 + 0.0597\log_{10}\left(\frac{[Fe^{3+}]}{[Fe^{2+}]}\right) \tag{6.65}$$

For the Fe/Fe^{2+} the couple, the Nernst equation is:

$$E_h = E_0 - 0.0597\log_{10}\left(\frac{1}{[Fe^{2+}]}\right) \tag{6.66}$$

A plot of the $\log_{10}\left(\frac{[Fe^{3+}]}{[Fe^{2+}]}\right)$ vs E_h can be generated by replacing the $E_0^{Fe^{3+}/Fe^{2+}}$ value in (6.65), and by selecting values for the concentration ratio over a range such as, for example, 0.01 to 10 in steps of 0.5. The plot of $\log_{10}\left(\frac{1}{[Fe^{2+}]}\right)$ against E_h can also be obtained by selecting a suitable range for $[Fe^{2+}]$ to apply in (6.66).

Figure 6.11 shows the resulting redox diagram for the Fe^{3+}/Fe^{2+}, and Fe/Fe^{2+} couples, where the lines/dots indicate potential ranges where different iron species predominate.

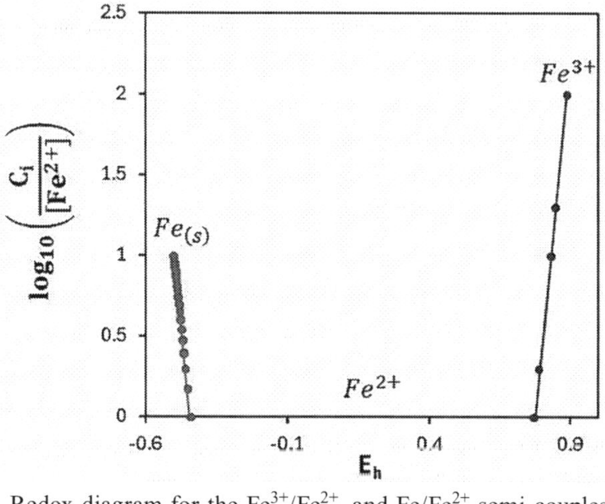

Figure 6.11 Redox diagram for the Fe^{3+}/Fe^{2+}, and Fe/Fe^{2+} semi-couples. The lines/dots indicate the E_h ranges where the different Iron species predominate.

6.18 THE REDOX DIAGRAMS: THE POURBAIX (E_h - pH) DIAGRAMS

When pH affects a redox reaction, the Pourbaix diagrams can be built up. These report the pE (or E_h) – pH stability ranges of the different chemical species, which are interrelated with each other via redox semi-reactions.

Let us consider the oxidation of water to oxygen and its reduction to hydrogen. The two key semi-reactions to account for are:

$$4H^+ + O_2 + 4e^- \rightarrow 2H_2O,\ E_h^{O_2/H_2O} = 1.229\ \text{V}$$

and

$$2H^+ + 2e^- \rightarrow H_2 \qquad E_h^{H^+/H_2} = 0.000\ \text{V}$$

These reactions define the boundaries of water stability in an E_h – pH diagram: below a certain potential, water tends to produce hydrogen gas, while above a certain potential, it tends to produce oxygen gas.

In a Pourbaix diagram, the water stability boundaries—defined by the oxidation of water to oxygen (O_2) and the reduction of water to hydrogen (H_2)—are essential reference lines. These boundaries set the redox limits of the system, marking the potential range within which water remains stable. Above the upper boundary, water tends to oxidize, releasing oxygen gas, while below the lower boundary, it tends to reduce, producing hydrogen gas. All other species in the diagram are stable only within this E_h – pH range, as they would otherwise drive the breakdown of water into its gaseous components. These lines help frame the stability zones for various species, providing a clear reference for redox behavior in aqueous environments.

Assuming that the pressure of hydrogen and oxygen is 1 atm, the Nernst equations for the two semi-reactions are:

$$E_h^{O_2/H_2O} = 1.229 - \frac{0.0597}{4}\log_{10}\left(\frac{1}{[H^+]^4}\right)$$

That, after simplification, becomes:

$$E_h^{O_2/H_2O} = 1.229 - 0.0597 \times pH$$

and

$$E_h^{H^+/H_2} = 0.000 - \frac{0.0597}{2}\log_{10}\left(\frac{1}{[H^+]^2}\right)$$

That, after simplifying, yields:

$$E_h^{H^+/H_2} = -0.0597 \times pH$$

Therefore, choosing a range of pH going from 0 to 14 step 0.5, the values of $E_h^{O_2/H_2O}$, and $E_h^{H^+/H_2}$ can be calculated, thereby leading to the graph reported in Figure 6.12.

This figure shows two lines. The dashed-dot one represents the boundary condition above which water transforms into molecular oxygen, while the solid one is the boundary condition below which water reduces to hydrogen. All the $E_h - pH$ values between the aforementioned lines represent the stability boundary conditions for water. It must be evidenced that water in nature does not oxidize to oxygen nor reduce to hydrogen unless very extreme conditions occur. The very majority of terrestrial waters lay in the $E_h - pH$ region in between the lines reported in Figure 6.12. Due to its electrochemical stability, water permitted the development of life as we know it.

The surface horizons of typical Mediterranean soils reveal a pH value ranging between 5.5 and 7.5 and an E_h value of around 1 V. This means that the surficial horizon lies in oxidizing conditions. That is the reason why, for example, iron artifacts tend to rust when they are buried. In fact, in those conditions, metallic iron oxidizes to Fe^{3+}, which is then transformed into rust (Fe_2O_3).

The procedure for building the Pourbaix diagrams applies similarly to all the electrochemical transformations. Therefore, once the key semi-reactions (as well as the key non-redox equilibria) involved in any transformation have been identified, the application of the Nernst equation combined with the expression for the equilibrium constants governing the non-redox steps allows the drawing of the $E_h - pH$ diagrams where the stability interval for each species implied in the transformation can be visualized. However, it must be kept in mind that only redox (i.e., electron-transfer) reactions influence the pE_h (or E_h) axis in a Pourbaix diagram. Non-redox equilibria, like precipitations/dissolutions, affect pH boundaries but do not shift pE_h directly.

Figure 6.13 reports the Pourbaix diagram for the different species of iron. Region A indicates the $E_h - pH$ values where the electrochemical reactions proceed in the direction of the reduction to metallic iron; region B represents the $E_h - pH$ interval where metallic iron is converted to aqueous Fe^{2+}; the $E_h - pH$ range represented by region C indicates that iron is predominantly present in the form of Fe^{3+}; when switching to region D, iron oxidizes to its maximum oxidation number (+6), thereby forming ferrate ion ($FeSO_4^{2-}$); as the $E_h - pH$ values lay in region E, precipitation of rust occurs; in region F, iron forms a solid Fe_3O_4 system; finally, in region G, a white solid iron(II) oxide is obtained.

Certain iron species are favored in specific pH and pE_h ranges due to their stability in either oxidizing or reducing conditions. For example, Fe^{3+}, a highly oxidized form of iron, is more stable in acidic and strongly oxidizing environments (high pE_h), where electron-donating (reducing) reactions are less likely to occur. On the other hand, FeO_4^{2-}, the ferrate ion exists in highly alkaline (basic) and strongly oxidizing conditions because it represents iron in its highest oxidation state (+6). The combination of high pH and high pE_h stabilizes FeO_4^{2-} by preventing both reduction and protonation, which would otherwise convert it to lower oxidation states or other forms.

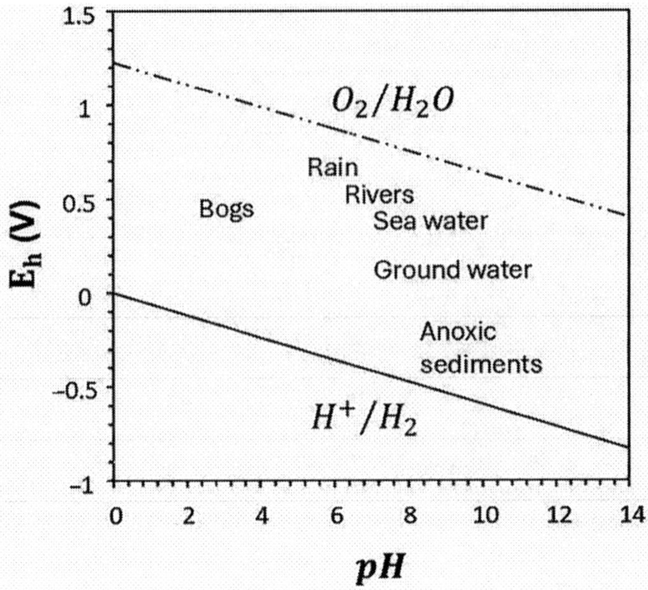

Figure 6.12 $E_h - pH$ diagram illustrating the oxidation of water to molecular oxygen and its reduction to molecular hydrogen. The dashed-dot curve represents the $E_h - pH$ boundary above which water oxidizes to oxygen, while the solid line shows the boundary below which water reduces to hydrogen. Typical $E_h - pH$ conditions for various natural systems are also indicated.

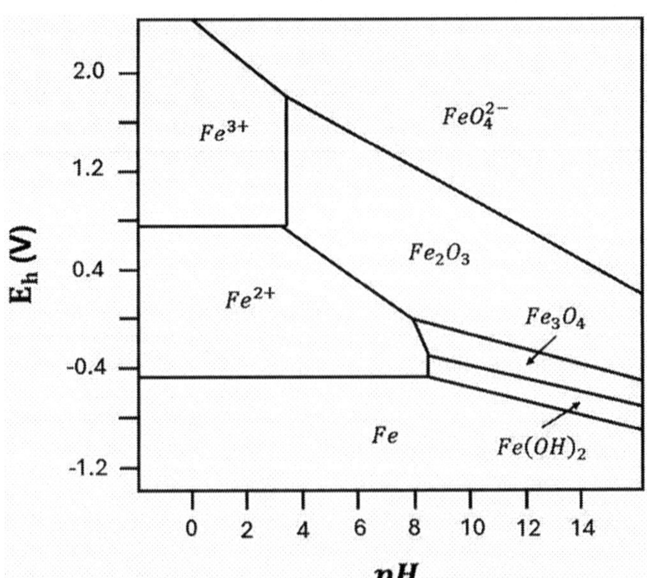

Figure 6.13 Pourbaix diagram for the iron. It was obtained by considering a constant temperature of 25°C. The regions delimitated by the lines are the $E_h - pH$ stability intervals for each iron species.

EXERCISES

1. Given the dissociation constant of acetic acid in water with $K_a = 1.8 \times 10^{-5}$, calculate the concentration of hydrogen ions in a 0.1 M solution of acetic acid.
2. A buffer solution is prepared by mixing 0.2 moles of acetic acid ($K_a = 1.8 \times 10^{-5}$) and 0.1 moles of sodium acetate in 1 liter of water. Using the Henderson-Hasselbalch equation, calculate the pH of this buffer solution.
3. Explain why adding a strong acid (like HCl) to a buffer solution made of acetic acid and sodium acetate affects the *pH* less than adding it to pure water. Calculate the *pH* change when 0.01 moles of HCl is added to 1 liter of the buffer solution from Exercise 2.
4. Given that the solubility product constant (K_{sp}) of calcium carbonate is 4.9×10^{-9}, determine the concentration of Ca^{2+} ions in a saturated solution of calcium carbonate at *pH* 7 and *pH* 5. Explain how the solubility changes with *pH* and why.
5. For the redox reaction $Zn^{2+} + 2e^- \rightarrow Zn$, the standard electrode potential is −0.76 V. Using the Nernst equation, calculate the electrode potential (E_h) when the concentration of Zn^{2+} is 0.01 M at 25°C.
6. Using the following standard electrode potentials:

$$Fe^{3+} + e^- \rightarrow Fe^{2+},\ E_0^{Fe^{3+}/Fe^{2+}} = 0.767 \text{ V}$$

$$Fe^{2+} + e^- \rightarrow Fe,\ E_0 = -0.440 \text{ V}$$

   Construct a simplified Pourbaix diagram for iron, showing the regions where Fe, Fe^{2+}, and Fe^{3+} are stable as a function of *pH* and *pE*.
7. Calculate the *pH* of a 0.1 M solution of sulfuric acid. Assume complete dissociation of the first proton ($K_{a1} \approx 10^3$) and partial dissociation of the second proton ($K_{a2} = 1.2 \times 10^{-2}$).
8. The solubility product constant (K_{sp}) of lead chloride is 1.6×10^{-5}. Calculate the solubility of this salt in pure water and a 0.1 M sodium chloride solution.
9. For the endothermic reaction $N_2O_4 \rightleftharpoons 2NO_2$, predict the shift in equilibrium position when the temperature is increased. Explain why the equilibrium constant increases with temperature for an endothermic reaction.

FURTHER READING

Atkins, P., J. de Paula and J. Keeler. 2018. Atkins' Physical Chemistry, 11th Ed. Oxford University Press.

Atkins, P. and L. Jones. 2010. Chemical Principles: The Quest for Insight, 5th Ed. W.H. Freeman and Company.

Bard, A.J. and L.R. Faulkner. 2001. Electrochemical Methods: Fundamentals and Applications, 2nd Ed. John Wiley and Sons.

Harris, D.C. 2015. Quantitative Chemical Analysis. W.H. Freeman and Company, 9th Ed.

Hites, R.A. and J.D. Raff. 2012. The Elements of Environmental Chemistry, 2nd Ed. John Wiley and Sons.

Manahan, S.E. 2017. Environmental Chemistry, 10th Ed. CRC Press.

Oxtoby, D.W., H.P. Gillis and L.J. Butler. 2015. Principles of Modern Chemistry, 8th Ed. Cengage Learning.

Stummm W. and J.J. Morgan. 1996. Aquatic Chemistry: Chemical Equilibria and Rates in Natural Waters, 3rd Ed. John Wiley and Sons.

Chapter 7

The Chemical Equilibria in Non-Ideal Conditions

7.1 FOUNDATIONS OF SOLUTION EQUILIBRIA: FROM IDEALIZED TO REAL-WORLD CONDITIONS

In previous chapters, we explored the unique molecular structure and properties of water—such as hydrogen bonding, high dielectric constant, and hydration capacity—which define its role as an effective solvent. Idealized chemical equilibria in aqueous solutions were analyzed under simplified assumptions, including negligible solute-solute interactions and uniform solute-solvent interactions. These models enabled a straightforward examination of acid-base reactions, solubility, and redox equilibria, providing a foundation for understanding water behavior in controlled environments.

This chapter now focuses on real-world non-ideal conditions where assumptions break down, and variables like ionic strength, temperature, and pressure introduce more complexity. For example, in soils, where diverse minerals, organic matter, and variable *pH* create a dynamic environment, water acts as a solvent and actively participates in intricate chemical equilibria and redox reactions. High ionic strength in these systems often leads to ion clustering and pairing, impacting the effective concentrations of ions. Consequently, activity coefficients must be applied to accurately describe ion behavior, as the simple concentration-based models from ideal solutions are insufficient. For instance, in solutions like seawater or saline soils, strong ion-dipole interactions decrease the activity coefficients of ions such as hydronium, resulting in *pH* values that deviate from those predicted under ideal conditions. Similarly, in soils with abundant calcium carbonate, interactions between calcium and carbonate reduce the free concentration of these ions, impacting nutrient availability and *pH* buffering capacity. Such real-world deviations highlight the need for advanced models that

DOI: 10.1201/9781003780212-7

incorporate activity coefficients and account for the heterogeneous and complex nature of non-ideal solutions in soils.

7.2 REAL-WORLD DEVIATIONS FROM IDEAL CONDITIONS

To understand the behavior of real-world systems, here are some examples of what happens when ideal conditions break down.

In an acetone-water mixture, the interactions between acetone and water molecules are weaker than in pure water (due to strong hydrogen bonding) and pure acetone[1]. Acetone disrupts the water hydrogen bonding network, making it easier for molecules to escape into the vapor phase. This leads to a higher vapor pressure than expected and a lower boiling point than predicted for an ideal solution. Similarly, the interactions between methanol and water are not as extensive as the hydrogen-bonding network in pure water[2]. Methanol molecules disrupt this network, leading to a reorganization of hydrogen bonds and overall weaker intermolecular cohesion in the mixture. As a result, the solution shows a higher vapor pressure than would be expected under ideal conditions.

When hydrochloric acid (HCl) dissolves in water, it ionizes completely, forming strong ionic interactions between H^+ and Cl^- ions and water molecules[3]. These ion-dipole interactions are significantly stronger than the hydrogen bonds in pure water, reducing the tendency of water molecules to escape into the vapor phase. This results in a lower vapor pressure and a higher boiling point than predicted by ideal conditions.

Glycerol, a highly polar molecule capable of strong hydrogen bonding, forms stronger interactions with water than pure water or pure glycerol alone[4]. As a result, fewer molecules escape into the vapor phase, leading to a lower vapor pressure than expected for an ideal mixture.

Sulfuric acid is also highly polar and dissociates in water[5], thereby resulting in strong ion-dipole interactions between H^+, HSO_4^- ions, and water molecules. These interactions are much stronger than hydrogen bonds in pure water, leading to a significantly reduced vapor pressure. In fact, concentrated sulfuric acid solutions have such low vapor pressures that they are often used as desiccants to absorb moisture from the air.

---

[1] Coutinho, K., N. Saavedra and S. Canuto. 1999. Theoretical analysis of the hydrogen bond interaction between acetone and water. Journal of Molecular Structure (Theochem). 466: 69–75.

[2] B. da Silva, et al. 2011. Hydrogen bond networks in water and methanol with varying interaction strengths. Physical Chemistry Chemical Physics. 13: 593–603.

[3] M.J. Packer and D.C. Clary. 1995. Interaction of HCl with water clusters: $(H_2O)_nHCl$, $n = 1$–3. Journal of Physical Chemistry. 99: 14323–14333.

[4] Nakagawa, H. and T. Oyama. 2019. Molecular basis of water activity in glycerol–water mixtures. Frontiers in Chemistry. 7: 731.

[5] Loshali, R. and N.D. Kandpal. 2016. Intermolecular interactions in sulphuric acid-water system. Indian Journal of Chemistry. 55A: 1080–1083.

In acetic acid and water mixtures, hydrogen bonding between acetic acid and water molecules is stronger than in pure acetic acid, where acetic acid molecules tend to dimerize[6]. The high affinity of acetic acid for water reduces the number of molecules that escape into the vapor phase, resulting in a lower vapor pressure than expected under ideal conditions.

7.3 UNDERSTANDING CHEMICAL EQUILIBRIA IN COMPLEX NON-IDEAL ENVIRONMENTS

In real-world systems, chemical equilibria rarely align with the idealized conditions often described in theoretical models. The latter assume minimal solute-solute interactions, uniform solute-solvent interactions, and solutes that can be considered point-like particles with negligible volume and independent behavior. In practice, these assumptions become unreliable due to factors like high solute concentrations, the presence of multiple interacting ions, and variations in temperature and pressure, all of which can significantly alter equilibrium behavior. Understanding these deviations is essential for accurately predicting the behavior of complex natural systems.

As outlined above, non-ideal behavior arises when molecules and ions in solution interact in ways not accounted for by ideal models. These interactions include electrostatic forces, hydrogen bonding, and van der Waals interactions, all of which impact the behavior of chemical species in solution. In non-ideal solutions, the "activity" of a chemical — that is, its effective concentration in a reaction/transformation—deviates from its actual concentration. This requires using activity coefficients, which adjust for these interactions and provide a more accurate representation of the system state.

High solute concentration is a primary cause of non-ideal behavior. In concentrated solutions, ions or molecules are closer together, leading to significant interparticle interactions that can alter the equilibrium position. For instance, in highly concentrated ionic solutions, electrostatic forces between ions cause ion pairing or clustering, effectively reducing the number of free ions available to participate in reactions and leading to deviations from predictions based purely on concentration. In this context, activity coefficients reflect the effective concentration of each species participating in a reaction or transformation.

Consequently, in non-ideal conditions, the equilibrium constant expression for a general reaction/transformation:

$$aA + bB \rightleftharpoons cC + dD$$

Must become:

$$K_{eq}^{T} = \frac{a_C^c a_D^d}{a_A^a a_B^b} \tag{7.1}$$

[6] Yamabe S. and N. Tsuchida. 2003. A computational study of interactions between acetic acid and water molecules. Journal of Computational Chemistry. 24(8): 939–947.

Here a_i^x represents the activity of the *i*-th chemical species raised to its stoichiometric coefficient. Activity coefficients deviate from the unity in non-ideal systems, representing the non-linear relationship between concentration and activity. Notably, activity coefficients approach unity only under conditions of extreme dilution or in pure substances.

Ionic strength, a measure of the total ion concentration, further complicates non-ideal systems by enhancing electrostatic interactions between charged species, which can either stabilize or destabilize ions, shifting, therefore, the equilibrium state. The Debye-Hückel theory provides a framework for calculating activity coefficients by accounting for these interactions (see Chapter 5). According to this theory, the activity coefficient typically decreases as ionic strength increases, indicating that ions become less reactive due to mutual electrostatic effects.

In non-ideal conditions, temperature and pressure also have a more complex impact on chemical equilibria. While temperature effects on equilibrium constants in ideal systems follow Le Chatelier's principle (favoring endothermic or exothermic reactions), non-ideal systems introduce additional complexities. Temperature changes can significantly affect solubility, activity coefficients, and gas behavior in solution, leading to equilibrium shifts that are not straightforwardly predictable. For instance, gases deviate from ideal behavior at high pressures, influencing their solubility in water according to a modified form of Henry's Law, where gas-solvent interactions require corrections. For a discussion of non-ideal gas behavior, refer to Chapter 12.

7.4 WATER ACTIVITY: THE COMPLEX INTERPLAY OF IONIC STRENGTH AND ION INTERACTIONS

In Chapter 5, the Debye-Hückel model was briefly discussed to understand the relationship between the activity coefficient and ionic strength, which reflects the total concentration of ionic charges in a solution. Numerous experimental studies have explored the relationship between water activity and ionic strength in real-world solutions. For instance, Figure 7.1 illustrates how water activity varies in acidic solutions containing different amounts of salts[7].

A visual inspection of the figure shows that as the cation size increases, the reduction in water activity becomes less pronounced. To illustrate this, consider the solutions involving HCl, NaCl, and KCl. These solutions differ only in their cations, with size increasing in the order $H^+ < Na^+ < K^+$. Consequently, the charge density (*d*) follows the order $d_{H^+} > d_{Na^+} > d_{K^+}$. This trend suggests that water molecules bind more strongly to smaller, highly charged ions, forming tightly

[7] The data in Figure 7.1 have been obtained from the tables reported in: Messnaoui, B., et al. 2020. Use of the Hygrometric Method for Determination of Water Activity of an Aqueous Hydrochloric Acid System at 298.15 K. Journal of Chemical and Engineering Data. 65: 3536–3545.
– Awakura, Y., et al. 1987. Activities of Water and HCl in Aqueous Solution Systems of $HCl\text{-}MCl_n$ including $CuCl_2$, $NiCl_2$, and $FeCl_3$. Hydrometallurgy. 19: 137–157.

bound hydration shells. The binding weakens as the charge density decreases, allowing for greater water mobility and higher water activity. Hence, H^+, with its high charge density, forms a tightly bound hydration shell that significantly reduces water mobility and lowers water activity. In contrast, monovalent ions with lower charge density, such as Na^+, and K^+, create weaker hydration shells, allowing for greater water mobility and higher water activity at the same ionic strength. The larger size of K^+ compared to Na^+, means that water molecules around the former are more flexible, leading to even higher water activity in potassium-containing solutions. However, this simple relationship between ion size and water activity does not fully apply when considering multivalent cations such as Fe^{3+}, Ni^{2+}, and Cu^{2+}. Although one might expect Fe^{3+}, with its high charge density, to produce the lowest water activity, this is not observed. The same observation can be applied to the water activity in the presence of Ni^{2+}, and Cu^{2+}, which should display the opposite trend, as reported in Figure 7.1. The explanation lies in the complex interactions between these ions and water molecules, influenced by hydrolysis, hydration shell characteristics, and specific ion effects.

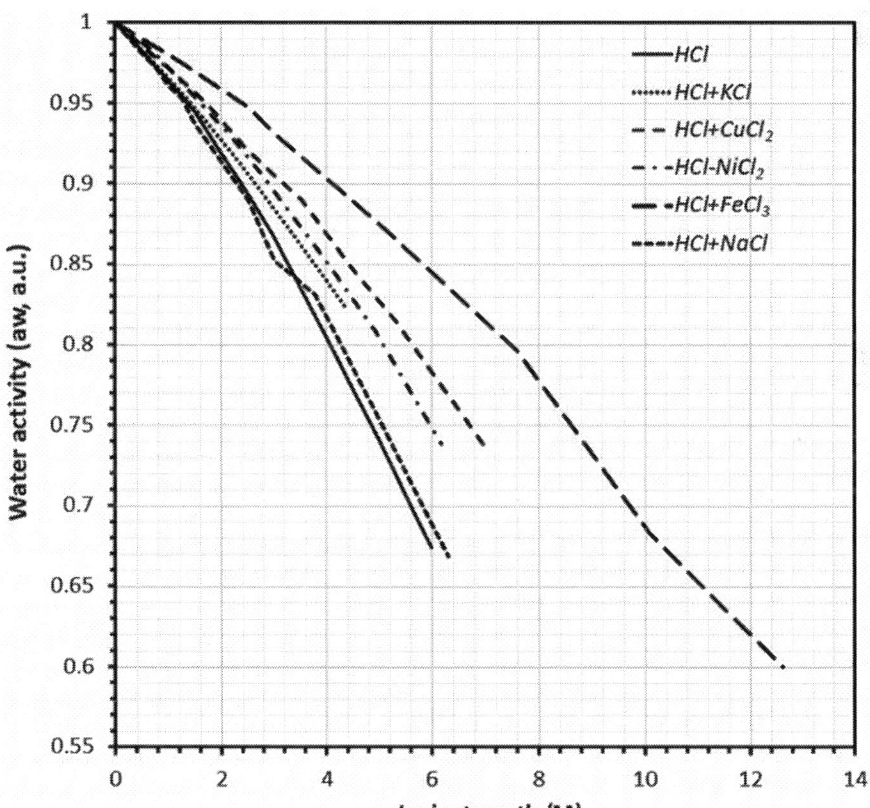

Figure 7.1 Variation of the water activity of HCl-solutions containing different amounts of salts.

7.5 COMPLEX INTERACTIONS BEYOND CHARGE DENSITY

Fe^{3+} has a pronounced tendency to undergo hydrolysis, forming stable species such as $[Fe(H_2O)_6]^{3+}$, $[Fe(H_2O)_5OH]^{2+}$, and other higher-order hydrolyzed complexes. These hydrolyzed complexes tightly bind water within their coordination sphere but have a limited impact on the bulk water structure beyond the immediate shell. Consequently, while water activity within the coordination sphere decreases, the effect on the surrounding water is less significant, allowing for relatively higher water activity compared to Ni^{2+} and Cu^{2+}. In contrast, the latter two ions, which have lower charge densities, can interact dynamically with water, extending their structuring influence into the bulk water. This broader interaction disrupts the hydrogen-bonding network more effectively, resulting in a more significant decrease in water activity.

To understand this behavior further, it is essential to introduce the concepts of kosmotropes and chaotropes.

Kosmotropes are ions that strongly interact with water molecules, enhancing the hydrogen-bonding network and increasing the structural order of water. They are often referred to as "water structure makers." Kosmotropic ions, such as Fe^{3+}, generate tightly bound hydration shells that are highly organized and stable. These shells limit the ions' ability to disrupt water structures beyond their immediate vicinity, which can result in higher water activity compared to other ions.

Chaotropes are ions that disrupt the hydrogen-bonding network of water, leading to a decrease in structural order. These ions are known as "water structure breakers" because they interact less strongly with water, allowing greater disruption and flexibility in the surrounding water molecules. While Ni^{2+} and Cu^{2+} can still be considered kosmotropic, their interactions with water can extend more dynamically into the bulk solution, exerting a broader structuring effect that disrupts the water network more effectively than Fe^{3+}.

The behavior of these ions can be understood through the Hofmeister series, which ranks ions based on their ability to structure or disrupt water structuring. Fe^{3+}, as a strong kosmotrope, tightly structures water within its primary hydration shell, but its influence on bulk water structure diminishes beyond this shell. In contrast, Ni^{2+}, and Cu^{2+} have interactions that extend further, leading to a greater decrease in water activity at similar ionic strengths.

In summary, the interaction of ions with water involves more than charge density and size; hydrolysis, hydration shell characteristics, and the kosmotropic or chaotropic nature of the ions must also be considered. These factors explain why Fe^{3+} solutions might exhibit higher water activity compared to Ni^{2+}, and Cu^{2+} solutions, even at the same ionic strength.

7.6 WATER ACTIVITY IN ACID SOLUTIONS: A BALANCE OF KOSMOTROPIC AND CHAOTROPIC EFFECTS

The explanation based on the Hofmeister effect can also be applied to understand the behavior of acids in aqueous solutions. Figure 7.2 illustrates the changes in water activity as different amounts of strong and weak acids are added[8].

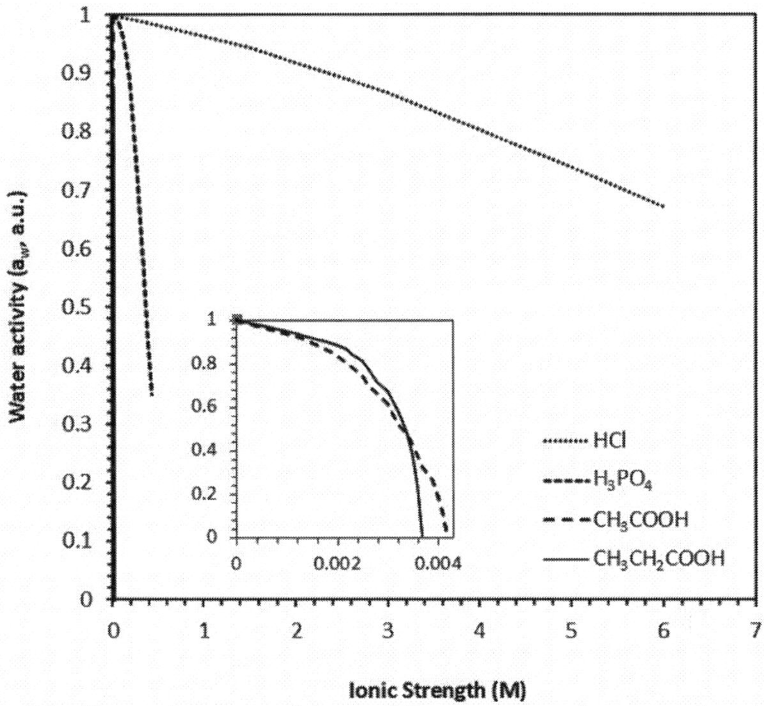

Figure 7.2 Changes in water activity as different amounts of strong and weak acids are added to the solution (**A**). The box (**B**) provides a detailed view of the ionic strength range ($0 \leq I \leq 0.043$ M) to highlight the behavior of water activity in acetic and propionic acid solutions.

HCl dissociates completely in water to form H^+ and Cl^- ions. The H^+ ion, being a proton, has an extremely high charge density and binds strongly to water molecules, forming a tightly bound hydration shell. Cl^-, being a relatively large monovalent anion, has a lower charge density but still exhibits kosmotropic characteristics. The high charge density of H^+significantly structures water by binding water molecules tightly, thereby decreasing the number of free water

[8] The data reported in Figure 7.2 have been obtained by using the tables reported in R.S. Hansen, F.A. Miller and S.D. Christian. 1955. Activity coefficients of components in the systems water–acetic acid, water–propionic acid, and Water–n-Butyric acid at 25°. The Journal of Physical Chemistry. 59(5): 391–395.
– Jiang, C. 1996. Thermodynamics of aqueous phosphoric acid solution at 25°C. Chemical Engineering Science. 51(5): 689–693.

molecules and reducing water activity gradually. The Cl^- ion, while less impactful, still contributes to the kosmotropic effect.

Phosphoric acid is a triprotic acid that partially dissociates in water to release H^+ ions and various phosphate anions (e.g., $H_2PO_4^-$, HPO_4^{2-}, and PO_4^{3-}). These anions have varying degrees of kosmotropic behavior due to their negative charges and ability to have strong H-bonding capability. The presence of multiple negatively charged species enhances water structuring by binding water molecules in an organized manner. This results in a sharp decrease in water activity at lower ionic strengths compared to HCl. This pronounced decrease is due to the formation of multiple hydration shells around the phosphate anions, which effectively sequester water molecules.

Acetic acid is a weak acid that partially dissociates in water to form CH_3COO^- (acetate) and H^+ ions. The acetate ion has a lower charge density and is more chaotropic compared to Cl^- and the phosphate anions. The CH_3COO^- ion interacts less strongly with water compared to more highly charged ions. Its presence disrupts the hydrogen-bonding network to a certain extent, leading to a decrease in water activity that is less pronounced than that observed for stronger acids like HCl or H_3PO_4. The effect of acetic acid on water activity reflects its more moderate kosmotropic properties.

Propionic acid is another weak acid, similar to acetic acid, that partially dissociates to form $CH_3CH_2COO^-$ (propionate) and H^+ ions. With its larger alkyl group, the propionate ion is slightly more hydrophobic and has a lower charge density than the acetate ion, making it even more chaotropic. The hydrophobic nature of propionate further disrupts the water structure, leading to a higher water activity at a given ionic strength than the other acids. This is because the weak interaction between propionate ions and water allows more water to remain free and unstructured.

In summary, the observed changes in water activity for different acids are due to the balance between the kosmotropic and chaotropic effects of their dissociated ions. HCl and H_3PO_4 exhibit stronger kosmotropic behavior and reduce water activity more effectively than CH_3COOH and CH_3CH_2COOH. In contrast, these latter two chemical species, with their more chaotropic anions, disrupt the water structure to a lesser extent, leading to higher water activity.

7.7 WATER ACTIVITY AND SOLUTE TRANSPORT DYNAMICS IN SOIL SYSTEMS

As discussed in previous chapters, water is fundamental to the movement of dissolved substances within the environment, playing a central role in various biogeochemical processes and influencing biological and ecological systems at both micro and macro scales. When a solute is dissolved in water, the process of hydration occurs, characterized by interactions between water molecules and the solute. Depending on the nature of the solute, either a kosmotropic or a chaotropic effect may dominate. Kosmotropic solutes promote an organized, ice-like structure in water, reducing water activity and mobility and forming low-density water

(LDW) micro-domains. These structured micro-domains restrict water movement due to stronger solute-water interactions. In contrast, chaotropic solutes disrupt the water hydrogen-bonding network, increasing water activity and mobility. This leads to the formation of high-density water (HDW) micro-domains, where water molecules are more densely packed and dynamic.

As explained in the context of colligative properties, water moves from less ordered, high-activity HDW regions to more ordered, low-activity LDW regions (HDW → LDW). Importantly, water molecules do not move alone; they transport solute particles with which they are associated. Consequently, solutes move in the same direction as water—toward regions of lower water activity.

Solid surfaces, such as soil particles and plant roots, interact with water through hydrogen bonds and other weak interactions (Figure 7.3A). These surfaces act as kosmotropes, forming a thin, structured LDW film. As water moves away from these surfaces, the chaotropic effect becomes more prominent, generating HDW regions farther from the surface. This water movement from HDW to LDW regions facilitates the transport of solutes initially bound to soil surfaces toward plant roots, where they can be absorbed (Figure 7.3B).

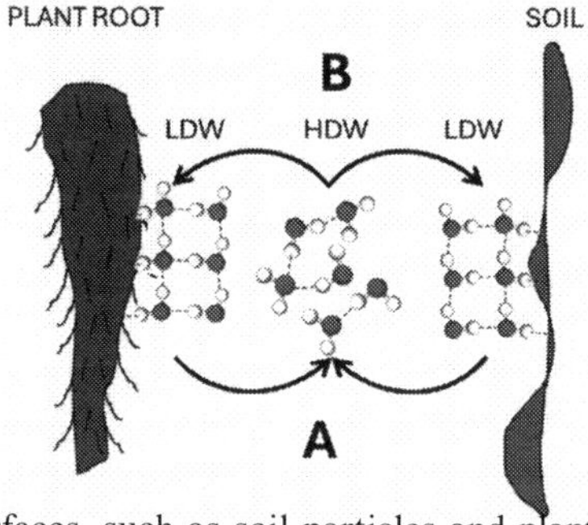

Figure 7.3 (**A**) Solid surfaces, such as soil particles and plant roots, interact with water molecules to promote the formation of structured, ice-like, low-density water (LDW) regions. As the distance from the surface increases, water molecules become more disordered, forming high-density water (HDW) regions. (**B**) Water movement occurs from HDW to LDW regions, transporting solutes in this direction and facilitating nutrient adsorption by plant roots. The image was generated by Napkin.ai.

Soil exchange capacity refers to the ability of soil surfaces to exchange solutes with bulk water. When the focus is on cations, this property is referred to as cation exchange capacity (CEC). Similarly, anion exchange capacity (AEC) is used for anions. This exchange process is crucial for maintaining nutrient availability and facilitating plant uptake from the soil solution, as it ensures that essential ions and molecules are continuously cycled between the soil matrix and the root zone. This dynamic process supports optimal nutrient absorption by enabling the transfer of nutrients from LDW regions on soil surfaces to the HDW regions in bulk solution, and ultimately toward the plant roots. The interaction between water activity and solute mobility helps maintain a balance in the rhizosphere, preventing nutrient depletion near the roots and ensuring that plants can access a steady supply of vital elements. This mechanism is fundamental to plant growth, productivity, and overall soil fertility, making it a key component of sustainable agricultural practices and effective soil management strategies.

7.8 EXPLORING COLLIGATIVE PROPERTIES THROUGH WATER ACTIVITY AND SOLUTE INTERACTIONS

Based on the principles related to the Hofmeister effect, it is now possible to understand the physicochemical mechanisms behind colligative properties.

Osmotic pressure is the pressure required to stop the flow of a solvent through a semi-permeable membrane during the osmosis process. The latter involves the movement of solvent through a semi-permeable membrane from a less concentrated solution to a more concentrated one until equilibrium is reached. This occurs because a concentrated solution has lower water activity, with strong solute-solvent interactions limiting the availability of free water molecules. In contrast, a more diluted solution contains water that is more mobile, with higher water activity due to weaker solute-solvent interactions. The difference in water activity drives the movement of the solvent from the more diluted solution to the concentrated one until equilibrium is achieved, where the rate of solvent movement in both directions across the membrane is equal.

Other colligative properties, such as boiling point elevation and freezing point depression, show similar dependencies on solute-solvent interactions. Boiling point elevation, or ebullioscopic elevation, refers to the increase in the boiling point of a solvent when a non-volatile solute is dissolved in it. This occurs because adding a solute decreases the activity of the solvent. The stronger solute-solvent interactions make solvent molecules more tightly associated, reducing their tendency to escape into the vapor phase. As a result, more energy (higher temperature) is needed for vaporization, leading to an elevated boiling point.

Freezing point depression, or cryoscopic lowering, refers to the decrease in the freezing point of a solvent when a non-volatile solute is added. This phenomenon can be explained by the reduction in solvent activity as solute particles disrupt the arrangement of solvent molecules needed for crystallization. In non-ideal conditions, where solute-solvent interactions are stronger, solvent molecules become less available for phase change, effectively lowering the energy of the liquid phase. This reduced energy means a lower temperature is required to transition to the solid phase, resulting in a depressed freezing point.

7.9 BEYOND IDEAL SOLUTIONS: UNDERSTANDING COLLIGATIVE PROPERTIES IN NON-IDEAL CONDITIONS

When studying colligative properties under ideal conditions, the general rule that they are independent of the chemical nature of the solute becomes apparent. In other words, colligative properties are determined by the concentration of the solute particles rather than their specific chemical identity. However, in non-ideal conditions, properties such as boiling point elevation, freezing point depression, and osmotic pressure no longer adhere strictly to this rule. Under these conditions, the interactions between solute and solvent lead to a significant dependence on the chemical nature of the solute, necessitating a more detailed evaluation of these interactions.

In non-ideal solutions, the nature of the solute becomes crucial due to specific solute-solvent interactions. Ions that form strong ion-dipole interactions with the solvent can significantly alter the behavior of the solution, affecting solvent's activity and changing the observed colligative properties compared to ideal predictions. Activity coefficients are used to modify the effective concentration of solute particles to account for these deviations. These coefficients are influenced by factors such as ionic strength, solute size, charge, and polarity, making colligative properties dependent not just on the number of particles but also on how these particles interact with the solvent and each other.

For aqueous solutions, the Hofmeister series demonstrates how different ions can have specific effects on water structure and solvent activity. Kosmotropic ions, which order water molecules, and chaotropic ions, which disrupt them, can impact properties like freezing point depression and boiling point elevation differently, even at the same molar concentration. Deviations from ideal behavior are particularly pronounced in highly concentrated solutions or those with complex solutes, such as macromolecules or salts with high charge density. For example, solutions containing calcium chloride ($CaCl_2$) or magnesium sulfate ($MgSO_4$) exhibit significant deviations due to strong solute-solvent interactions that alter the chemical potential and activity of the solvent.

In summary, colligative properties in non-ideal conditions are influenced by the chemical nature of the solute. The specific interactions between solute particles and the solvent affect the solvent activity and lead to deviations from ideal behavior. Thus, while the number of solute particles remains a key factor, the type of solute and its interactions with the solvent must also be considered for an accurate prediction of colligative properties in non-ideal solutions.

7.10 A MATHEMATICAL APPROACH TO COLLIGATIVE PROPERTIES AND NON-IDEAL CONDITIONS

As outlined above, colligative properties—including osmotic pressure, boiling point elevation, and freezing point depression—are influenced by solute concentration under ideal conditions, where their behavior is independent of the chemical nature of the solute. However, non-ideal conditions introduce complexities due to solute-solvent interactions that require activity corrections and detailed mathematical treatment.

Osmotic pressure, under ideal conditions, is described by the van't Hoff equation:

$$\Pi = i \cdot c \cdot R \cdot T$$

Here, Π represents osmotic pressure, i is the van't Hoff factor, c is the molar concentration of the solute, R is the gas constant, and T is the absolute temperature. However, real solutions often show significant deviations due to interactions between solute and solvent molecules. Correction factors like activity coefficients are used to predict osmotic pressure accurately in non-ideal conditions.

The study of macromolecular solutions, such as those involving proteins, highlights these deviations. The osmotic pressure for such solutions can be represented with virial expansions:

$$\Pi = R \cdot T \cdot (A_1 \cdot c + A_2 \cdot c^2 + A_3 \cdot c^3 + \cdots)$$

where A_1, A_2, and A_3 are the virial coefficients, reflecting the impact of solute-solvent interactions on osmotic behavior. In particular, A_1 represents the ideal behavior, while A_2, A_3, and subsequent terms account for non-ideal interactions. These higher-order coefficients, such as A_2 (the second virial coefficient), indicate the extent of solute-solvent and solute-solute interactions with $A_2 > 0$ suggesting dominant solute-solvent interactions and $A_2 < 0$ indicating stronger solute-solute attractions.

Under ideal conditions, colligative properties such as boiling point elevation and freezing point depression are determined by the simple equations:

$$\Delta T_b = i \cdot K_b \cdot m$$

$$\Delta T_f = i \cdot K_f \cdot m$$

where ΔT_b and ΔT_f represent the temperature changes for boiling point elevation and freezing point depression, respectively, i is the van't Hoff factor, K_b, and K_f are the ebullioscopic and cryoscopic constants of the solvent, and m is the molality of the solution. These equations assume that the solution behaves ideally, meaning that the properties depend solely on the number of solute particles and not on their interactions with the solvent.

However, solute-solvent and solute-solute interactions become significant in non-ideal solutions, necessitating modifications to these basic equations. One way to account for non-ideal behavior is to incorporate an activity coefficient (γ), modifying the expressions as follows:

$$\Delta T_b = i \cdot K_b \cdot m \cdot \gamma$$

$$\Delta T_f = i \cdot K_f \cdot m \cdot \gamma$$

These equations account for the effective concentration of solute particles by adjusting for the activity coefficient, which varies based on the strength of solute-solvent interactions and the solution's ionic strength.

Alternatively, non-ideal behavior can be represented by using the virial expansions that include higher-order correction terms:

$$\Delta T_b = i \cdot K_b \cdot m \cdot (1 + B_1 \cdot m + B_2 \cdot m^2)$$

$$\Delta T_f = i \cdot K_f \cdot m \cdot (1 + C_1 \cdot m + C_2 \cdot m^2)$$

here, B_1, B_2, C_1, and C_2 are the coefficients representing first-order and second-order corrections due to solute-solvent and solute-solute interactions. These coefficients can be determined experimentally or estimated through theoretical models that account for the chemical nature of the solute and its interaction with the solvent.

The inclusion of higher-order virial terms (e.g., B_1, B_2 for boiling point elevation and C_1, C_2 for freezing point depression) indicates that the solution behavior deviates from ideality due to complex molecular interactions. Such deviations are particularly significant in solutions with high ionic strength, macromolecules, or strong polar solutes, where non-ideal effects become more pronounced. Understanding these corrections is essential for accurately modeling real-world processes, such as those found in environmental chemistry, where the behavior of pollutants and natural solutes in water can be complex and influenced by non-ideal solution dynamics[9].

7.11 THE *pH* IN NON-IDEAL CONDITIONS

In non-ideal systems, the calculation and interpretation of *pH* become more complex due to interactions that deviate from the assumptions of idealized models. In ideal solutions, *pH* is obtained directly from the concentration of hydrogen ions. However, in real-world systems—such as soils, natural waters, and concentrated salt solutions—high ionic strength, ion pairing, and electrostatic interactions require adjustments using activity coefficients to accurately reflect the effective concentration of H_3O^+ ions. In other words, *pH* must be redefined as follows:

$$pH = \log_{10} \frac{1}{\{H_3O^+\}}$$

where the curly brackets represent another way to indicate the activity of a chemical species.

In solutions with high ionic strength, like seawater, electrostatic interactions between ions reduce the effective concentration of free hydrogen ions by promoting ion clustering or ion pairing. This clustering lowers the activity coefficient of hydrogen ions, impacting the observed *pH*. For instance, in seawater, which contains high concentrations of salts such as sodium, magnesium, calcium, potassium, and chloride ions, the activity coefficient of H_3O^+ is reduced, altering the *pH* relative to ideal conditions. Pure water at 25°C has a *pH* of exactly 7, but seawater typically has a *pH* around 8. This alkalinity is due to buffering by carbonate and bicarbonate ions, which stabilize *pH* despite changes in H_3O^+ concentration. In non-ideal conditions, such dissolved ions can significantly influence *pH*, unlike in ideal solutions where such effects are negligible.

---

[9] This paragraph is based on the following references: Nagy, E. 2019. Chapter 21— Pressure-Retarded Osmosis (PRO) process. *In*: Nagy, E. (ed.). Basic Equations of Mass Transport Through a Membrane Layer, 2nd Ed. 505–531. Elsevier. 9780128137222.

– Fullerton, G.D., R.J. Zimmerman, C. Antu III and I.L. Cameron. 1992. New expressions to describe solution nonideal osmotic pressure, freezing point depression, and vapor pressure. Biochemistry and Cell Biology. 70: 1325–1331.

– Chang, Q. 2016. Osmotic pressure. *In*: Chang, Q. (ed.). Colloid and Interface Chemistry for Water Quality Control. 37–47. Academic Press. 9780128093153.

The Debye-Hückel theory provides a framework for estimating activity coefficients in these non-ideal conditions, showing that as ionic strength increases, activity coefficients decrease, making ions less "free" in solution. This is relevant in systems like soils and natural waters with elevated ionic strength, where *pH* calculations based solely on hydrogen ion concentration may be misleading without incorporating activity corrections.

Temperature further complicates *pH* in non-ideal systems. While temperature affects the dissociation constant of water in ideal conditions, in non-ideal environments, it also influences activity coefficients and solubility. For example, rising temperatures in natural waters like lakes and rivers can increase carbon dioxide solubility, thereby forming carbonic acid and subsequently lowering *pH*. Temperature changes can also alter interaction energy between ions, modifying their activity coefficients and potentially leading to under- or overestimations of *pH* in non-ideal settings.

7.12 EXAMPLES OF *pH* IN NON-IDEAL CONDITIONS

As discussed in the previous sections, ionic strength significantly impacts water activity and, consequently, the activity of H_3O^+ ions, thereby influencing the *pH* of aqueous solutions. Ionic strength plays a pivotal role in influencing the activity of water and, consequently, the activity of hydronium ions, directly affecting *pH*. This effect is particularly pronounced in solutions with higher ionic concentrations, where electrostatic interactions between ions alter water activity and *pH* behavior. For instance, while simple salts like KCl and NaCl may have minimal impacts, complex solutions with varying ionic strengths show more significant shifts in *pH*. Such conditions necessitate the consideration of activity coefficients to accurately calculate *pH*, especially when dealing with environmental systems like seawater or soil solutions, where a diverse range of ions contribute to non-ideal behavior.

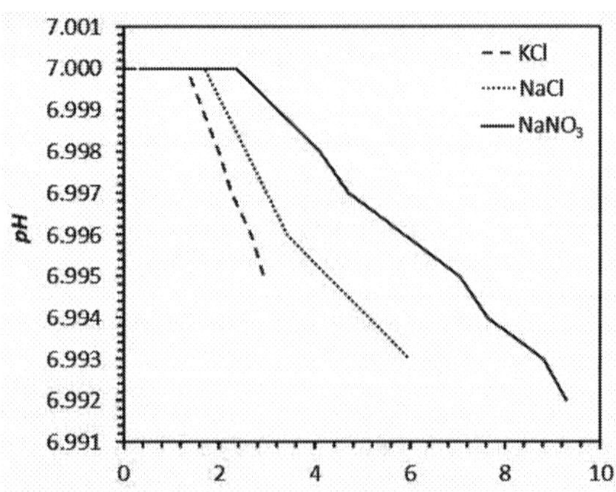

Figure 7.4 Effect of ionic strength (i.e., ionic concentration) on the *pH* of a water solution. The impact is observed only at the third decimal place of the *pH* value, indicating the minimal influence of non-H_3O^+ producing systems on *pH*.

This effect is illustrated in Figure 7.4, where the *pH* impact of neutral salts like KCl, NaCl, and $NaNO_3$ is observed only at the third decimal place, indicating their minimal influence under typical lab conditions.

Although the impact of ionic strength on *pH* may appear negligible in simple solutions, its importance becomes apparent in more complex systems, such as natural waters and soils. In seawater, for instance, dissolved salts contribute to a buffering effect that modifies the activity of hydronium ions, resulting in a *pH* of around 8 despite similar H_3O^+ concentrations as pure water at 25°C. This difference arises from the activity coefficient adjustments and the buffering effects of carbonate and bicarbonate ions[10].

Buffers prepared with high salt concentrations also exemplify non-ideal behavior. For example, a phosphate buffer might exhibit a *pH* of around 7 under dilute conditions, but with added sodium chloride, the effective *pH* can decrease due to a reduced activity coefficient of hydronium ions. In soil solutions, ionic strength variations caused by the presence of organic matter, clay particles, and dissolved ions like calcium and magnesium further demonstrate how non-ideal conditions influence *pH* and nutrient availability, underscoring the importance of considering activity corrections.

7.13 THE ACID-BASE EQUILIBRIA IN NON-IDEAL CONDITIONS

In non-ideal systems, acid-base equilibria are influenced by factors such as high ionic strength, variable activity coefficients, complexation, and the heterogeneous nature of the medium. These conditions require adjustments to the traditional acid-base models. In fact, chemical interactions in non-ideal environments such as soils and saline waters impact both ionization and *pH* buffering capacity.

In high ionic strength solutions, electrostatic interactions lead to ion clustering or ion pairing, which reduces the free energy and alters the chemical activity of ions. Consequently, the dissociation constants of acids and bases depend not only on the concentration but also on the activity coefficients. For example, in high ionic strength environments, the apparent dissociation constant of a weak acid, such as acetic acid (CH_3COOH), deviates from its value in pure water. This is because the activity coefficients of acetate (CH_3COO^-) and hydronium ions (H_3O^+) decrease with increasing ionic strength, thereby resulting in a lower degree of dissociation and a different *pH* than expected under ideal conditions.

Just as a numerical example, let us consider the calculation of the *pH* of a 0.1 M acetic acid solution. All the students who have passed the general

[10] Millero, F.J., D.R. Schreiber and R.N. Roy. 1972. The effect of pressure on the ionization of water at various temperatures from molal-volume data. Journal of Solution Chemistry. 1(3): 239–256.
– Souza, V., A.P. Ordine, I.C.S. Fraga, M.A. Getrouw, P.P. Borges, J.C. Damasceno, et al. 2006. Effect of NaCl and HCl concentrations on primary pH measurement for the certification of standard materials. Brazilian Archives of Biology and Technology. 49 (Special): 79–85.

chemistry exam know that the *pH* of the aforementioned solution is ≈ 2.87. Now, let us assume that the ionic strength of the acetic acid solution is varied (e.g., by adding a certain amount of sodium chloride) so that the activity of the hydronium ion is lowered to 0.75. This effective concentration of H_3O^+ must be used in the equilibrium constant expression to calculate the *pH,* which results, therefore, ≈ 2.79.

Complexation also significantly influences acid-base equilibria in non-ideal conditions. As examples, we must state that metal ions such as calcium, magnesium, and iron can form stable complexes with anions, thereby effectively reducing the free concentration of these ions available for acid-base reactions. Therefore, when considering calcareous soils, calcium ions from calcium carbonate ($CaCO_3$) form complexes with carbonate ions. The formation of such complexes affects the availability of free carbonate in solution and modifies soil *pH* buffering capacity as well as the solubility of carbonate minerals. Similarly, soluble complexes between metal ions and organic ligands in acidic environments can stabilize certain protonated or deprotonated species, further shifting acid-base equilibria. A numerical example to show the effect of the ion in common on the buffering capacity and solubility of calcium carbonate is left to the students.

Ion pairing, where oppositely charged ions associate in solution, is another factor that impacts acid-base behavior in non-ideal conditions. Common in saline and brackish waters with high multivalent ion concentrations, ion pairing reduces the concentration of free ions in solution, affecting the values used in the equilibrium expressions. This can cause shifts in the acid-base equilibria, leading to deviations in *pH* from ideal predictions as the system compensates for decreased activity of dissociated species.

7.14 TYPES OF SOIL ACIDITY AND THEIR ROLE IN SOIL *pH* MANAGEMENT

As discussed throughout this book, soil is a highly complex system influenced by numerous factors that can affect its *pH.* Specifically, soil *pH* is the result of multiple equilibria involving various types of acids and bases present within the system. While calculating the *pH* of a solution in controlled laboratory conditions is relatively straightforward, determining the *pH* of soil requires direct measurement to capture accurate information.

Traditionally, soil *pH* is measured by suspending a known amount of soil in deionized water or a 1 M KCl solution, using a soil-to-solution ratio of 1:2.5. The choice of solution depends on the type of acidity (i.e., the concentration of H_3O^+ ions) being assessed. In soil science, soil acidity can be categorized into actual acidity, exchangeable acidity, residual acidity, and virtual (or potential) acidity.

Actual acidity refers to the active acidity in the soil solution, represented by the concentration of hydronium ions (H_3O^+) freely available in the aqueous phase of the soil. This type of acidity is what is directly measured when determining soil *pH* using deionized water or another neutral solution. Actual acidity reflects the immediate acidic conditions that plants and soil organisms experience and influences nutrient availability, microbial activity, and overall soil health. It is a dynamic measure that can change with environmental factors such as rainfall, fertilizer application, and biological processes within the soil. Unlike residual or exchangeable acidity, actual acidity is not bound to soil particles and is readily influenced by the soil solution's composition.

Exchangeable acidity is associated with hydrogen (H^+) and aluminum (Al^{3+}) ions that are loosely bound to the surface of soil particles, such as clay minerals and organic matter, and can be readily exchanged with cations in the soil solution. This type of acidity is measured by treating the soil with a neutral salt solution, commonly potassium chloride (KCl), which displaces these ions from the soil particles into the solution. Exchangeable acidity contributes to the overall potential acidity of the soil and impacts its ability to buffer *pH* changes. It is particularly significant in acidic soils, where aluminum ions can reach levels toxic to plants. Understanding exchangeable acidity is crucial for assessing soil fertility and determining effective soil management practices, such as liming, neutralizing acidity, and enhancing soil health.

Residual acidity refers to the acidity that remains in the soil and cannot be directly measured in the soil solution or through standard extraction methods. It is primarily associated with hydrogen and aluminum ions bound to soil particles, particularly clay minerals and organic matter, and is not readily exchangeable. This type of acidity contributes to the overall potential acidity of the soil and can affect its long-term pH balance. Residual acidity becomes significant when soil conditions change, such as through liming, which can release these bound ions and result in *pH* adjustments over time. Understanding residual acidity is important for managing soil fertility and determining the appropriate amendments to achieve optimal soil *pH* for plant growth.

Virtual (or potential) acidity is the total acidity that could be released into the soil solution under specific conditions but is not currently active or measurable. It represents the acidity that may be generated by processes such as the decomposition of organic matter, oxidation of certain soil minerals (e.g., sulfides), or chemical reactions that release hydrogen (H^+) and aluminum (Al^{3+}) ions over time. Unlike actual acidity, which is directly measurable in the soil solution, or exchangeable acidity, which involves ions loosely held on soil particles, virtual acidity encompasses latent sources of acidity that have not yet been activated or released.

Virtual acidity becomes particularly relevant when soil management practices, such as deep tillage or drainage, expose previously stable soil layers to oxygen and moisture. These conditions can trigger chemical reactions that increase active acidity in the soil. Understanding virtual acidity is essential for predicting long-term changes in soil *pH* and assessing potential impacts on soil fertility and plant growth.

7.15 BUFFERING CAPACITY IN NON-IDEAL SYSTEMS

Non-ideal conditions significantly influence the buffering capacity of real world systems due to the presence of weak acids, weak bases, and complex interactions between ions and the surrounding matrix. In soils, buffering capacity arises from dissolved ions in the soil solution, and ions are adsorbed onto the surfaces of clay minerals and organic matter. These interactions alter the availability of reactive ions, thereby affecting the soil's ability to neutralize *pH* changes.

Under non-ideal conditions, buffering effectiveness is often reduced by processes such as the formation of complexes and ion pairs, which lower the concentration of free ions available for acid-base reactions. For example, in calcareous soils rich in calcium carbonate, buffering capacity is influenced by the interaction of calcium ions with carbonate, forming stable complexes. While calcium carbonate contributes to buffering, such interactions can limit its capacity to effectively neutralize added acids or bases, reducing its overall buffering efficiency.

In these scenarios, soil *pH* does not simply reflect hydrogen and hydroxide ion concentrations. Instead, it represents a dynamic equilibrium involving multiple factors, including water dissociation, ion exchange processes, precipitation of insoluble compounds, and complexation reactions. These interconnected processes shape the chemical environment of the soil, illustrating the complexity of *pH* regulation in non-ideal systems.

7.16 CASE STUDY: SICILIAN SOIL BUFFERING PROPERTIES

An example of soil buffering capacity under non-ideal conditions is provided by a soil sample from the experimental station in western Sicily, approximately 100 km south of Palermo (37°38' 12.48"N, 13°45' 57.71"E). This region experiences a semi-arid Mediterranean climate characterized by mild, wet winters and hot, dry summers[11]. The soil has a clayey texture (62.1% clay, 21.8% silt, and 16.1% sand; see Chapter 8 for a definition of soil texture) and minimal gravel content. During winter, the saturated soil appears dense and swollen, but as it dries by late spring or early summer, it forms a polygonal pattern of surface shrinkage cracks.

The initial soil *pH*, measured in deionized water using a 2.5:1 water-to-soil suspension, was 8.6. To study its buffering capacity, the soil was treated as a weak base and titrated with 1 M HCl, yielding the titration curve shown in Figure 7.5A. The same soil was then treated as a weak acid and titrated with 1 M NaOH, producing the titration curve presented in Figure 7.5B. These two curves were combined in Figure 7.5C by assigning negative values to the volume of added HCl, thereby producing a single curve similar to those reported in Figure 6.6 in Chapter 6.

7.16.1 Buffer Capacity Curve Analysis

The combined titration curve in Figure 7.5C was analyzed by calculating its first derivative, resulting in the maxima shown in Figure 7.5D. These maxima correspond to the inflection points of the titration curve, where $pH = pKa$. Using

[11] Peel, M.C., B.L. Finlayson and T.A. McMahon. 2007. Updated world map of the Köppen-Geiger climate classification. Hydrology and Earth System Sciences. 11: 1633–1644.

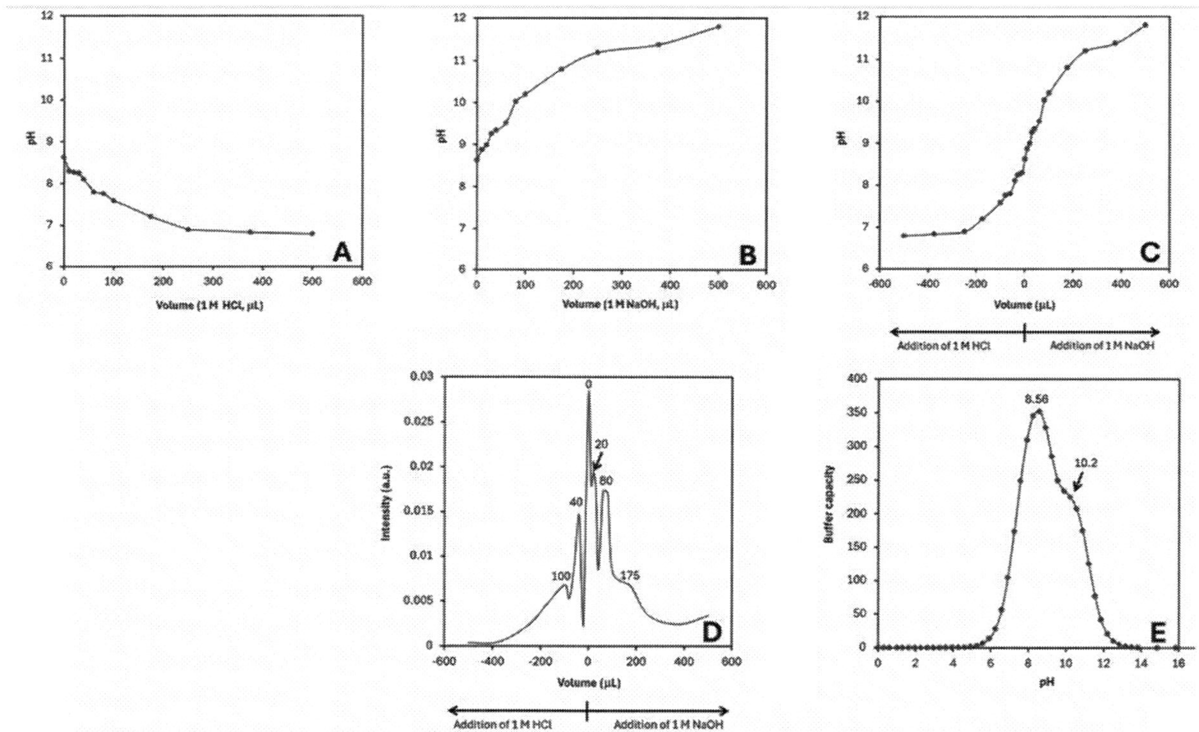

Figure 7.5 Titration curve of Sicilian soil (see the main text for soil details) treated as a weak base: titration with 1 M HCl (**A**). Titration curve of the same soil treated as a weak acid: titration with 1 M NaOH (**B**). The two titration curves were combined by artificially assigning negative values to the volumes of added HCl (**C**). The first derivative of the combined titration curve shown in (**C**) was calculated, and the resulting maxima are presented in (**D**). These maxima correspond to the inflection points of the titration curve where $pH = pKa$. The pKa values were used to generate the buffer capacity curve, providing an assessment of the soil's buffering properties.

these *pKa* values, equation (6.47) was applied to generate the buffer capacity curve, as shown in Figure 7.5E.

Inspection of the buffer capacity curve (Figure 7.5E) reveals two distinct features. The first is a pronounced maximum at *pH* 8.56, representing the highest buffering capacity, where the soil can maintain its functionalities within the *pH* range of 8.23–8.89. The second feature is a shoulder at *pH* 10.2, which is not an effective buffering point. This is due to the asymmetrical shape of the curve around this *pH*, indicating instability. As a result, even small *pH* variations near *pH* 10.2 can cause significant changes in soil properties.

7.17 KEY FACTORS SHAPING SOIL BUFFERING CAPACITY

As discussed throughout this chapter, soil buffering capacity, particularly under non-ideal conditions, is governed by a network of chemical equilibria. These equilibria involve chemical species mediating acid-base balance, ion exchange, and nutrient availability. Understanding these interactions is essential for predicting soil behavior and managing its chemical stability in diverse environmental conditions. Below are some of the primary mechanisms influencing soil buffering capacity.

7.17.1 The Carbonate Buffer System: A Primary Soil Stabilizer

The carbonate system, comprising carbonic acid and its conjugate bases (bicarbonate and carbonate), is a major buffering mechanism in soils, especially those rich in carbonate minerals like calcium carbonate ($CaCO_3$). The buffering reactions are as follows:

$$CO_{2(g)} + H_2O \rightleftharpoons H_2CO_{3(aq)}$$

$$H_2CO_{3(aq)} + H_2O \rightleftharpoons HCO_3^- + H_3O^+$$

$$HCO_3^- + H_2O \rightleftharpoons CO_3^{2-} + H_3O^+$$

In non-ideal conditions, the effectiveness of this system is influenced by the ionic strength of the soil solution and the presence of competing ions. These factors can form complexes with bicarbonate and carbonate, reducing their activity and overall buffering capacity. As a result, the carbonate system in real-world soils often performs less effectively than in idealized scenarios.

7.17.2 Aluminum Hydrolysis: A Contributor to Soil Acidification

In acidic soils, aluminum significantly influences buffering capacity through hydrolysis, which releases protons into the soil solution. The key hydrolysis reactions are:

$$Al^{3+}_{(aq)} + 2H_2O \rightleftharpoons Al(OH)^{2+}_{(aq)} + H_3O^+$$

$$Al(OH)^{2+}_{(aq)} + 2H_2O \rightleftharpoons Al(OH)^{+}_{2(aq)} + H_3O^+$$

$$Al(OH)^{+}_{2(aq)} + 2H_2O \rightleftharpoons Al(OH)_{3(aq)} + H_3O^+$$

In highly basic conditions, aluminum undergoes an additional reaction:

$$Al(OH)_{3(aq)} + OH^- \rightleftharpoons Al(OH)^{-}_{4(aq)}$$

These reactions are strongly affected by soil *pH*, organic matter content, and interactions with competing ions. In acidic soils, aluminum hydrolysis intensifies acidification by releasing additional protons, highlighting its critical role in shaping soil chemistry.

7.17.3 Cation Exchange: Dynamic Ion Balancing

Cation exchange between soil particles (e.g., clay minerals and organic matter) and the soil solution is another vital process affecting buffering capacity. Representative reactions include:

$$2H^+_{(soil)} + Ca^{2+}_{(aq)} \rightleftharpoons 2H^+_{(aq)} + Ca^{2+}_{(soil)}$$

$$Na^+_{(soil)} + K^+_{(aq)} \rightleftharpoons Na^+_{(aq)} + K^+_{(soil)}$$

These equilibria regulate nutrient availability and buffer *pH* changes in soils. The extent of ion exchange depends on factors such as soil cation exchange capacity (CEC), the ionic strength of the solution, and the affinity of soil particles for specific cations. Together, these factors maintain the dynamic chemical balance essential for soil stability.

7.17.4 Organic Acids: The Role of Humic and Fulvic Systems

Organic matter, composed of acids, bases, and amphiphilic compounds like humic and fulvic acids (see Chapter 11), contributes to soil buffering through acid-base equilibria. Representative reactions include:

$$RCOOH + H_2O \rightleftharpoons RCOO^- + H_3O^+$$

$$ROH + H_2O \rightleftharpoons RO^- + H_3O^+$$

$$RNH_2 + H_2O \rightleftharpoons RNH_3^+ + OH^-$$

In non-ideal conditions, interactions between organic acids and metal ions form complexes, reducing the free concentration of protons and affecting soil buffering capacity.

7.17.5 Ammonium and Nitrate Equilibria: Agricultural Soil Dynamics

In agricultural soils, ammonium (NH_4^+) and nitrate (NO_3^-) play crucial roles in nutrient availability and *pH* regulation. Key reactions include ammonium dissociation under basic conditions:

$$NH_4^+ + H_2O \rightleftharpoons NH_3 + H_3O^+$$

and nitrification, i.e., the oxidation of ammonium to nitrate:

$$NH^+_{4(aq)} + 2O_2 \rightleftharpoons NO^-_{3(aq)} + 2H^+_{(aq)} + H_2O$$

Nitrification releases protons into the soil, contributing to acidification over time. The availability of ammonium and nitrate is influenced by factors such as pH, temperature, moisture levels, and microbial activity, underscoring their importance in soil chemical processes.

7.17.6 Iron and Manganese: Redox Processes in Soil Buffering

Redox reactions involving iron and manganese are critical for soil buffering capacity under non-ideal conditions. These elements exist in multiple oxidation states and participate in dynamic equilibria that influence both soil *pH* and redox conditions. The key reactions include the iron redox equilibrium:

$$Fe^{3+}_{(aq)} + 1e^- \rightleftharpoons Fe^{2+}_{(aq)}$$

and the manganese redox equilibrium:

$$MnO_{2(s)} + 4H^+ + 2e^- \rightleftharpoons Mn^{2+}_{(aq)} + H_2O$$

Under reducing conditions, these reactions shift iron and manganese to their more soluble reduced forms, $Fe^{2+}_{(aq)}$, and $Mn^{2+}_{(aq)}$, respectively. This transition not only increases their mobility in the soil but also impacts soil acidity and its capacity to buffer *pH* changes. The interplay of redox and acid-base equilibria highlights the importance of these elements in maintaining the chemical stability of soils under varying environmental conditions.

7.18 REDOX REACTIONS IN SOIL: KEY DRIVERS OF SOIL CHEMISTRY

Redox reactions are fundamental processes in soil systems, shaping nutrient availability, soil health, and plant growth. These reactions are influenced by oxygen availability, water content, organic matter, microbial activity, and soil structure. The redox potential of soils varies significantly depending on their degree of oxygenation: in well-aerated soils, high redox potential favors oxidation reactions, while in waterlogged or poorly oxygenated soils, reducing conditions dominate. Below are the primary redox-driven cycles in soils.

7.18.1 Iron and Manganese: Cycling Between Oxidation States

Iron and manganese undergo redox reactions that significantly impact soil chemistry under varying oxygen levels. Under aerobic conditions, iron is predominantly found as Fe(III) in insoluble oxides, and manganese exists as Mn(IV) in similar forms. However, in anaerobic or poorly oxygenated soils, these metals are reduced to their more soluble forms, Fe(II) and Mn(II), according to the semi-reactions reported in subsection 7.17.6. As already highlighted, the reduction of Fe(III) and Mn(IV) increases their mobility in soil, making them more available for uptake by plants and microorganisms. However, prolonged waterlogging can lead to excessive concentrations of Fe(II) and Mn(II), which may become toxic to plants.

7.18.2 Nitrogen: Balancing Nitrification and Denitrification

Nitrogen redox reactions are integral to soil fertility and nutrient cycling. In aerobic soils, nitrifying bacteria oxidize ammonium (NH_4^+) to nitrate (NO_3^-) through intermediate steps such as nitrite (NO_2^-) formation. Conversely, in anaerobic conditions, denitrifying bacteria reduce nitrate to nitrogen gas (N_2) or nitrous oxide (N_2O):

- Nitrification (aerobic): $NH_4^+ \rightleftharpoons NO_2^- \rightleftharpoons NO_3^-$
- Denitrification (anaerobic): $NO_3^- \rightleftharpoons NO_2^- \rightleftharpoons N_2 + N_2O$

While nitrification increases nitrogen availability for plants in aerobic soils, denitrification under anaerobic conditions leads to nitrogen loss from soils, particularly in wetlands and flooded areas.

7.18.3 Sulfur: Oxidation and Reduction in Anaerobic Soils

The sulfur cycle is heavily influenced by redox processes, especially in anaerobic environments. In aerobic soils, sulfur typically exists as sulfate (SO_4^{2-}), a form readily absorbed by plants. Under anaerobic conditions, sulfate-reducing bacteria convert sulfate into hydrogen sulfide (H_2S):

- Sulfate Reduction: $SO^{2-}_{4(aq)} + 8e^- + 10H^+ \rightleftharpoons H_2S + 4H_2O$

Hydrogen sulfide, a toxic gas, can accumulate in waterlogged soils, affecting plant and microbial life. Additionally, sulfide reacts with Fe(II) to form insoluble iron sulfide (FeS), thereby immobilizing iron in anaerobic environments and altering soil metal dynamics.

7.19 REDOX REACTIONS AND ENVIRONMENTAL CONTAMINANTS

Redox reactions play a crucial role in the fate and transport of environmental contaminants, particularly metals and organic pollutants. The redox state of the soil or aquatic environment can determine whether certain contaminants are immobilized or mobilized, affecting their bioavailability and potential toxicity.

7.19.1 Heavy Metals

The redox state of soils and sediments influences the mobility and speciation of heavy metals such as arsenic (As), chromium (Cr), and mercury (Hg). For example, arsenic exists primarily as arsenate (As^{5+}) under oxidizing conditions and as arsenite (As^{3+}) under reducing conditions. Arsenite is more soluble and toxic than arsenate, and its mobility increases under anaerobic conditions commonly found in wetlands and groundwater:

$$AsO^{3-}_{4(aq)} + 2e^{-} \rightleftharpoons AsO^{3-}_{2(aq)}$$

Similarly, chromium can exist in two oxidation states, Cr(III) and Cr(VI). The latter is highly toxic and mobile, while the former is less toxic and tends to form insoluble compounds that are less bioavailable. Reducing conditions in soils can convert Cr(VI) to Cr(III), thereby mitigating the environmental and health risks associated with chromium contamination.

7.19.2 Organic Pollutants

Many organic pollutants, such as hydrocarbons and chlorinated solvents, are also subject to redox transformations in soils and sediments. Microorganisms play a key role in these processes by using organic pollutants as electron donors or acceptors, thereby breaking down the contaminants into less harmful forms. For example, under anaerobic conditions, certain bacteria can reductively dechlorinate tetrachloroethylene to trichloroethylene and eventually to ethylene, a non-toxic compound:

$$C_2Cl_4 + H^{+} + 2e^{-} \rightleftharpoons C_2Cl_3H + Cl^{-}$$

These redox reactions are critical for the bioremediation of contaminated soils and groundwater, where microorganisms are harnessed to detoxify pollutants through natural or engineered processes.

7.20 FUTURE RESEARCH AND PERSPECTIVES

Advancing our understanding of soil chemistry requires a deeper exploration of how non-ideal conditions impact soil processes at molecular and macro scales.

Future research should focus on real-time monitoring of soil chemical dynamics under diverse environmental conditions, such as extreme weather events, prolonged droughts, and flooding. Additionally, integrating advanced analytical techniques like spectroscopy, isotope tracing, and molecular modeling could provide new insights into the mechanisms underlying ion exchange, metal-organic interactions, and redox equilibria.

Another promising avenue lies in studying the interactions between soil chemistry and emerging contaminants, such as microplastics and pharmaceuticals, which may disrupt natural equilibria. Enhancing predictive models to account for non-ideal behaviors in managed and natural soils will support sustainable agriculture, carbon sequestration, and environmental remediation efforts. By bridging soil science with cutting-edge technologies, future research can better address global food security, climate change, and soil health challenges.

EXERCISES

1. Given the titration data of a Sicilian soil sample ($pH = 8.6$), calculate the buffer capacity at pH 8.56 using the titration curve in Figure 7.5C and the buffer capacity equation $\beta = \frac{\Delta C}{\Delta pH}$. Assume that 5 mL of 1 M HCl is required to lower the pH from 8.56 to 8.23.

   Calculate the buffer capacity (β) at pH 8.56.

   Discuss the factors influencing the observed buffering capacity in this system.
2. A 0.1 M acetic acid solution at high ionic strength (ionic strength = 0.1 M) shows an effective hydronium ion activity ($a_{H_3O^+}$) of 0.75. The dissociation constant Ka for acetic acid in ideal conditions is 1.8×10^{-5}.

   Recalculate the pH of the solution considering the non-ideal conditions.

   Compare the result with the ideal solution pH and discuss the implications of ionic strength on acid-base equilibria.
3. Using the iron redox equilibrium $Fe^{3+} + 1e^- \rightleftharpoons Fe^{2+}$, calculate the standard redox potential (E_0) for this reaction at pH 5 in waterlogged soil. Assume that the activity of Fe^{2+} is 10^{-3} M and that of Fe^{3+} is 10^{-4} M. Use the Nernst equation:

$$E = E_0 - \frac{0.059}{n} \times \log_{10}\left(\frac{[Red]}{[Ox]}\right)$$

   where $E_0 = +0.77$ V and $n = 1$.
4. For a clay-rich soil sample, the following cation exchange reactions are observed:

$$2H^+_{soil} + Ca^{2+}_{aq} \rightleftharpoons 2H^+_{aq} + Ca^{2+}_{soil}$$

If the initial concentration of Ca^{2+}_{aq} is 0.01 M and decreases to 0.005 M, calculate the amount of Ca^{2+} adsorbed onto the soil. Assume the soil sample contains 1 g of clay with a cation exchange capacity (CEC) of 50 mmol/100 g.

5. Consider the reduction of arsenate ($AsO^{3-}_{4(aq)}$) to arsenite ($AsO^{3-}_{2(aq)}$) in anaerobic conditions:

$$AsO^{3-}_{4(aq)} + 2e^{-} \rightleftharpoons AsO^{3-}_{2(aq)}$$

Write the Nernst equation for this redox couple.

Discuss the factors that might influence arsenic mobility and toxicity in soils under reducing conditions.

FURTHER READING

Conte, P. and P. Lo Meo. 2020. Nuclear Magnetic Resonance with Fast Field-Cycling Setup: A Valid Tool for Soil Quality Investigation. Agronomy. 10(7): 1040.

Conte, P. 2021. Fast Field-Cycling NMR Relaxometry: A Technique to Study Water and Ion Dynamics in Soil Systems. Annual Reports on NMR Spectroscopy. 104: 141–188.

Collins, K.D. and M.W. Washabaugh. 1985. The Hofmeister effect and the behaviour of water at interfaces. Quarterly Reviews of Biophysics. 18(04): 323–422.

Henderson, M.A. 2002. The interaction of water with solid surfaces: fundamental aspects revisited. Surface Science Reports. 46: 1–308.

Lo Nostro, P. and B.W. Ninham. 2012. Hofmeister phenomena: an update on ion specificity in biology. Chemical Reviews. 112: 2286–2322.

Marcus, Y. 1985. Ion Solvation. Wiley-Interscience.

Nightingale, E.R. Jr. 1959. Phenomenological theory of ion solvation. Effective radii of hydrated ions. The Journal of Physical Chemistry. 63: 1381–1387.

Zhang, Y. and P.S. Cremer. 2006. Interactions between macromolecules and ions: the Hofmeister series. Current Opinion in Chemical Biology. 10: 658–663.

Chapter **8**

The Physical Science of Soil Colloids

8.1 THE HIDDEN WORLD OF COLLOIDS: A BRIEF INTRODUCTION

The physical chemistry of colloids explores a remarkable class of ubiquitous and diverse materials. Colloidal particles, found extensively in the environment, include pollen, dust, fibers, metals, bacteria, and essential soil components. In soils, colloids primarily consist of clay minerals and organic matter (see Chapters 10 and 11, respectively), with sizes ranging from approximately 2 nm to 2 μm. These particles are large enough to exhibit supramolecular properties but small enough to remain dispersed in a liquid medium, such as water, for significant durations (at least hours). Within this size range, surface properties dominate over bulk characteristics, leading to distinctive behaviors and specific interactions with molecules and ions in their surroundings.

The term "soil colloids" was introduced in the late 19th century, drawing parallels between soil behavior and known colloidal systems. Scientific interest in the colloidal properties of soil components has intensified since the early 21st century, particularly in the clay fraction (<2 μm) and, in some cases, fine silt (2–10 μm). Colloidal suspensions, which involve dispersing colloidal particles in another phase, most commonly consist of solids in a liquid medium. These systems may include macromolecules like proteins or small oxide particles, where the key distinction from solutions lies in phase boundaries and solubility. While solutions are thermodynamically stable, colloidal suspensions exhibit kinetic stability, sustained by energy barriers (e.g., electrostatic repulsion) that prevent aggregation or sedimentation. However, altering conditions, like ionic strength, can destabilize these systems.

DOI: 10.1201/9781003780212-8

Colloids are categorized as lyophobic (low affinity for the dispersing medium) or lyophilic (high affinity). In aqueous systems, the terms hydrophobic and hydrophilic are used. Lyophobic suspensions, such as sols, are common for materials like AgI, Au, and oxides, and they exhibit properties like light scattering and, in some cases, distinctive coloration.

8.2 COLLOIDS DEFINED: SIZE AND BEHAVIOR

Table 8.1 highlights the distinctions between true solutions, colloidal dispersions, and suspensions based on particle size and behavior.

Table 8.1 Differentiation between true solutions, colloidal dispersions, and suspensions based on solute size and molecular behavior

Type of System	Solute Size	Visibility	Type of Motion
True solution	<2 nm	Invisible	Random molecular motion (diffusion)
Colloidal dispersion	2 nm–2 mm	Visible under ultramicroscopy	Brownian motion
Suspensions	>2 mm	Visible under standard optical microscopy	Gravitational settling and convective motion

True solutions consist of solutes smaller than 2 nm, molecularly dispersed, and invisible to the naked eye and advanced microscopy. Their complete dissolution in the solvent ensures homogeneity without phase boundaries.

Colloidal dispersions contain solutes sized between 2 nm and 2 μm. Although too small for optical microscopes, these particles are detectable using ultramicroscopy. Surface properties dominate their behavior including Brownian motion, the random movement induced by solvent collisions, which helps maintain their stability by counteracting gravitational forces.

Suspensions comprise particles larger than 2 μm. Visible under standard optical microscopy, these particles are subject to convective and gravitational forces, causing sedimentation over time if undisturbed. This contrasts sharply with the kinetic stability of colloidal dispersions.

Overall, particle size profoundly influences the physical and dynamic properties of these systems, determining their stability, visibility, and interactions with external forces. Understanding these distinctions is vital for applications ranging from pharmaceuticals to environmental science and materials engineering.

8.3 THE TYNDALL EFFECT: AN INSIGHT INTO COLLOIDAL BEHAVIOR

The Tyndall effect, named after the 19th-century scientist John Tyndall, refers to the scattering of light by particles in a colloidal dispersion. This phenomenon is a defining feature of colloidal systems and distinguishes them from true solutions (Figure 8.1). When a beam of light passes through a colloidal system, the dispersed particles scatter light in multiple directions, making the light beam visible from the side. This scattering occurs because the particle sizes in colloidal systems (2 nm to 2 μm) are comparable to the wavelength of visible light, allowing effective interaction with light waves. In contrast, particles in true solutions are smaller than the wavelength of light and do not scatter it, making the light beam invisible in such systems.

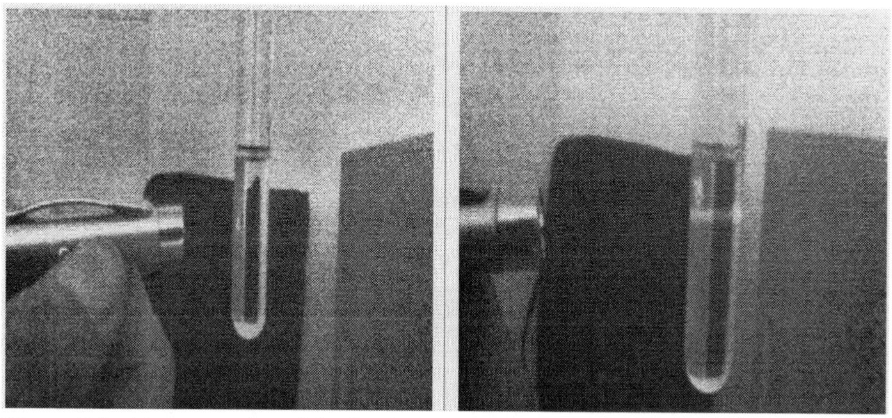

Figure 8.1 Demonstration of the Tyndall effect distinguishing a solution from a colloidal dispersion. The left-hand image shows a light beam passing through a phosphoric acid solution without visible scattering, indicating a true solution. The right-hand image shows a light beam passing through a dispersion of cellulose in phosphoric acid, where the scattered light is visible due to the larger particle size in the colloidal range. This highlights the ability of colloidal particles to scatter light, making the beam visible. Both images belong to the author of the book.

8.3.1 The Mechanism of Light Scattering

The scattering of light in the Tyndall effect arises from the interaction between light waves and dispersed colloidal particles. As light strikes a colloidal particle, the particle absorbs and re-emits the light, scattering it in multiple directions (Figure 8.2). The intensity and pattern of scattering depend on the particle size and the wavelength of the incident light. Shorter wavelengths, such as blue light, scatter more effectively than longer wavelengths, such as red light. This wavelength dependency is why the scattered light often appears bluish.

The relationship between particle size and scattering intensity is also important: smaller colloidal particles scatter light less intensely and in a more uniform pattern, whereas larger colloidal particles produce stronger and more directional scattering. This property makes the Tyndall effect a valuable tool for studying particle size distributions in colloidal systems.

8.3.2 Applications of the Tyndall Effect

The Tyndall effect has significant implications in soil science and environmental monitoring. It provides a straightforward method for distinguishing colloidal systems from true solutions, as colloidal particles scatter light while dissolved molecules in solutions do not (as illustrated in Figure 8.1). Beyond this fundamental distinction, the intensity and pattern of light scattering can offer valuable insights into the size, concentration, and distribution of colloidal particles.

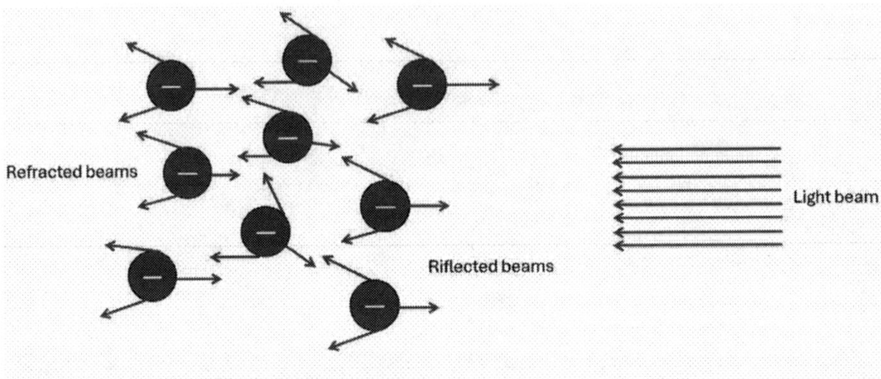

Figure 8.2 A light beam passes through a colloidal dispersion, where part of the light is scattered in multiple directions by the colloidal particles. This scattering, characteristic of the Tyndall effect, makes the light beam visible from the side.

In soil science, the Tyndall effect is used to study colloidal clay minerals, which are critical in determining soil structure, water retention, and nutrient availability. The scattering intensity allows researchers to assess particle size and dispersion, thereby helping to understand soil properties and erosion dynamics. Similarly, in environmental applications, the Tyndall effect measures water turbidity, an essential parameter for monitoring sediment transport in rivers, lakes, and runoff. Quantifying the concentration of suspended colloidal particles makes it possible to evaluate the impacts of soil erosion and sediment deposition on ecosystems.

8.4 THE DYNAMIC EQUILIBRIUM OF COLLOIDS: FORCES, CLASSIFICATIONS, AND INTERACTIONS

Colloidal systems, characterized by the dispersion of particles within a continuous medium, are classified based on the physical states of the dispersed phase and the medium. These combinations give rise to diverse types of colloids, such as gas-liquid systems (e.g., foams like cappuccino froth), liquid-liquid emulsions (e.g., mayonnaise), and solid-liquid dispersions (e.g., soil colloids). Other notable examples include gas-solid systems (e.g., aerogels) and liquid-solid systems (e.g., gels) (Figure 8.3).

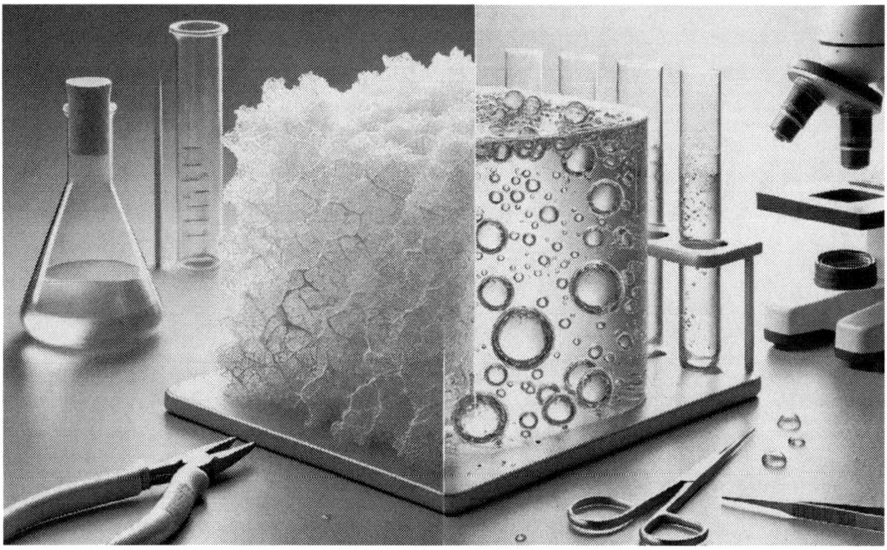

Figure 8.3 The image on the left-hand side depicts a silica aerogel, highlighting its translucent, porous, and ethereal structure. The image on the right-hand side showcases a gel in a transparent container with a viscous texture and visible bubbles. The image was generated with the aid of ChatGPT.

This classification underscores the ubiquity and versatility of colloids in both natural and industrial contexts, bridging disciplines such as chemistry, materials science, and environmental science.

A critical feature of colloidal systems is their stability, which arises from the balance between dispersive and gravitational forces. Dispersive forces, including Brownian motion and electrostatic repulsions, counteract the downward pull of gravity on colloidal particles, maintaining their suspension (Figure 8.4). Brownian motion, the random movement of colloidal particles caused by collisions with solvent molecules, prevents sedimentation by keeping the particles in constant motion. Electrostatic repulsions between similarly charged particles create an energetic barrier that hinders aggregation and coalescence. The system achieves equilibrium when these dispersive forces are sufficiently strong to counterbalance gravitational forces, ensuring the colloidal particles remain uniformly dispersed for extended periods. A colloidal system is typically considered stable when the dispersion maintains its uniformity for at least 4 hours, as this timeframe reflects a sufficient resistance to processes such as sedimentation or aggregation.

Colloidal systems can also be classified based on the nature of the interactions between the dispersed particles and the surrounding medium, leading to two categories: electrocratic and solventocratic colloids. The former are systems where particle stability is predominantly governed by electrostatic interactions, such as those arising from surface charges or adsorbed ions. These systems often exhibit sensitivity to changes in ionic strength, *pH*, or the addition of electrolytes, which can alter the charge distribution and destabilize the colloid by promoting flocculation or sedimentation. Examples include sols of metallic particles, such as Agl or Au, and oxides like hematite and gibbsite. Conversely, solventocratic colloids derive their stability primarily from interactions with the solvent, such as hydration layers or steric effects provided by adsorbed polymers or surfactants. These systems tend to be less sensitive to ionic changes and rely on the affinity of the dispersed particles for the solvent. Hydrophilic colloids in water, such as protein solutions or polysaccharide dispersions, are common examples of solventocratic colloids.

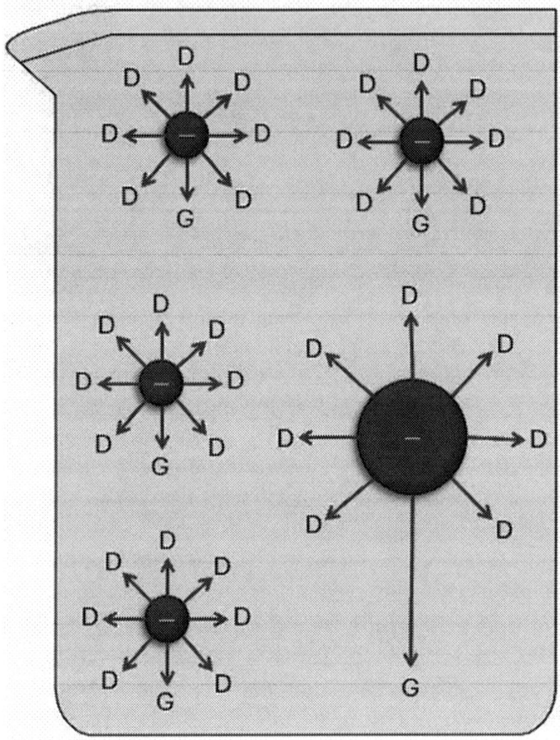

Figure 8.4 Interaction of dispersive forces (D) and gravitational force (G) in determining the stability of colloidal particles. Smaller particles are stabilized by dispersive forces, which counteract gravitational pull, allowing them to remain suspended in the medium. Larger particles experience stronger gravitational force relative to dispersive forces, thereby causing flocculation and sedimentation.

Among colloids, fulvic acids—described in detail in Chapter 11—stand out as particularly stable. In aqueous systems, fulvic acids resist flocculation for much longer than the typical stability threshold of 4 hours, with flocculation not beginning until at least two days. This exceptional stability is attributed to their chemical structure and interactions with water molecules. In contrast, humic acids and humin, discussed thoroughly in Chapter 11, exhibit less stability than fulvic acids. These colloids start flocculating within shorter timeframes, highlighting their relatively weaker dispersive forces or stronger susceptibility to destabilizing factors such as ionic strength or gravitational effects.

8.5 THE GOUY-CHAPMAN-STERN MODEL: THE THEORY OF THE DOUBLE LAYER

As discussed in the previous paragraph, colloidal systems can be broadly categorized into electrocratic and solventocratic types based on their stabilization mechanisms. In the context of soil science, only electrocratic systems are relevant. This distinction arises because soil colloids—comprising clay minerals and organic substances, which will be explored in Chapters 10 and 11, respectively—carry surface charges that interact with surrounding ions in aqueous systems. Consequently, natural aqueous systems involving soil colloids are inherently electrocratic, with their behavior governed by electrostatic interactions and the dynamics of the electrical double layer (EDL).

8.5.1 The Structure of the Electrical Double Layer

When a soil colloidal particle is dispersed in an aqueous medium, its surface is typically charged due to isomorphic substitution, ionization, adsorption, or dissociation processes. This surface charge attracts counterions (oppositely charged ions) from the surrounding solution, forming a distinct region near the particle known as the electrical double layer. The EDL consists of two main components:

1. Stern Layer. Closest to the particle surface, this layer includes counterions strongly adsorbed to the charged surface. These ions are immobilized due to electrostatic attraction and form a compact layer. This portion of the double layer was first proposed by Stern to improve the original Gouy-Chapman model, which did not account for the finite size of ions and their direct adsorption.
2. Diffuse Layer. Beyond the Stern layer lies the diffuse layer, where counterions and co-ions (similarly charged ions to the particle) are distributed according to a balance between electrostatic forces and thermal motion. The ion concentration decreases with increasing distance from the surface, gradually approaching the bulk ionic composition of the solution.

The combination of these two regions forms the electrical double layer, significantly impacting colloidal stability and interaction forces.

8.5.2 The Gouy-Chapman Model

The Gouy-Chapman model was the first theoretical approach to describe the diffuse layer mathematically. It assumes that ions are point charges distributed according to the Boltzmann distribution, where their concentration depends on the electric potential and thermal energy.

The electric potential ($\psi(x)$) within the diffuse layer is governed by the Poisson equation:

$$\frac{d^2\psi(x)}{dx^2} = -\frac{\rho(x)}{\varepsilon\varepsilon_0} \tag{8.1}$$

where $\psi(x)$ is the electric potential at a distance x from the charged surface; $\rho(x)$ is the local charge density; ε is the relative permittivity of the medium; ε_0 is the permittivity in the void [Note: 48 in chapter 5].

The charge density arises from the ionic distribution in the solution and is given by the Boltzmann distribution:

$$\rho(x) = \sum_i z_i e c_i \exp\left(-\frac{z_i e\psi(x)}{k_B T}\right) \tag{8.2}$$

Here, z_i is the valence of the ions; e is the elementary charge; c_i is the bulk concentration of the i-th ion; k_B is the Boltzmann constant; T is the temperature of the Kelvin system. This equation implies that ions of opposite charge to the surface (counterions) are concentrated closer to the surface, while similarly charged ions (co-ions) are repelled. The concentration of ions decreases exponentially with distance from the surface, as modulated by the electric potential.

The solution to the Poisson-Boltzmann equation introduces the Debye length (κ^{-1}), which characterizes the thickness of the diffuse layer and the extent of the ionic cloud around a colloidal particle:

$$\kappa^{-1} = \sqrt{\frac{\varepsilon\varepsilon_0 k_B T}{2e^2 \sum_i c_i z_i^2}} \tag{8.3}$$

Since the ionic strength (I) is defined as in equation (5.19), the Debye length can be simplified and redefined as:

$$\kappa^{-1} = \sqrt{\frac{\varepsilon\varepsilon_0 k_B T}{4e^2 I}} \tag{8.4}$$

This relationship highlights how environmental factors influence the size of the ionic cloud surrounding a colloidal particle. Specifically, the Debye length decreases as ionic strength (I) increases, compressing the diffuse layer and reducing the range of electrostatic interactions (Figure 8.5A). Conversely, the Debye length increases with temperature as higher thermal energy promotes broader ionic distributions around the particle (Figure 8.5B).

The limits of the Gouy-Chapman model lie in the following:

1. Ions are considered point charges with no finite size, neglecting steric effects at high concentrations.
2. The model assumes that ions interact purely through electrostatics and do not bind specifically to the charged surface.
3. Under conditions of strong surface charge, the exponential approximation in the Boltzmann distribution becomes invalid, leading to inaccuracies.

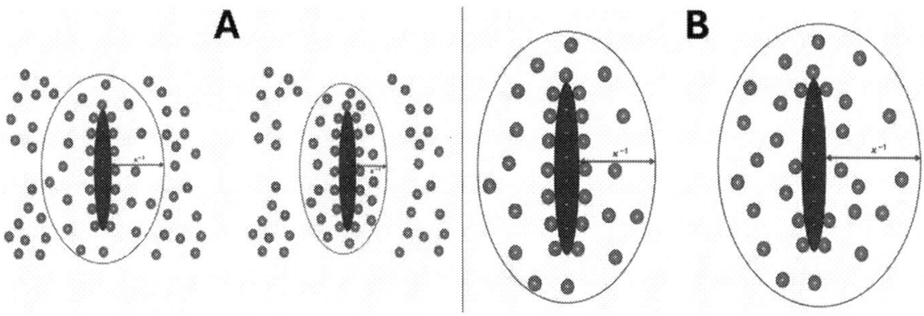

Figure 8.5 Effect of the ionic strength (**A**) and temperature (**B**) on the Debye length (κ^{-1}). In panel (**A**), an increase in ionic strength compresses the electrical double layer, reducing the Debye length due to enhanced electrostatic screening by the surrounding ions. In panel (**B**), an increase in temperature enhances the thermal motion of ions, reducing electrostatic screening and expanding the Debye length.

8.5.3 Stern's Contribution

Stern refined the Gouy-Chapman model by introducing the Stern layer, a compact region where counterions are tightly bound to the charged surface. Unlike ions in the diffuse layer, they do not follow the Boltzmann distribution due to immobilization. This modification accounts for the finite size of ions and their specific adsorption, which the Gouy-Chapman model neglects.

The total electric potential drop ψ_0 across the electrical double layer (EDL) is expressed as the sum of the potentials in the Stern layer (ψ_s) and the diffuse layer (ψ_d):

$$\psi_0 = \psi_s + \psi_d \tag{8.5}$$

The Stern layer behaves like a parallel-plate capacitor, with the potential drop (ψ_s) given by:

$$\psi_s = \frac{\sigma}{C_s} \tag{8.6}$$

where σ is the surface charge density and C_s the Stern layer capacitance is calculated as follows:

$$C_s = \frac{\varepsilon\varepsilon_0}{d_s} \tag{8.7}$$

here, d_s is the thickness of the Stern layer.

The diffuse layer potential (i.e., ψ_d) retains the Gouy-Chapman form:

$$\psi_d = \frac{2k_BT}{e}\sin h^{-1}\left(\frac{\sigma}{\sqrt{8\varepsilon\varepsilon_0 k_BTI}}\right) \tag{8.8}$$

By accounting for these two distinct regions, Stern's model resolves discrepancies observed in high ionic strength systems, where the diffuse layer compresses, and the Stern layer dominates.

8.5.4 Capacitance and Stability

Stern's model also introduces the concept of capacitance[1], combining the Stern and diffuse layer capacitances (C_s, and C_d, respectively) as capacitors in series:

$$C_{tot}^{-1} = C_s^{-1} + C_d^{-1} \tag{8.9}$$

The diffuse capacitance is related to the inverse of the Debye length by equation (8.10):

$$C_d = \varepsilon\varepsilon_0\kappa \tag{8.10}$$

This dual-layer structure accurately predicts electrostatic interactions, which are critical for colloidal stability, adsorption processes, and electrokinetic properties.

8.6 STABILITY FACTORS AND FLOCCULATION MECHANISMS

Flocculation refers to the process where colloidal particles aggregate into larger clusters, known as flocs or floccules, through a variety of physical and chemical interactions. These flocs can either float to the surface or settle at the bottom, depending on their density relative to the surrounding medium, as well as the viscosity and flow conditions of the system. The process begins when the stabilizing forces, primarily electrostatic repulsions, that maintain colloidal particles in a dispersed state are disrupted or weakened. This destabilization allows attractive forces, such as van der Waals interactions or polymer bridging, to dominate, pulling the particles together.

Flocculation is driven by changes in environmental or chemical factors, such as *pH*, ionic strength, or the presence of flocculants like multivalent cations or polymers. When flocculation occurs, the system transitions from a stable colloidal suspension to a phase where larger aggregates are formed. These aggregates grow in size as smaller clusters collide and merge, often assisted by thermal motion or external stirring. The behavior of the resulting flocs—whether they settle or float—depends on the balance of gravitational, buoyant, and hydrodynamic forces acting on them.

8.6.1 The Effect of Electrostatic Neutralization on Colloidal Stability

Colloidal particles typically carry a surface charge, which can be either positive or negative, depending on their composition and the surrounding medium. This

[1] Capacitance is a scalar physical quantity that quantifies the ability of a conductor or system of conductors to store electric charge when subjected to an electric potential relative to its surroundings or a potential difference relative to other conductors. It is defined as the ratio of the charge stored (Q) to the potential difference (V) across the conductor: C = Q/V, where C is the capacitance expressed in Farad (F).

surface charge forms an electrical double layer around each particle, consisting of a compact Stern layer of tightly bound counterions and a diffuse layer of more loosely associated ions. The overlap of these electrical double layers between adjacent particles generates electrostatic repulsive forces that help stabilize the colloidal system, preventing particle aggregation. This stability is particularly important in natural systems where colloidal behavior influences key processes, such as nutrient transport in soils.

Electrostatic neutralization occurs when counterions (ions of opposite charge) are introduced into the system, reducing or completely neutralizing the surface charge of the particles. This diminishes the repulsive forces generated by the EDL, thereby lowering the energy barrier that prevents particles from approaching one another. As the surface charge is neutralized, particles become more susceptible to attractive forces such as van der Waals interactions. These forces act over short distances and promote aggregation when the repulsive barrier is removed (Figure 8.6).

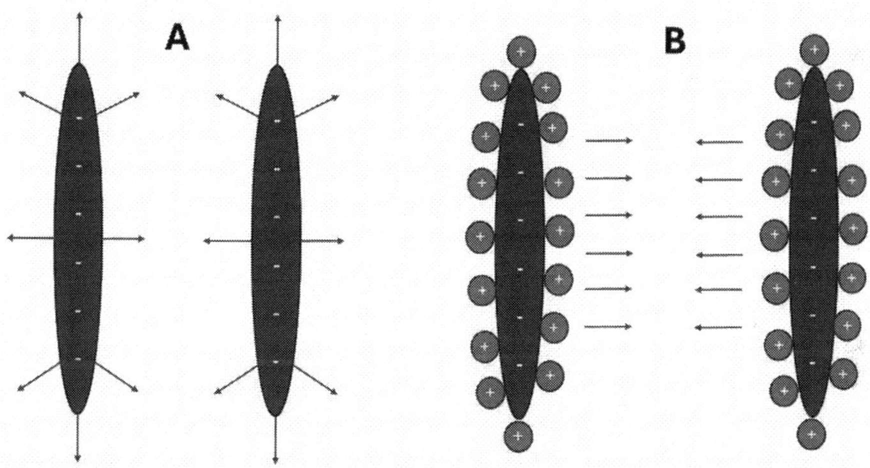

Figure 8.6 Dispersive and attractive forces between colloidal particles. In panel **A**, dispersive forces dominate between negatively charged colloidal particles, maintaining stability and preventing aggregation. In panel **B**, introducing positive counterions neutralizes the negative surface charges, reducing repulsive forces. This neutralization facilitates attractive interactions, initiating flocculation as the particles aggregate due to van der Waals forces and diminished electrostatic repulsion.

The degree of neutralization and subsequent aggregation depends on several factors, including the concentration and valence of counterions. Multivalent ions (e.g., Ca^{2+}, Mg^{2+}, and Al^{3+}) are particularly effective in neutralizing surface charges because they bind more strongly to the charged surface than monovalent ions. As the concentration of such ions increases, the colloidal particles approach the isoelectric point (IEP), the *pH* or ionic condition at which the net surface charge is zero. The repulsive electrostatic forces are minimized at this point, and particles aggregate readily, forming flocs or floccules.

The IEP is highly system-dependent and varies based on the chemical composition of the colloidal particles and the surrounding medium. For example, metal oxides typically have an IEP in the acidic *pH* range, while clay minerals and organic matter have IEPs in neutral to slightly alkaline ranges. Adjusting the ionic strength or *pH* of the system allows control over the aggregation process, which can be advantageous in applications like coagulation-flocculation during water treatment, where colloidal particles are intentionally aggregated for easier removal.

However, excessive neutralization can lead to over-flocculation, where large flocs become unstable and break apart under mechanical stress or turbulent flow conditions. Additionally, re-stabilization may occur if an excess of counterions reverses the particle surface charge, reintroducing electrostatic repulsion.

8.6.2 Colloidal Stability and the Effect of *pH*

The *pH* of the surrounding medium plays a crucial role in determining the stability of colloidal systems, as it directly influences the surface charge of colloidal particles. Many colloidal particles have ionizable surface groups that respond to changes in *pH* by gaining or losing protons. These changes in surface charge modulate the electrostatic interactions between particles, ultimately affecting the balance between stability and aggregation.

For colloidal particles with negatively charged surfaces, such as organic colloids, lowering the *pH* leads to the protonation of surface functional groups such as carboxyl and hydroxyl ones. On the one hand, protonation reduces the negative surface charge, thereby diminishing the repulsive forces that prevent aggregation. As the repulsion weakens, attractive forces such as van der Waals interactions dominate, leading to particle aggregation (Figure 8.7). On the other hand, raising the *pH*, increases the negative charge on such surfaces due to deprotonation, enhancing electrostatic repulsion and stabilizing the colloidal system.

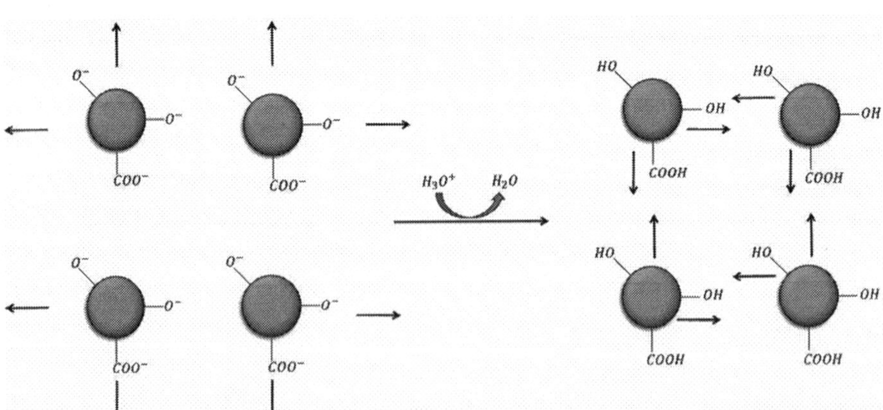

Figure 8.7 Effect of pH on colloid flocculation. Under basic conditions (left), colloidal particles remain dispersed due to strong electrostatic repulsion arising from negatively charged surfaces. As the pH decreases, the negative charges are progressively protonated, weakening the electrostatic dispersive forces. In this environment, hydrogen bonding and van der Waals interactions dominate, forming flocs (right).

For positively charged colloids, the behavior is reversed. Increasing the *pH* causes deprotonation of positively charged surface groups, reducing the overall positive charge and weakening repulsive interactions. This destabilization promotes aggregation, especially at intermediate *pH* levels, where electrostatic stabilization is minimal. At low *pH*, protonation enhances the positive surface charge, strengthening repulsive forces and stabilizing the system.

While *pH* is an effective tool for controlling colloidal stability, extreme *pH* conditions may introduce additional complexities. For instance, excessively acidic or alkaline environments may alter the chemistry of the medium or dissolve certain colloidal materials, thereby changing the overall behavior of the system.

8.6.3 The *pH* Sensitivity of Colloidal Stability in Soils and the Chemistry of Acid and Basic Soils

The sensitivity of colloidal stability to *pH* is a critical factor in natural systems, particularly in soils, where it governs the aggregation and dispersion of clay particles. These processes profoundly affect soil structure, water retention, nutrient transport, and overall soil fertility. In acidic soils, reducing electrostatic repulsion between colloidal particles often promotes aggregation, which can improve soil structure by enhancing particle cohesion. However, excessive aggregation may restrict water infiltration and aeration, leading to potential challenges for plant growth. Conversely, in alkaline soils, the greater dispersion of clay particles can decrease soil permeability and lead to a breakdown of soil structure, resulting in compacted layers and reduced water retention.

8.6.4 Chemical Composition of Acid Soils

Acidic soils are characterized by a low *pH*, typically below 6.5. From a chemical standpoint, these soils exhibit a dominance of hydrogen (H^+) and aluminum (Al^{3+}) ions in the soil solution and on exchange sites within the soil matrix. The high concentration of H^+ ions is responsible for the low *pH*, as they displace other cations from the soil colloids and contribute to the acidification of the soil. Aluminum becomes soluble in acidic conditions, releasing Al^{3+} ions, which further acidify the soil through hydrolysis reactions:

$$[Al(H_2O)_6]^{3+} + H_2O \rightleftharpoons [Al(H_2O)_5OH]^{2+} + H_3O^+$$

This reaction produces additional hydrogen ions, thereby exacerbating soil acidity. Acidic soils often lack essential base cations such as calcium (Ca^{2+}), magnesium (Mg^{2+}), and potassium (K^+), which are replaced by H^+ and Al^{3+} on the cation exchange complex. These soils are common in regions with high rainfall, where leaching of base cations is prevalent, and in areas with parent materials low in base-forming minerals.

8.6.5 Chemical Composition of Basic (Alkaline) Soils

Basic or alkaline soils are defined by a *pH* greater than 7.5 and are often associated with high concentrations of carbonate (CO_3^{2-}) and bicarbonate (HCO_3^-) ions. These soils frequently occur in arid and semi-arid regions, where limited rainfall reduces the leaching of soluble salts. Instead, salts accumulate, leading to high concentrations of sodium (Na^+), calcium (Ca^{2+}), and magnesium (Mg^{2+}) in the soil solution. Sodium-dominated soils, known as sodic soils, are particularly problematic due to the dispersive effect of Na^+ on soil particles, which disrupts soil structure.

In alkaline soils, the presence of carbonate and bicarbonate ions can lead to the precipitation of calcium and magnesium carbonates, thereby reducing the availability of these essential nutrients. Additionally, high *pH* conditions decrease the solubility of micronutrients such as iron, manganese, and zinc, leading to deficiencies that can impact plant growth.

8.6.6 Implications of Soil *pH* on Colloidal Behavior

Clay particles are more likely to aggregate in acidic soils due to the reduced negative charge on their surfaces, caused by the protonation of functional groups like carboxyl and hydroxyl groups. This reduced charge weakens the electrostatic repulsion between particles, allowing van der Waals forces and H-bonds to promote aggregation. While aggregation improves soil structure, it can limit water infiltration and restrict root access to oxygen.

In alkaline soils, the high concentrations of Na^+ ions and the low protonation of surface groups lead to the dispersion of clay particles, destabilizing soil structure. Sodium ions, being monovalent, exhibit a weak ability to neutralize the negative surface charges of clay particles compared to divalent (e.g., Ca^{2+} and Mg^{2+}) or trivalent (e.g., Al^{3+} and Fe^{3+}) cations. This is because monovalent ions create a less compressed electrical double layer characterized by a larger Debye length (κ^{-1}). In such conditions, the repulsive forces between negatively charged clay particles dominate, promoting dispersion (Figure 8.8A).

During rainfall, the ionic strength of the soil solution decreases as dissolved salts are diluted. This reduction in ionic strength further increases the Debye length, extending the range of electrostatic repulsion between clay particles. As a result, dispersed clay particles remain suspended in the soil solution, reducing aggregate stability and clogging soil pores. This phenomenon leads to soil compaction and waterlogging, where water infiltrates poorly and remains stagnant, submerging the soil. Once rainfall ends and water begins to evaporate, the ionic strength of the soil solution increases as salts concentrate. This increased ionic strength shortens the Debye length, compressing the electrical double layer and reducing the range of repulsive forces. The clay particles then come into closer proximity, forming aggregates. However, in sodium-dominated soils, this aggregation often leads to the formation of dense, hard crusts as the soil dries, a condition known as cracking. These crusts impede root penetration and hinder water infiltration, exacerbating soil degradation over time.

In contrast, soils rich in divalent cations like calcium exhibit the opposite behavior (Figure 8B). Due to their higher charge density, divalent cations are more effective at neutralizing the negative charges on clay surfaces. This leads to a much more compressed electrical double layer (shorter Debye length) and stronger attractive forces between clay particles. These forces promote aggregation and the formation of stable soil structures, improving permeability, water retention, and aeration. Calcium, in particular, encourages the formation of aggregates that resist compaction and allow for better water and air movement through the soil profile.

The contrasting effects of sodium and calcium highlight the critical role of cation type and ionic strength in determining soil structure and behavior. Sodium-dominated soils are prone to dispersion, waterlogging, and compaction during rainfall, whereas calcium-rich soils maintain a more stable and productive structure under similar conditions.

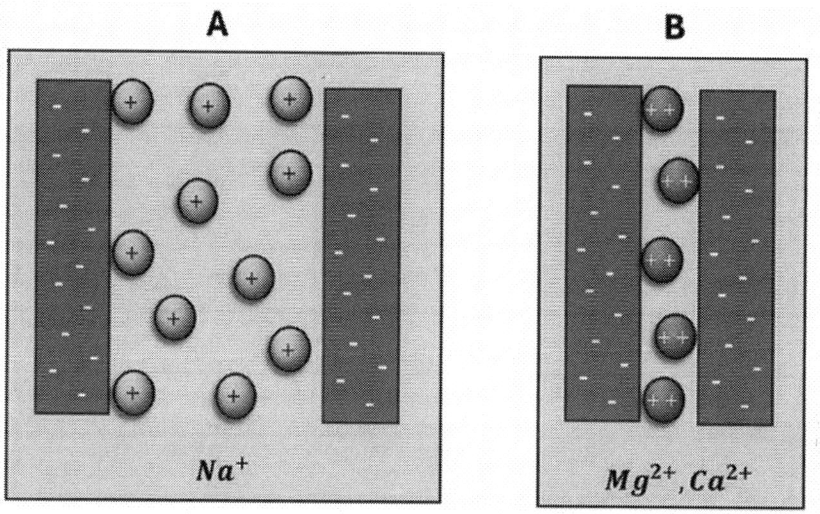

Figure 8.8 Effect of cation type on soil colloidal behavior. In panel **A**, sodium-dominated soils exhibit dispersion due to the weaker neutralizing effect of monovalent Na^+ ions, resulting in a less compressed electrical double layer and extended electrostatic repulsion between negatively charged clay particles. In panel **B**, calcium- and magnesium-dominated soils exhibit aggregation due to the stronger neutralizing effect of divalent Ca^{2+} and Mg^{2+} ions, leading to a more compressed electrical double layer and enhanced attractive forces. This contrast highlights the role of cation charge density in determining soil structure and stability.

8.7 ADSORPTION PROPERTIES OF SOIL COLLOIDS

Due to their surface area and charge density, soil colloids have a remarkable ability to adsorb ions, molecules, and other substances from the surrounding environment. This property is pivotal in nutrient retention, pollutant immobilization, and soil fertility. The behavior of adsorption on colloidal surfaces can be mathematically described by adsorption isotherms, of which the Langmuir and Freundlich equations are the most widely used. These models offer a foundation for understanding the adsorption capacity of soil colloids and their interactions with adsorbed species[2].

8.7.1 The Freundlich Model

Let us consider A a generic compound dissolved in a liquid solvent such as water in natural systems. When a colloidal phase, such as those contained in soils, is introduced, the following equilibrium can be established:

$$A_l \rightleftharpoons A_s \tag{8.11}$$

Here, A_l represents the compound in the liquid phase at the equilibrium, while A_s is the same compound adsorbed onto the colloidal surface at equilibrium. The equilibrium described by equation (8.11) is governed by an equilibrium constant, referred to as the Freundlich constant (K_F), which can be expressed as:

$$K_F = \frac{q_s}{C_l}$$

From this relationship, the amount of compound adsorbed can be written as:

$$q_s = K_F \times C_l \tag{8.12}$$

Here, q_s (mol kg^{-1}) represents the amount of the compound adsorbed per unit mass of the colloidal phase while C_l (mol L^{-1}) is the equilibrium concentration in the liquid phase. The linear form of equation (8.12) applies only under ideal conditions, where no changes in surface properties occur and where no maximum adsorption capacity can be defined. However, such conditions are rarely encountered in real systems. Therefore, a more empirical form of the equation is commonly used:

$$q_s = K_F \times C_l^N \tag{8.13}$$

Here, N is the linearity parameter, indicating adsorption intensity and describing deviations from ideality.

Equation (8.13) is well-suited for describing the adsorption of inorganic compounds, such as heavy metals. However, it often fails to accurately model adsorption when dealing with aromatic organic pollutants, particularly in soils rich in natural organic matter[3]. This is because interactions between soil organic

---

[2] Paragraphs 8.7.1 to 8.7.5 contain mathematical elaborations adapted from Conte, et al. 2021. Recent developments in understanding biochar's physical–chemistry. Agronomy. 11(4): 615.

[3] Zhu, D. and J.J. Pignatello. 2005. Characterization of aromatic compound sorptive interactions with black carbon (charcoal) assisted by graphite as a model. Environmental Science and Technology. 39: 2033–2041. doi:10.1021/es0491376.
– Lattao, C., X. Cao, J. Mao, K. Schmidt-Rohr and J.J. Pignatello. 2014. Influence of molecular structure and adsorbent properties on sorption of organic compounds to a temperature series of wood chars. Environmental Science and Technology 48: 4790–4798.
– Xiao, F. and J.J. Pignatello. 2015. Interactions of triazine herbicides with biochar: Steric and electronic effects. Water Research. 80: 179–188.

fractions and other organic compounds—such as those having anthropogenic origin—are influenced by several factors, including:

- **Cavity effects (cav):** These involve the structural arrangement of solvent molecules (e.g., water) at the solid-liquid interface, facilitating the transport of the solvated solute to specific adsorption sites on the surface.
- **Dispersion forces (disp):** Also referred to as London dispersion forces, these arise from induced dipole interactions and play a significant role in solute adsorption.
- **Dipole-dipole interactions (dip):** Resulting from permanent dipoles, these interactions include solute-solvent, solute-solute, solvent-substrate, and solute-substrate interactions.
- **Hydrogen bonds (H-b):** Similar to dipole-dipole interactions, hydrogen bonds contribute to solvent-solute, solute-solute, substrate-solvent, and substrate-solute interactions.
- **π-π interactions (π-π):** These interactions involve aromatic rings and occur between solutes, solvents, substrates, or combinations of these, influencing adsorption at multiple levels.

To account for these additional complexities, some authors[3] have proposed a modified version of the Freundlich model, expressed as:

$$q_s = K_F^* \times C_l \tag{8.14}$$

In this formulation K^*_F incorporates information about hydrophobic and electrostatic interactions between the adsorbate and the adsorbent. It is expressed as:

$$K_F^* = K_{ow} \times e^{(b_1 - E)}$$

Here, K_{ow} represents the octanol-water partition coefficient[4], reflecting hydrophobic interactions, while E accounts for the electrostatic contribution. The term b_1 is an empirical parameter dependent on the solute concentration.

---

[4] The octanol-water partition coefficient is a measure used in chemistry and environmental science to describe how a substance distributes itself between water and a fat-like substance referred to as octanol. The latter is used as a stand-in for lipids (fats) because it behaves similarly to biological membranes. Water, of course, represents the aqueous environment. When a chemical is added to a mixture of water and octanol, it will naturally "*choose*" where it prefers to stay according to the partition equilibrium: $(\text{chemical})_w \rightleftarrows (\text{chemical})_o$, where the subscripts w, and o lie for water, and octanol, respectively. If the chemical dissolves more in water, it is hydrophilic (water-loving). If it dissolves more in octanol, it is lipophilic (fat-loving). The octanol-water partition coefficient is calculated as: Concentration in octanol/Concentration in water. The importance of the octanol-water partition constant lies in the fact that, depending on the field of application, it can provide highly valuable information. For example, it indicates how easily a substance can pass through cell membranes (relevant in biology and pharmacology); it helps predict how a pollutant may move through the environment (e.g., whether it remains in water or accumulates in organisms); and it offers insights into a compound's bioaccumulation potential and toxicity.

This modified Freundlich model allows for better characterization of complex adsorption behaviors in soils, especially when aromatic organic pollutants are involved.

8.7.2 The Kinetics of Adsorption

A brief introduction to chemical kinetics is necessary to understand how the Langmuir model describes the adsorption properties of soil colloidal systems differently from the Freundlich model.

When considering the equilibrium described in equation (8.11), where A represents the adsorbate distributed between the solution and the soil colloidal phase, the adsorption kinetics can be expressed as:

$$\frac{dQ}{dt} = k_x(Q_e - Q)^x \tag{8.15}$$

Here, Q denotes the amount of adsorbate (in moles) per unit mass (kg) of soil colloid, while Q_e represents the amount of adsorbate in solution at sorption equilibrium (expressed in moles per kilogram of solvent). The parameter x indicates the reaction order ($x \geq 0$), and k_x is the kinetic constant. This constant serves as an experimental measure of the affinity between the adsorbate and the adsorbent: higher k_x values signify stronger adsorbate-adsorbent interactions.

The reaction order is determined experimentally by integrating equation (8.15), which reveals how the reaction rate depends on the concentration of the reactants. Assuming $x = 1$, integration, and rearrangement yield equation (8.16):

$$Q_t = Q_e(1 - e^{-K_1 t}) \tag{8.16}$$

This equation indicates that at a given time t, the quantity of adsorbate on the soil colloidal surface is governed by an exponential relationship modulated by the concentration of the adsorbate in solution. For any $x \neq 1$, the resolution of equation (8.15) gives:

$$Q_t = Q_e - \left[Q_e^{(1-x)} - (1-x)k_x t\right]^{\frac{1}{(1-x)}} \tag{8.17}$$

The reaction order can only be determined experimentally. Specifically, by plotting Q_t-vs-t, the graphs shown in Figure 8.9 are obtained for $x = 1$, 2, and 3, respectively.

If the experimental data align with one of the simulated curves shown in Figure 8.9, then the corresponding reaction order is identified as first (Figure 8.9A), second (Figure 8.9B), or third (Figure 8.9C) order kinetics on soil colloids. The interpolation of the experimental data provides the kinetic constant, which describes the affinity between the adsorbent and the adsorbate.

Due to the complexity of adsorption phenomena, the aforementioned kinetics are often referred to as pseudo-first-order, pseudo-second-order, and pseudo-third-order kinetics, respectively. As a general observation, since the actual order of naturally occurring kinetics is typically unknown in advance, these processes are most often simplified to either first-order or, at most, second-order kinetics. This simplification results in the practical consideration of only two types of adsorption kinetics: pseudo-first-order and pseudo-second-order kinetics.

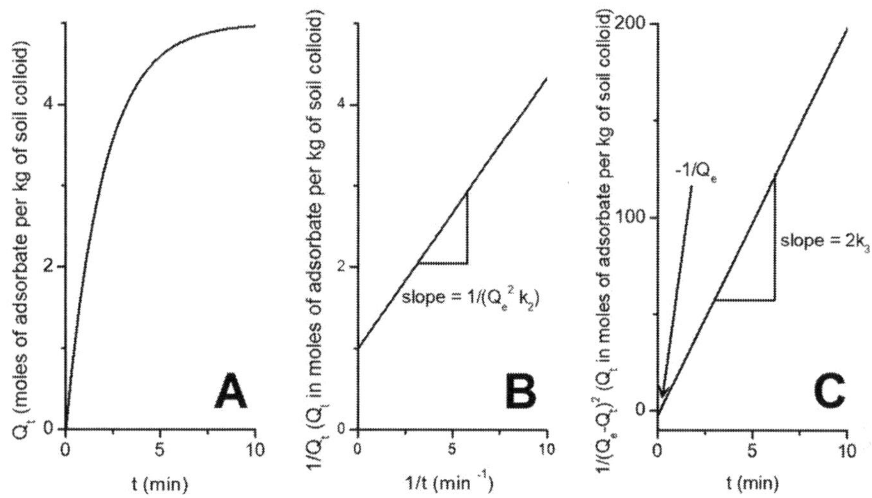

Figure 8.9 Simulated curves for first-order (**A**), second-order (**B**), and third-order (**C**) absorption kinetics on soil colloids.

8.7.3 The Meaning of the Reaction Order

The reaction order is often associated with the concept of molecularity, which describes the number of reactant molecules that come together to form an "activated complex" in the rate-limiting step of the reaction. This activated complex subsequently leads to the formation of reaction products. In a first-order reaction, only one molecule among the reactants participates in the formation of the activated complex. Conversely, the reaction is classified as second-order when two molecules are required to form the collision complex. Reactions occurring on the surface of a solid-phase system, such as a catalyst, can also be described as third-order. In these cases, the collision complex involves all reactants interacting with the surface of the catalyst.

It is important to note that the relationship between reaction order and molecularity is a simplification. Nonetheless, this approach provides a useful framework for developing simplified adsorption theories, such as the one proposed by Langmuir in the early 20th century.

8.7.4 The Langmuir Model

Let us denote the soil colloid by the letter S, the adsorbate by the letter A, and the $A - S$ complex containing a molar amount of adsorbate interacting with the soil colloidal surface by S_1. The equilibrium can be expressed as follows:

$$S + A \rightleftharpoons S_1 \tag{8.18}$$

In this equation, the forward reaction represents the adsorption of A onto S, forming the complex S_1, while the reverse reaction represents the desorption of A from the surface of S.

Assume the following conditions:

- The soil's colloidal surface is flat.
- Each molecule of A interacts with only one functional group on the soil colloidal surface.
- The adsorbate molecules form only a monolayer on the soil's colloidal surface.
- The interactions between A molecules are isotropic.

Under these assumptions, the steady-state condition can be described by the following relationship:

$$r_d = r_i \tag{8.19}$$

Here,

- $r_d = k_d[A][S]$ represents the second-order adsorption rate, where k_d is the adsorption rate constant,
- $[A]$ is the molar concentration of the adsorbate in solution at equilibrium and $[S]$ is the molar concentration of available adsorption sites on the soil colloidal surface at equilibrium.
- $r_i = k_i\,[S_1]$ represents the first-order desorption rate, where k_i is the desorption rate constant, and $[S_1]$ is the molar concentration of the adsorbate bound to the soil colloidal surface at equilibrium.

It is important to note that the expressions for r_d, and r_i are equivalent to equation (8.15).

Given that the total amount of adsorbing sites is defined as $[S_0] = [S] + [S_1]$, the equilibrium constant for equation (8.19) is expressed as $k_e = k_d/k_i$. The fraction of sites covered by the adsorbate is represented as $\vartheta_A = [S_1]/[S_2]$. From this, equation (8.20), the simplest form of the Langmuir adsorption isotherm, can be derived:

$$\vartheta_A = \frac{k_e[A]}{k_e[A]+1} \tag{8.20}$$

Figure 8.10A illustrates simulations of the Langmuir equation (8.20) for varying k_e values. As k_e increases toward 1, the behavior changes from the quasi-linear trend observed at low k_e values (e.g., $k_e = 0.025$) to an L-shaped curve. However, equation (8.20) is not suitable for fitting S-shaped experimental data. This limitation arises from the inherent constraints of the approach, which assumes ideal conditions such as monodentate adsorbents in dilute solutions and a completely flattened soil colloidal surface—conditions that are rarely achievable in real scenarios.

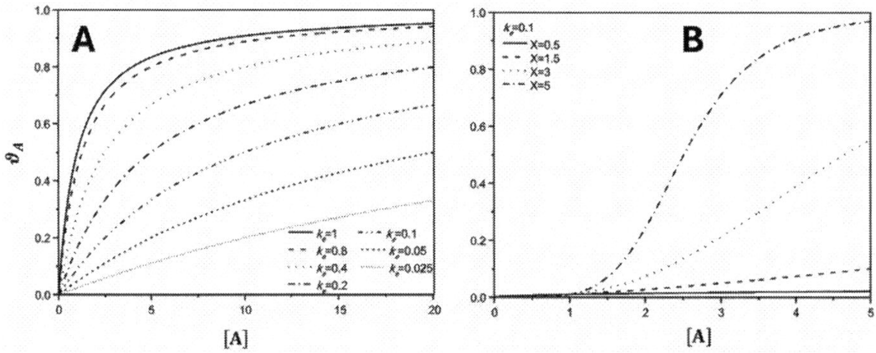

Figure 8.10 (**A**) Langmuir isotherms are simulated using equation (8.20). The curves were generated using the equilibrium constant (k_e) values indicated in the graph. (**B**) Langmuir isotherms are simulated using equation (8.21). The curves were obtained by fixing the k_e value at 0.1 and using the X values specified in the graph.

A modified Langmuir isotherm that accounts for non-ideal conditions is given by equation (8.21):

$$\vartheta_A = \frac{k_e[A]^X}{k_e[A]^X + 1} \tag{8.21}$$

where X is an empirical parameter associated with the effects of the adsorbate concentration. Equation (8.21) serves as a suitable alternative to the model described in equation (8.20) when S-shaped trends are observed (Figure 8.10B).

A slightly modified version of equation (8.21) is known as the Sips equation, which is expressed as equation (8.22):

$$q_e = \frac{q_{max}(k_e[A])^X}{1+(k_e[A])^X} \tag{8.22}$$

Equation (8.22) offers the advantage of expressing k_e consistently in units of L mol^{-1}, regardless of the value of X. Additionally, it simplifies the Freundlich equation when $1 \gg (k_e[A])^X$.

8.7.5 Combining Freundlich and Langmuir Models for Practical Insights

The Freundlich and Langmuir models provide valuable insights into adsorption processes but address different aspects of adsorption behavior. The Freundlich model captures the non-linear adsorption dynamics often observed at low concentrations, where surface heterogeneity is dominant. Conversely, the Langmuir model describes adsorption as a monolayer process on a uniform surface, emphasizing saturation effects as binding sites are progressively occupied. In natural systems, adsorption often involves a combination of these behaviors, necessitating the integration of both models for a more comprehensive understanding.

Hybrid models like the Sips equation merge the strengths of Freundlich and Langmuir's approaches. By incorporating an empirical parameter for surface heterogeneity, the Sips equation provides a versatile framework that accommodates non-linear adsorption at low concentrations and saturation at higher concentrations. This hybridization is particularly useful for describing adsorption processes on soil colloids, where the surface properties are highly variable and dynamic.

Experimental studies demonstrate that both the adsorbate characteristics and the surface chemistry of the colloid influence the adsorption behavior of soil colloids. For instance, soil colloids functionalized with polar groups such as COOH, OH, and C=O tend to favor Freundlich-type adsorption for hydrophilic ions due to their affinity for these functional groups. In contrast, non-polar or hydrophobic adsorbates, such as organic pollutants or hydrocarbons, often exhibit Langmuir-like behavior, where interactions are dominated by hydrophobic forces and the formation of a monolayer on the soil surface.

Applications of these integrated models extend across various fields. In agriculture, understanding the adsorption of nutrients like ammonium and phosphate on soil colloids informs fertilizer application strategies to minimize leaching and maximize efficiency. In environmental remediation, hybrid models are used to predict the adsorption of heavy metals, such as lead and cadmium, on modified clay minerals or biochar[5] surfaces, enabling the design of effective pollutant removal strategies. Similarly, in wastewater treatment, the adsorption of dyes or pharmaceuticals on engineered adsorbents like activated carbon is often described using these combined models.

Additionally, the transition between Freundlich and Langmuir regimes during adsorption can provide insights into the saturation dynamics of soils and sediments. For example, adsorption may follow Freundlich's behavior at low concentrations of pollutants due to the availability of a wide range of binding sites with different affinities. As concentrations increase, saturation occurs, and Langmuir dynamics dominate, highlighting the finite capacity of the soil's colloidal surface.

Future advancements in this area may include the development of even more sophisticated hybrid models that incorporate kinetic factors and multi-component adsorption systems. Such models could address real-world scenarios where multiple adsorbates compete for binding sites on complex surfaces. By combining theoretical insights with experimental data, researchers can refine our understanding of soil colloid interactions and enhance applications in agriculture, environmental management, and industrial processes.

[5]The meaning of the term biochar and its physical-chemical characteristics are discussed in paragraph 14.4.3.

EXERCISES

1. Classify the following systems based on their particle size:
   (a) Salt solution with particles <2 nm.
   (b) Soil suspension with particles between 2 nm and 2 μm.
   (c) Muddy water with particles >2 μm.
2. Explain why the Tyndall effect is observed in colloidal systems but not in true solutions.
3. For a Langmuir adsorption process, given $k_e = 0.1\,\mathrm{L\ mol^{-1}}$ and $[A] = 0.5\,\mathrm{mol\ L^{-1}}$, calculate the surface coverage (ϑ_A).
4. Discuss how lowering *pH* affects the stability of negatively charged soil colloids.
5. Why is the Sips equation considered a bridge between the Freundlich and Langmuir models?

FURTHER READING

Conklin, A.R. 2005. Introduction to Soil Chemistry: Analysis and Instrumentation, 1st Ed. Wiley-Interscience.

Essington, M.E. 2004. Soil and Water Chemistry: An Integrative Approach, 1st Ed. CRC Press.

Molina, F.V. 2014. Soil Colloids: Properties and Ion Binding, 1st Ed. CRC Press.

Sparks, D.L. 2003. Environmental Soil Chemistry, 2nd Ed. Academic Press.

Sposito, G. 2008. The Chemistry of Soils, 2nd Ed. Oxford University Press.

Chapter 9

Fundamental Soil and Colloidal Properties

9.1 SOIL TEXTURE: THE FOUNDATION OF SOIL PROPERTIES AND FUNCTIONALITY

Soil texture is among the most fundamental physical properties of soils. It affects soil behavior and functionality in both natural and managed ecosystems. While the coarsest texture components form the bulk of the soil structure, the colloidal fraction—primarily composed of clay particles—shows a paramount influence due to its tiny size, high surface area, and charge density. These properties govern critical processes such as water retention, permeability, nutrient availability, aeration, and biological activity, thereby making soil texture central to agricultural productivity and environmental conservation. Due to these considerations, we can state that soil texture is more than a mere physical classification. By analyzing the distribution of particle sizes and their interactions, we uncover the essential role soil texture plays in sustaining soil functionality and health.

9.2 UNDERSTANDING SOIL TEXTURAL CLASSES: SAND, SILT, AND CLAY

Soil texture refers to the relative proportion of mineral particles of varying sizes within the soil. These particles are classified into three major categories based on their diameters: sand, silt, and clay. Sand, the coarsest fraction, is subdivided into coarse (0.2–2 mm) and fine sand (0.02–0.2 mm), while silt particles range from 0.002 to 0.02 mm, and clay particles measure less than 0.002 mm. Figure 9.1 illustrates the relative sizes of these particle classes, emphasizing the substantial differences in their dimensions.

DOI: 10.1201/9781003780212-9

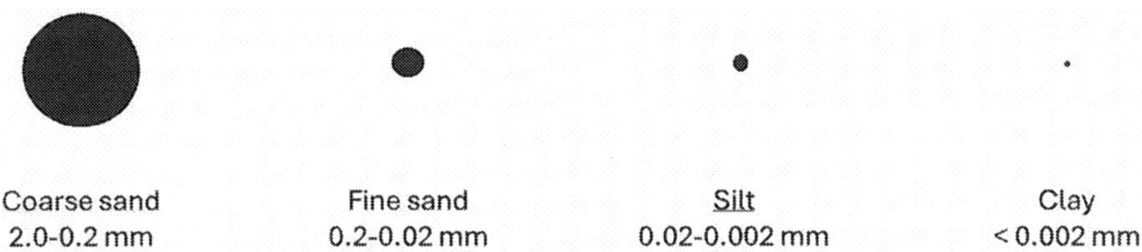

Figure 9.1 Relative particle sizes of soil texture classes. The circles are proportional to each other, thereby showing that coarse sand is the largest-sized particle, whereas clay is the smallest one.

Particles larger than 2 mm, such as gravel and stones, are excluded from the definition of soil texture and are classified as the "skeleton" or coarse fragments. These coarse fragments do not contribute to the finer-textured fraction of the soil but can significantly impact soil structure and behavior.

The distribution of sand, silt, and clay particles defines the soil textural class, which can be determined using a soil texture triangle (Figure 9.2). For instance, soil with 30% sand, 40% silt, and 30% clay is classified as a clay loam. Understanding soil texture is crucial as it affects physical, chemical, and biological processes, influencing soil behavior and ecosystem functionality.

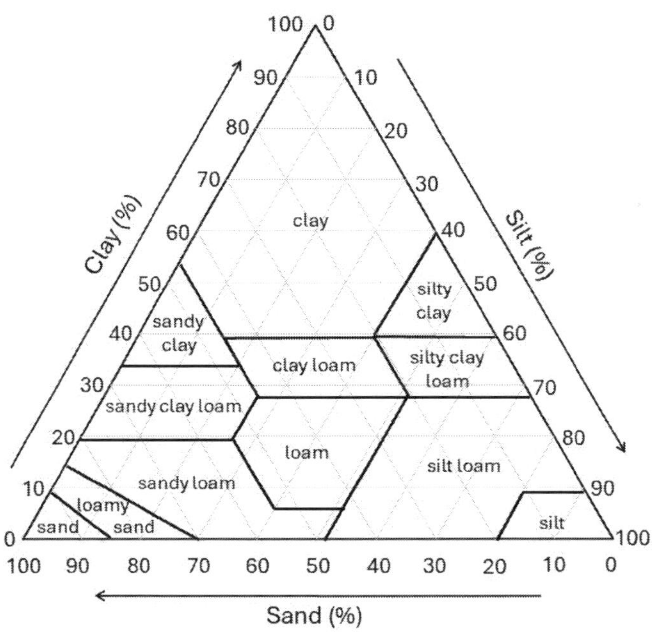

Figure 9.2 A soil texture diagram is used to recognize the textural class of soil.

9.2.1 Stony Soils: Challenges and Environmental Impacts

Soils with a high proportion of coarse fragments (larger than 2 mm) are called stony soils. These soils pose mechanical challenges for cultivation due to the interference of coarse particles with tilling and planting. Additionally, stony soils exhibit low water retention, as the skeleton reduces the proportion of fine particles capable of absorbing and holding water. While this property enhances drainage, it can lead to water stress for plants in arid regions.

Nutrient availability in stony soils is typically limited because coarse fragments have minimal surface area and lack the capacity for cation exchange or nutrient retention. Furthermore, the physical barriers created by these fragments can restrict root penetration, reducing access to deeper water and nutrients. Environmentally, stony soils provide excellent drainage but are prone to leaching contaminants, increasing the risk of groundwater pollution.

9.2.2 Sandy Soils: Permeability and Nutrient Challenges

Sandy soils are dominated by particles with large diameters (0.2–2 mm), resulting in a coarse texture. The large particle size means sandy soils have low surface area per unit mass (for a deeper understanding of the relationship between texture size and surface area, the reader is addressed below), limiting their ability to retain nutrients and water. Water infiltrates rapidly, reducing water availability for plants and increasing the risk of leaching nutrients and contaminants into groundwater.

Despite their low nutrient content, sandy soils are highly permeable, allowing for excellent drainage and aeration. This property supports aerobic microbial activity, which accelerates the decomposition of organic matter. As a result, sandy soils are often light and easy to work with, making them ideal for crops requiring well-drained conditions, such as root vegetables.

However, the rapid drainage and lack of water retention make sandy soils less suitable for crops that require consistent moisture levels. From an environmental perspective, the high permeability can exacerbate the downward movement of pesticides and pollutants, increasing the vulnerability of aquifers to contamination.

9.2.3 Silty Soils: Balancing Water Retention and Fertility

Silty soils have particle sizes ranging from 0.002 to 0.02 mm, placing them between sand and clay in terms of texture. Silty soils are known for their smooth, flour-like texture, which makes them prone to compaction. When compacted, these soils often exhibit low permeability to both air and water, leading to conditions that mimic those of clay soils, such as poor drainage and reduced aeration.

Silty soils are generally more fertile than sandy soils because their smaller particles have a moderate surface area for nutrient and water retention. However, due to their small particle size and limited cohesion, they are highly susceptible to erosion, particularly in regions with high rainfall or strong winds.

Environmentally, silty soils are less permeable to contaminants than sandy soils, reducing the likelihood of groundwater contamination. However, their susceptibility to compaction and erosion can negatively affect plant growth and soil health, necessitating conservation measures in agricultural and environmental management.

9.2.4 Clay Soils: High Retention and Low Permeability

Clay soils are composed predominantly of particles smaller than 0.002 mm. Due to their fine size, clay particles have a very high surface area relative to their mass, which enables them to retain significant amounts of water and nutrients. This property is particularly important in agriculture, as clay-rich soils provide essential nutrients through cation exchange and adsorption processes.

However, the fine texture of clay generates poor drainage and aeration, leading to waterlogging in wet periods. This can result in anaerobic conditions that inhibit plant root respiration and promote the activity of anaerobic microbes. Conversely, during dry periods, clay soils tend to shrink and crack, damaging plant roots and disrupting soil structure.

From an environmental perspective, clay soils are impermeable to contaminants, preventing the infiltration of harmful substances into deeper soil layers or groundwater. This makes them effective natural barriers against pollution but also creates challenges for waste disposal or remediation projects that involve contaminant migration through the soil.

9.2.5 Why Soil Texture Matters: Impacts on Soil Health and Management

Understanding soil texture is critical because it influences a wide range of physical, chemical, and biological processes that determine soil behavior and functionality in natural and managed ecosystems. These processes include water movement and retention, nutrient availability, aeration, and biological activity, each of which plays a pivotal role in soil health and productivity.

9.2.5.1 Soil Texture and Water Dynamics

Soil texture profoundly influences water retention and movement. Coarse-textured soils, such as sandy soils, have larger pores facilitating rapid water infiltration and drainage. However, their limited surface area and poor capacity to hold water often lead to low water availability for plants. In contrast, fine-textured soils, like clayey soils, possess small pores that retain water due to capillary forces, but their low permeability can result in slow infiltration and waterlogging.

Silty soils, with particle sizes between sand and clay, exhibit intermediate behavior, balancing water retention and drainage. Understanding these dynamics is crucial for managing irrigation in agricultural systems and mitigating issues such as runoff or drought stress. Furthermore, texture influences the soil's ability to store plant-available water, which is the difference between field capacity and permanent wilting point.

9.2.5.2 Soil Texture and Nutrient Availability

Soil texture also impacts nutrient availability, primarily through its influence on cation exchange capacity (CEC) and nutrient retention. Clay particles and organic matter contribute significantly to CEC due to their high surface area and negative charge, allowing them to attract and retain positively charged nutrient ions such as potassium (K^+), calcium (Ca^{2+}), and ammonium (NH_4^+).

Sandy soils, with low surface area and negligible charge, have minimal nutrient-holding capacity, making them more susceptible to nutrient leaching. Conversely, clay-rich soils often retain nutrients effectively but may exhibit poor nutrient availability due to strong adsorption, which makes nutrients less accessible to plants. Silty soils again demonstrate intermediate behavior, with moderate CEC and nutrient retention capacity.

9.2.5.3 Influence of Texture on Soil Aeration

Texture is critical in determining soil aeration and essential for root respiration and microbial activity. With their abundant huge pores, coarse-textured soils provide excellent aeration but may struggle to retain sufficient moisture. While retaining water efficiently, fine-textured soils often have poor aeration due to restricted air movement in micropores.

Aeration also affects redox potential, which influences the availability of nutrients like iron and manganese. For example, poorly aerated clay soils are more likely to experience anaerobic conditions, which can result in the reduction of Fe^{3+} to Fe^{2+}, a form more readily absorbed by plants. Managing soil aeration is, therefore, critical for balancing oxygen availability with water retention.

9.2.5.4 Soil Texture and Root Penetration

The physical properties of soil texture dictate the ease of root penetration, which is essential for plant growth and stability. Sandy soils, with their loose structure, allow roots to penetrate easily, but their low nutrient and water retention capacity can limit plant development. In contrast, clay soils can pose a physical barrier to root growth due to compaction and high bulk density.

The moisture content of soils also influences root penetration. For example, dry clay soils often form hard clods that resist penetration, whereas well-hydrated clay soils are more pliable, enabling roots to grow. Silty soils provide a balance, offering moderate resistance to penetration while retaining adequate moisture and nutrients.

9.2.5.5 Practical Applications of Soil Texture Analysis

Understanding soil texture is fundamental for optimizing land use, crop selection, and soil management practices. Farmers and soil scientists use texture analysis to determine irrigation strategies, select appropriate crops, and apply soil amendments. For instance, coarse-textured soils benefit from frequent, light irrigation, while fine-textured soils require careful management to avoid waterlogging.

In environmental science, texture analysis helps predict soil erosion risks, with sandy soils being more prone to erosion due to low cohesion and clayey soils offering greater resistance. Moreover, textures inform strategies for contaminant management, as finer particles with high adsorption capacities are more effective at immobilizing pollutants.

9.2.6 Measuring Soil Texture: The Hydrometer Method and the Stokes' Law

Hydrometer analysis is a reliable method to measure soil texture[1], a technique based on the sedimentation of soil particles in a suspension. Using basic physics and simple laboratory tools, this method provides a practical way to determine the proportions of sand, silt, and clay in a soil sample.

9.2.6.1 What is a Hydrometer, and Why is it Used?

A hydrometer is an instrument used to measure the density of a liquid or suspension. It consists of a calibrated stem and a weighted bulb that allows it to float vertically in a liquid (Figure 9.3). In soil texture analysis, the hydrometer measures the density of a soil-water suspension at different time intervals. These measurements reflect the concentration of soil particles still suspended in the liquid, as larger particles settle more quickly than smaller ones. The technique is grounded in Stokes' law, which states that the settling velocity of particles depends on their size, density, and the fluid's viscosity. Thus, the hydrometer helps quantify the sand, silt, and clay proportions in a sample.

9.2.6.2 How Does the Hydrometer Method Work?

The process begins with soil sample preparation. A representative soil sample (e.g., 50 g) is dried, crushed, and sieved to remove coarse fragments. The sample is then treated with a dispersing agent, such as sodium hexametaphosphate, to break apart soil aggregates. The dispersed sample is placed in a graduated cylinder with water to create a suspension, which is thoroughly mixed to ensure uniform dispersion. Once the suspension is prepared, a hydrometer is used to measure the density of the liquid at specific time intervals, following Stokes' Law, which governs the settling rates of particles based on their size. The first reading, taken at 40 seconds, reflects the proportion of sand, as sand particles settle quickly due to their larger size. The second reading, taken at 2 hours, primarily represents the clay fraction, as most silt particles will have settled by this time. The silt content is determined indirectly by subtracting the sand and clay fractions from the total sample. The silt proportion is calculated by subtracting the clay reading from the initial sand and clay reading as silt particles (0.002–0.02 mm) settle between these two intervals. For a more detailed explanation of the different steps of the hydrometer method, the reader is addressed to the paragraphs below.

9.2.6.3 Advantages of the Hydrometer Method

The hydrometer method is efficient and effective, especially for fine-textured soils where traditional mechanical sieving methods may fail. It provides quantitative data on soil particle sizes, enabling precise classification of soil texture. Furthermore, it is relatively simple, requiring minimal equipment and technical expertise, making it ideal for educational and agricultural applications.

[1] This method has been described based on the information reported at the following link: https://civilblog.org/2015/11/18/hydrometer-analysis-of-soil-what-why-how/

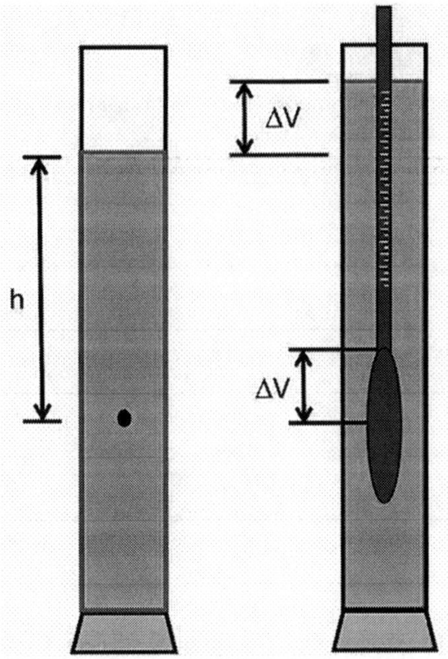

Figure 9.3 The hydrometer consists of a bulb containing a weight (e.g., lead or steel) to keep it in a vertical position, while the upper part of the sealed bulb contains air to regulate buoyancy. The soil sample is dispersed in a liquid containing a dispersing agent (e.g., sodium hexametaphosphate) and placed in a graduated cylinder. During sedimentation, soil particles settle at different rates based on their size, causing density variations in the suspension at a specific height (*h*). The hydrometer is immersed in the cylinder to a predefined depth, displacing a volume of liquid (ΔV) proportional to the local density. Readings are taken as follows: 1. After 40 seconds, determine the density corresponding to sand particles that settle the fastest. 2. After 2 hours, determine the density corresponding to clay particles, which remain in suspension for longer. The amount of silt is calculated by subtracting the contributions of sand and clay from the initial total density. These data allow for determining the soil's particle size composition (sand, silt, and clay) using Stokes' law.

9.2.6.4 Limitations of the Method

Despite its advantages, the hydrometer method has some limitations. It may not perform well with soils containing high organic matter, as organic particles can float and interfere with readings. Similarly, coarse-textured soils with a high sand content may settle too quickly for accurate measurements. Environmental factors such as temperature can also affect the fluid's viscosity, requiring careful calibration.

9.2.6.5 The Stokes' Law

Stokes' Law describes the settling velocity (v) of a spherical particle in a fluid. It is related to the particle radius (r), the density difference between the particle and the fluid ($\Delta\rho$), the dynamic viscosity of the fluid (μ), and the acceleration due to gravity (g). The formula is expressed as:

$$v = \frac{2 \cdot r^2 \cdot g \cdot \Delta\rho}{9 \cdot \mu} \tag{9.1}$$

Here, v is the settling velocity of the particle (m s^{-1}), r is the radius of the particle (m), g is the acceleration due to gravity (9.81 ms^{-2}), $\Delta\rho$ is the density difference between the particle and the fluid (kg m^{-3}), and μ is the dynamic viscosity of the fluid (Pa·s).

This equation shows that larger particles settle faster because their settling velocity is proportional to the square of their radius (r^2). Conversely, smaller particles settle more slowly, allowing them to remain suspended in the fluid for longer periods.

9.2.6.6 Applying Stokes' Law in the Hydrometer Method

Stokes' Law provides the theoretical foundation for the hydrometer method, allowing the settling velocity of soil particles to be estimated based on their size. In soils, particle sizes are categorized as coarse sand (0.2–2 mm), fine sand (0.02–0.2 mm), silt (0.002–0.02 mm), and clay (less than 0.002 mm). Larger particles, such as coarse and fine sand, settle rapidly—typically within seconds—because of their higher settling velocity. Silt particles take significantly longer, ranging from minutes to hours, while the smallest clay particles can remain suspended for hours or even days due to their low settling velocity.

The time required for particles of a given size to settle through a fluid to a specified depth can be calculated using Stokes' Law:

$$t = \frac{h}{v} = \frac{9 \cdot \mu \cdot h}{2 \cdot r^2 \cdot g \cdot \Delta\rho} \tag{9.2}$$

where t is the settling time in seconds, and h is the depth of measurement in meters. All the other parameters have already been defined above. For example, clay particles with a radius of 0.001 mm (half the size of the smallest silt particles) settling through a 0.1-meter column of water take significantly longer to settle than silt particles with a radius of 0.01 mm due to the r^2 term in the denominator of the Stokes' Law.

9.2.6.7 Hydrometer Readings

The hydrometer method measures the density of the suspension at specific time intervals to determine the proportions of sand, silt, and clay in the soil. The first reading, taken at 40 seconds, reflects the sand fraction, as coarse and fine sand particles will have settled by this time. A second reading at 2 hours reflects the clay fraction, as only these smallest particles remain suspended. The silt fraction is calculated as the difference between these two readings, as silt particles settle during the intermediate period.

9.2.6.8 Steps in the Hydrometer Method

The process begins with soil sample preparation. A representative soil sample (e.g., 50 g) is dried, crushed, and sieved to remove coarse fragments. The sample is then treated with a dispersing agent, such as sodium hexametaphosphate, to separate soil aggregates. The dispersed sample is placed in a graduated cylinder with water to create a suspension, which is mixed thoroughly to ensure uniform dispersion.

Once the suspension is prepared, a hydrometer is used to measure the density of the liquid at specific time intervals. The first reading (40 s) indicates the proportion of sand, while the second reading (2 h) measures the clay content. The silt fraction is calculated by subtracting the clay content from the total particles measured after 40 s. The percentages of sand, silt, and clay are determined using the following formulas:

— Percentage of sand:

$$\frac{R_1}{M}\cdot 100 \tag{9.3}$$

— Percentage of silt:

$$\frac{R_1 - R_2}{M}\cdot 100 \tag{9.4}$$

— Percentage of clay:

$$\frac{R_2}{M}\cdot 100 \tag{9.5}$$

where, R_1 is the hydrometer reading at 40 s (i.e., sand fraction), R_2 is the hydrometer reading at 2 h (i.e., clay fraction), and M is the initial mass of the soil sample (in g).

9.3 THE SURFACE AREA: A GENERAL OVERVIEW

Surface area refers to the total area that the surface of an object occupies, encompassing both its external and, in some cases, internal surfaces if the material contains pores or cavities. It is a crucial property that influences how materials interact with their environment, which plays a significant role in chemical reactions, adsorption, and heat or mass transfer. In porous materials, the concept of surface area becomes even more important, as the internal surfaces within pores or fissures can greatly exceed the external surface area. This is reflected in the different types of surface area used to characterize materials. The total surface area includes both the external and internal surfaces of a material, capturing the full extent of its interface with the surrounding environment. On the other hand, the specific surface area expresses the surface area per unit of mass or volume, commonly measured in square meters per gram ($m^2\ g^{-1}$). This measure is especially relevant for materials with high porosity, as their extensive internal pore networks contribute significantly to their overall surface area. For instance, clays and activated carbon exhibit high specific surface areas due to their fine particles and complex internal structures, making them highly effective in adsorption and retention processes. The geometrical surface area refers specifically to the mathematically defined area of the external shape of a material, which is often used for regular shapes like spheres, cubes, or flat surfaces. However, the geometrical surface area significantly underestimates the true surface area for irregularly shaped or porous materials, as it neglects the contributions from internal pores and surface roughness. These distinctions are vital in understanding how materials behave, particularly in applications such as catalysis, filtration, and soil science, where surface area directly impacts performance by determining the extent of interaction with fluids, gases, or other materials.

9.3.1 The Relationship Between Particle Size and Surface Area

The preceding paragraphs mentioned the relationship between particle size and surface area but did not explain it in detail. A practical demonstration is provided to illustrate the direct correlation between these two parameters.

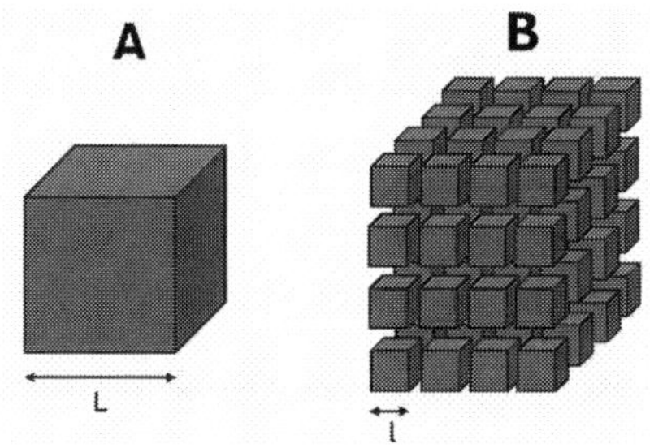

Figure 9.4 Demonstration of the relationship between particle size and surface area. (**A**) A single cube with side length L and surface area $6L^2$. (**B**) The same cube is divided into smaller cubes with side length l, where $l < L$. The total surface area of the smaller cubes (A_{tot}) is greater than the surface area of the original cube, thereby illustrating that smaller particle sizes result in increased surface area.

Let us consider the cube having side length L, as indicated in Figure 9.4A. Its surface area (A) is given by $6 \cdot L^2$. Now, let us divide this cube into n smaller cubes, each with a side length l that $l < L$ as shown in Figure 9.4B. The surface area of each smaller cube is $6 \cdot l^2$. If the total number of smaller cubes is n, so that $L = n \cdot l$, the total surface area (A_{tot}) of all the smaller can be expressed as:

$$A_{tot} = 6 \cdot n \cdot l^2$$

Now, dividing the surface area of the original cube by the total surface area of all the smaller cubes, we obtain:

$$\frac{A}{A_{tot}} = \frac{6 \cdot L^2}{6 \cdot n \cdot l^2}$$

Simplifying:

$$\frac{A}{A_{tot}} = \frac{1}{n}$$

Rearranging, we find:

$$A_{tot} = n \cdot A$$

In other words, the total surface area of the n smaller cubes is n times greater than the surface area of the original cube.

From this simple demonstration, it follows that when a soil particle is broken into several smaller parts, the total surface area of the fragments becomes larger than that of the original particle. Therefore, the surface area of soil particles increases as their size decreases. Bearing this in mind, the surface area of sand, silt, and clay particles follows the order:

$$\text{coarse sand} < \text{fine sand} < \text{silt} < \text{clay}$$

9.3.2 The Importance of Surface Area in Soil Physical Chemistry

Surface area is a critical factor influencing the physical, chemical, and biological behavior of soils. As mentioned earlier, when the size of soil particles decreases, their surface area increases significantly, making the colloidal fraction the most impactful in various soil processes. This high surface area provides numerous interaction sites for water, nutrients, and microbial activity, playing a central role in soil functionality. These interaction sites often consist of hydrogen donor or acceptor groups, which facilitate the formation of hydrogen bonds (see Chapter 5), enabling strong interactions between high-surface-area systems and chemicals of various natures. Additionally, high surface area promotes van der Waals interactions (see Chapter 5), particularly with chemicals that cannot form hydrogen bonds.

The significant influence of surface area is particularly evident in the soil's ability to retain water. Smaller particles, such as clay, possess a vast surface area relative to their volume, allowing them to adsorb and retain water effectively. This property is critical in arid regions where water availability is limited. In contrast, soils dominated by larger particles, such as sand, exhibit lower surface areas, resulting in reduced water retention and rapid drainage. The surface area also governs capillary forces[2], which control water movement through the soil profile and directly affect water availability for plants.

[2] Capillary forces refer to the intermolecular forces that arise at the interface between a liquid and a solid surface, causing the liquid to move or rise within narrow spaces (capillaries) without the assistance of external forces like gravity. These forces are the result of a combination of cohesive and adhesive forces as already defined in paragraph 5.16. Chemically, capillary forces can be defined as the net effect of surface tension (due to cohesive) and the wetting properties of the liquid on the solid surface (due to adhesive forces). When adhesive forces are stronger than cohesive forces, the liquid climbs the surface (e.g., water in glass), while if cohesive forces dominate, the liquid is depressed (e.g., mercury in glass). From a mathematical point of view, the height h to which a liquid rises in a capillary tube is described by the Jurin's law: $h = \frac{2\gamma \cos\theta}{\rho g r}$, where γ is the surface tension of the liquid, θ is the contact angle – see paragraph 5.16 – ρ is the liquid density, g is the gravitational acceleration, and r is the radius of the capillary provided that it shows a cylindrical shape.

Nutrient retention and exchange are also closely linked to surface area. Colloidal particles may carry negative charges on their surfaces, enabling them to attract and hold positively charged ions (cations) like potassium, calcium, and ammonium[3]. This process, known as cation exchange, is essential for maintaining soil fertility. Soils with high surface area, such as clay-rich soils, exhibit higher cation exchange capacity (CEC), enabling them to store and release nutrients more efficiently. Conversely, sandy soils, with minimal surface area, retain fewer nutrients and are consequently less fertile.

The surface area also plays a crucial role in soil aggregation and stability. Fine particles act as binding agents, promoting the formation of aggregates that enhance soil structure, porosity, aeration, and resistance to erosion. However, excessive aggregation in clay-rich soils can lead to compaction, impeding root penetration and water infiltration. This underscores the importance of balancing soil texture to achieve optimal soil structure and functionality.

Another vital aspect of surface area is its ability to interact with organic and inorganic compounds, including pollutants. The large surface area of colloidal particles makes them effective at adsorbing harmful substances like heavy metals and pesticides, thereby mitigating contamination. Moreover, the adsorption of organic molecules on clay surfaces stabilizes soil organic matter and contributes to carbon sequestration, which is essential for maintaining soil health and combating climate change. The high surface area of soil colloids also fosters microbial colonization and activity, providing extensive surfaces for microorganisms to access nutrients and thrive. This promotes key biological processes such as organic matter decomposition, nutrient cycling, and disease suppression, all of which are critical for soil health and fertility.

In conclusion, surface area is a cornerstone of soil physical chemistry. It governs essential processes like water retention, nutrient exchange, aggregation, pollutant mitigation, and biological activity, making it a central concept for understanding and managing soil behavior. Recognizing the importance of surface area is fundamental to optimizing agricultural practices and promoting sustainable soil management.

9.4 THE POROSITY: A BRIEF INTRODUCTION

The three-dimensional organization of matter, from the molecular to the macroscopic level, allows for forming aggregates containing spaces known as "pores." The number, shape, interconnection, and distribution of these pores—collectively called "porosity"—are fundamental to various properties of materials, such as their mechanical resistance and the transport of matter and energy.

In soil, porosity originates from the size and spatial arrangement of organic and inorganic particles. This differs from other materials, such as sedimentary rocks, where porosity results from imperfect grain packing, or effusive rocks, where it

[3] If the colloidal particle is organic matter (OM), it is important to consider that OM can also carry positive charges, allowing it to adsorb anions, depending on the soil *pH*.

forms through gas evolution during magma cooling. In ceramics, porosity depends on the size of the particles used in their production.

Soil porosity uniquely reflects the dynamic interplay of particle size, composition, and structural organization, influencing soil capacity to support water movement, air exchange, and biological activity. The size of the soil particles determines how closely they can pack, influencing the void spaces between them. For example, finer particles create more closely spaced pores, while coarser particles result in larger, less frequent voids.

The soil composition, including its mineral and organic matter content, further shapes porosity. Organic matter contributes to aggregate formation, enhancing pore stability and soil structural resilience. The spatial arrangement of particles—dictated by natural processes such as drying and wetting cycles, root penetration, and biological activity—determines how interconnected these pores are, ultimately affecting soil's ability to store and transmit water and air effectively.

Soil porosity is not static. It evolves in response to environmental conditions and human interventions. External factors, such as compaction, erosion, or tillage, can alter the spatial organization of particles and reduce the connectivity of pores, impacting soil function. Understanding this intricate balance between particle size, composition, and structure is critical for assessing and managing soil health, as porosity directly influences its physical and ecological roles.

9.4.1 Quantitative Assessment of Soil Porosity

Theoretically, porosity is defined as the ratio between the volume of pores (V_p) and the total volume of the material (V_t), such as the soil:

$$\Phi = \frac{V_p}{V_t} \tag{9.6}$$

It can be classified as absolute (or total) when all the pores are accounted for or effective when only the interconnected pores, allowing fluid flow, are considered. The latter is theoretically calculated as the ratio of the volume of the interconnected voids to the total volume of the soil.

Porosity can also be described as primary when it results directly from the process of material formation or secondary when it arises from chemical or mechanical changes occurring after the material has formed.

Figure 9.5 shows the different pore types contributing to soil porosity.

Closed or isolated pores (labeled "a") are voids entirely enclosed within the solid material and do not connect to the external environment. These pores are isolated and play no direct role in transporting fluids or gases, although they influence soil bulk density (see below) and overall porosity.

Blind pores (labeled "b") are partially open voids connecting to the external surface or the pore network at only one end, while the other remains closed. These pores can temporarily store water or air but contribute minimally to soil permeability because they do not form continuous fluid or gas flow pathways.

Interparticle pores (labeled "c" and "d") are spaces between soil particles or aggregates, forming the primary network responsible for water and air transport.

The larger, elongated pores (labeled "c") represent fissures that allow for rapid movement of fluids and gases, while the smaller but interconnected pores (labeled "d") support steady flow and play a crucial role in drainage and aeration.

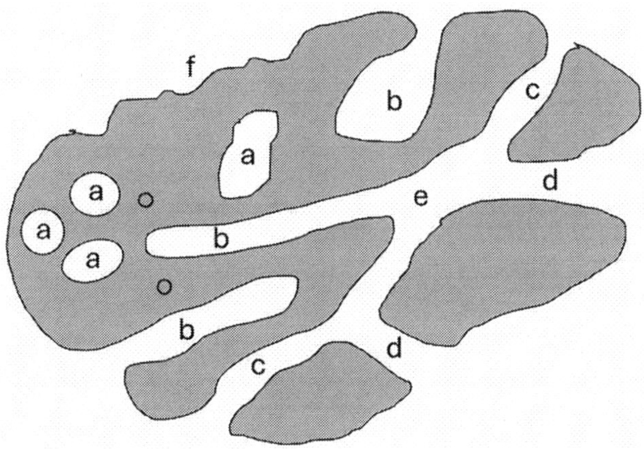

Figure 9.5 The pores indicated with the letter "a" are closed or isolated, while those labeled "b" are called blind pores. Interparticle pores are indicated by "c" and "d" (for a more detailed distinction between these, refer to the main text). Interconnected pores are labeled "e," and surface cracks are identified with "f."

Interconnected or open pores (labeled "e") allow fluid flow from the surface through the medium bulk volume, thereby contributing significantly to permeability and exchange processes.

The surface cracks (labeled "f") represent deformations on solid particle surfaces where fluids are not trapped and are easily exchanged with bulk volumes.

Finally, intraparticle pores (represented by the circles in Figure 9.5) are voids located within individual particles or aggregates. The solid material encloses these pores and, while not contributing to macroscopic transport, is essential for retaining water and nutrients, thereby influencing soil moisture and adsorption properties.

Each of these pore types contributes to the overall porosity of the soil in unique ways. Interparticle pores (c and d) dominate the processes of permeability and transport, while intraparticle pores (red circles) regulate water retention. Meanwhile, blind pores (b) and closed pores (a), though less involved in transport, impact soil storage and structural properties. These pore types define the dynamic and multifaceted nature of soil porosity, as depicted in Figure 9.5.

9.4.2 The Shape of the Pores

Figure 9.6 illustrates the various shapes that soil pores can exhibit. It is important to note that the geometric structure of pores is highly variable, with only intraparticle pores being adequately analyzed in detail. Experimental observations, such as those obtained through molecular adsorption techniques or electron microscopy, have shown that intraparticle pores can take on a range of geometries, including cylindrical shapes (open at one or both ends, Figure 9.6A), bottle-neck shapes

(Figure 9.6B), conical shapes (Figure 9.6C), spherical geometries, and slit-shaped configurations (Figure 9.6D).

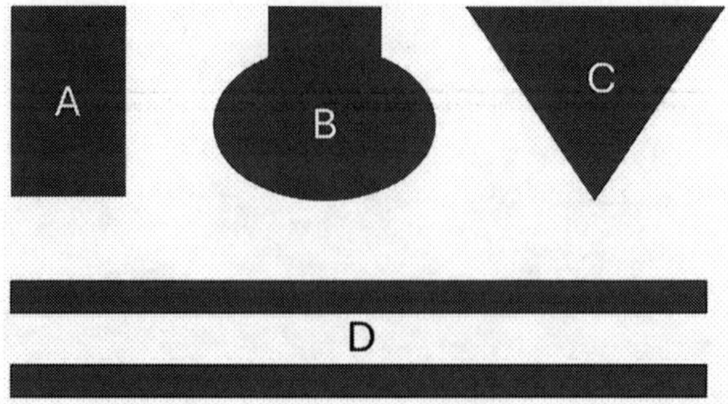

Figure 9.6 Geometric shapes of intraparticle pores. Experimental analyses, such as molecular adsorption and electron microscopy, have revealed that intraparticle pores can exhibit various geometries, including cylindrical shapes (**A**), bottle-neck shapes (**B**), conical shapes (**C**), and slit-shaped configurations (**D**).

Recognizing pore shape is far more than a mere intellectual exercise. As will become clear below, the choice of a specific pore geometry in the various models used to measure surface area and porosity can significantly influence the quantitative results for these parameters.

9.4.3 Pore Size

The International Union of Pure and Applied Chemistry (IUPAC) classifies pore sizes into the following categories:

- **Macropores:** pores with a diameter greater than 50 nm.
- **Mesopores:** pores with diameters ranging between 2 nm and 50 nm.
- **Micropores:** pores with diameters smaller than 2 nm.

Within the micropores, the IUPAC further distinguishes two subcategories:

- **Supermicropores:** pores with diameters between 0.7 nm and 2 nm.
- **Ultramicropores:** pores with diameters smaller than 0.7 nm.

This classification assumes that pores exhibit a spherical geometry. However, while the IUPAC classification provides a standardized framework for categorizing pore sizes based on physical dimensions, its applicability to soil science is limited. For example, if we approximate a water molecule as a sphere with a radius of 0.3 nm, a 2 nm pore—classified as a micropore under the IUPAC system—can accommodate only around 37 water molecules, while a supermicropore can fit just one or two, depending on its size. Ultramicropores, on the other hand, are too small to contain even a single water molecule if their diameter is less than 0.6 nm. This calculation underscores the limitation of IUPAC's geometric focus, which does not account for how water and nutrients interact dynamically with soil pores.

In soil science, the emphasis shifts from abstract geometric dimensions to the functional behavior of pores. As a result, classifications are based on the ability of pores to retain water, facilitate nutrient exchange, and support processes critical for plant growth and ecosystem functioning. Consequently, the following classification is applied:

- **Bonding pores**, with a diameter <5 nm. These pores are important to stabilize primary particle aggregates, enhancing soil structure and resistance to erosion, which is critical for maintaining soil integrity during agricultural practices.
- **Residual pores**, with a diameter <500 nm. These are significant for chemical interactions between the soil and plant nutrients at the molecular level.
- **Storage pores**, with sizes ranging from 500 and 50,000 nm. These pores "store" the water essential for plants and the survival of microorganisms.
- **Transmission pores**, with sizes between 50,000 and 500,000 nm. These allow the movement of water within the soil and, consequently, support plant root growth.

Figure 9.7 clearly illustrates the differences between IUPAC and soil science pore classifications. From this figure, it is evident that IUPAC micropores and part of the IUPAC mesopores correspond to the soil science bonding pores. The remaining portion of the IUPAC mesopores aligns with the soil science residual pores, while all IUPAC macropores fall within the soil science storage and transmission pores. By focusing on water availability and nutrient transport, the soil science classification offers a practical framework for optimizing soil management in agriculture and ecosystem conservation.

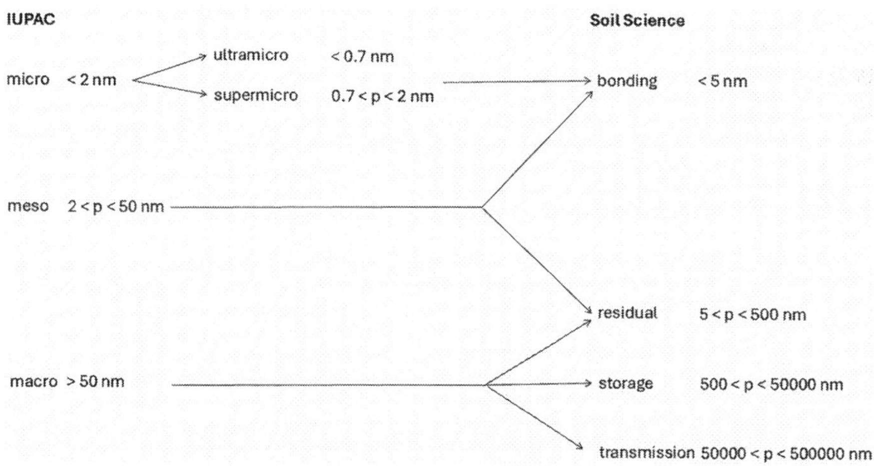

Figure 9.7 Comparison between IUPAC and soil science pore size classifications. The IUPAC categories—micropores (<2 nm), mesopores (2–50 nm), and macropores (>50 nm)—are shown alongside their corresponding classifications in soil science: bonding pores (<5 nm), residual pores (5–500 nm), storage pores (500–50,000 nm), and transmission pores (50,000–500,000 nm). This figure highlights the functional differences in how pores are categorized, with soil science classifications focusing on water availability and nutrient transport. The letter "p" lays for "pore."

9.5 HOW TO MEASURE SURFACE AREA, POROSITY, AND PORE SIZES: A BRIEF INTRODUCTION TO ANALYTICAL INSTRUMENTS

The instrument commonly used to measure surface area, porosity, and pore size distribution is often referred to as a "porosimeter." However, this term is actually lab slang, adopted for convenience when referring to such analyses. The term "porosimeter" is somewhat limiting, as it implies a focus solely on pore investigation. In reality, these instruments provide comprehensive insights into surface area and pore structure. The correct terminology depends on the technique employed: when gases are used for adsorption studies, the instrument is more accurately called a "gas adsorption analyzer," whereas for methods involving liquid mercury, it is referred to as a "mercury intrusion porosimeter."

The difference between the two types of instruments lies in their sensitivity to materials' physical characteristics. Specifically, the gas adsorption analyzer is highly sensitive to smaller pores, including micropores and mesopores (<50 nm, as classified by IUPAC), and provides detailed information on specific surface area and pore size distribution in the nanometer range. However, it cannot effectively detect larger macropores (>50 nm) because adsorption processes saturate beyond this range. Consequently, the gas adsorption analyzer is preferred for materials with intricate, small pores, such as zeolites, activated carbons, and catalysts. It is widely used in material science due to its high precision in measuring surface area and pore properties, which is essential for designing high-performance materials.

In contrast, the mercury intrusion porosimeter is more sensitive to larger pores (>50 nm) and can measure pore sizes up to several micrometers. This method involves forcing liquid mercury into pores under increasing pressure, making it effective for detecting larger-scale porosity. However, it struggles with smaller pores (<2 nm), as the high pressures required can damage delicate structures. Additionally, it does not provide surface area measurements. For this instrument, the term "porosimeter" is appropriate and does not represent lab slang.

Despite its apparent suitability for soil science, the mercury intrusion porosimeter is rarely used in this field. Its limitations in detecting smaller pores and the environmental and ethical concerns associated with mercury toxicity make it impractical for soil studies. Soil pores span a broad size range, from micropores to macropores, but research in this field focuses heavily on functional properties such as water retention, nutrient availability, and aeration. Since gas adsorption analyzers excel at fine-scale measurements and mercury intrusion is not well-suited for environmental samples, soil scientists often favor gas adsorption or alternative techniques like water retention curves or imaging methods. Together, gas adsorption analyzers and mercury intrusion porosimeters provide complementary insights into pore structures, with the choice of instrument depending on the material's pore size range and the intended application.

9.5.1 The Principles of the Surface Area Measurement

Whatever the type of instrument used (see the paragraph above), the principle of the measurement lies in the interaction between a fluid and the pore boundaries. In gas adsorption methods, a fluid such as liquid nitrogen is introduced into the pores of the material, where the molecules interact with the pore walls according to their distance from the boundary. Molecules closer to the pore boundary experience stronger binding forces, while those farther from the surface are held less tightly. These interactions, governed by molecular forces, form the basis for measuring surface area, pore size, and porosity.

The measurement of surface area is based on the amount of fluid that interacts with the material's surface. Imagine carefully pouring liquid nitrogen into a pore, like filling a glass. As the nitrogen fills the pore, it coats the walls through adsorption, a process where molecules adhere to the surface. The more intricate the surface (e.g., with ridges or curves), the more nitrogen molecules are required to coat it thoroughly. In a gas adsorption analyzer, this principle is applied by gradually introducing nitrogen at controlled pressures. The total amount of gas adsorbed is measured and used to calculate the specific surface area of the material using mathematical models like the Brunauer-Emmett-Teller (BET) theory (see below).

Liquid nitrogen is used in this process for several important reasons. Its low boiling point of –196°C (77 K) ensures a stable cryogenic temperature, reducing the kinetic energy of nitrogen molecules and allowing them to adsorb predictably onto the surface. Furthermore, nitrogen is a small, inert molecule that can penetrate even the smallest pores (e.g., micropores) and does not chemically interact with most materials, ensuring precise and consistent results. Additionally, nitrogen is widely available, cost-effective, and practical for routine laboratory use.

While gas adsorption focuses on fine-scale porosity, the same principles can be applied to liquid-based methods like mercury intrusion porosimetry. In this case, instead of adsorption, mercury is forced into the pores under increasing pressure. The relationship between the applied pressure and the pore size is governed by similar physical theories, with mercury interaction providing information on pore volume and size distribution. Unlike gas adsorption, mercury intrusion is better suited for analyzing macropores and does not measure surface area. However, the theory of fluid interaction with pore boundaries remains consistent across both methods.

In summary, nitrogen molecules coat the surfaces of small, intricate pores, while mercury fills larger pores under pressure, allowing these complementary techniques to cover a broad range of pore sizes. This dual approach provides a comprehensive understanding of porosity, from micropores to macropores.

9.5.2 The Basis for the Measurement of Pore Size Distribution

The measurement of pore size distribution relies on the principle of desorption, where the gas that was previously adsorbed onto the pore surfaces is gradually removed as pressure is reduced. To understand this, imagine a pore as a glass with a narrow neck and a wider base. When the glass is filled with liquid, the fluid first coats the walls and then fills the broader base as the pressure increases. During desorption, as the pressure decreases, the fluid begins to leave the pore, starting with the largest spaces—just like liquid would first exit the broad base of the glass—while the narrower neck retains the fluid for longer due to stronger molecular forces. This process mimics the behavior of gas molecules in real pores, where smaller pores require lower pressures to release the adsorbed gas due to their confined spaces and higher capillary forces (see below for the meaning of capillary force).

By analyzing the pressure at which gas desorbs, the pore size distribution can be calculated. Larger pores release gas at higher pressures, while smaller pores desorb gas more gradually as the pressure approaches zero. This stepwise release of gas, captured in the desorption isotherm, provides detailed insights into the range of pore sizes within the material. The glass analogy helps visualize how gas desorption reflects the unique geometry and connectivity of pores, forming the foundation for pore size distribution analysis.

9.5.3 The Brunauer-Emmett-Teller (BET) Theory

The BET model to describe surface area measurement was introduced in 1938 by Stephen Brunauer, Paul Hugh Emmett, and Edward Teller[4]. These authors outlined the concept of multilayer adsorption, where multiple layers of gas molecules accumulate on the surface of a material without undergoing chemical reactions.

BET theory focuses on the behavior of a non-reactive gas, known as the adsorbate, as it interacts with a solid material, referred to as the adsorptive. As discussed earlier, nitrogen is the most commonly employed adsorbate for these analyses due to its inert nature and well-characterized adsorption properties. Standard BET measurements are conducted at nitrogen boiling point (77 K), which ensures stable and predictable adsorption behavior, providing reliable and consistent data. In specific applications, other gases, such as argon, carbon dioxide, and water, can also be used to allow measurements at different temperatures or to target unique surface properties.

It is important to note that specific surface area is a scale-dependent property, meaning there is no single absolute value for a material surface area. Measurements obtained using BET theory depend on factors such as the choice of adsorbate and the adsorption cross-section of the gas molecules. This variability highlights the complexity of quantifying the surface area and underscores the need for standardized conditions to ensure accurate and comparable results.

The BET theory builds upon the Langmuir model, as discussed in Section 8.7.4. While the Langmuir model describes monolayer molecular adsorption, the BET theory extends this concept to account for multilayer adsorption. In BET theory, the Langmuir model is applied to each individual layer of adsorbed molecules,

[4] Brunauer, S., P.H. Emmett and E. Teller. 1938. Adsorption of Gases in Multimolecular Layers. Journal of the American Chemical Society. 60(2): 309–319.

enabling a more comprehensive understanding of adsorption behavior on surfaces. The resulting equation is[4]:

$$\frac{P}{W(P_0 - P)} = \frac{1}{C \cdot W_m} + \frac{C-1}{C \cdot W_m} \cdot \frac{P}{P_0} \tag{9.7}$$

This is a specific form of the Langmuir equation, where W represents the weight of gas adsorbed at a given relative pressure (P/P_0), and W_m denotes the weight of the adsorbate required to form a monolayer covering the surface. The constant C, known as the BET constant, is associated with the energy of adsorption in the first adsorbed layer. Its value reflects the strength of the interactions between the adsorbent and the adsorbate.

The plotting of the $\frac{P}{W(P_0 - P)}$ term against P/P_0 gives a straight line whose intercept is $\frac{1}{C \cdot W_m}$, while the slope is $\frac{C-1}{C \cdot W_m}$. Therefore, from the intercept and the slope, C, and W_m can be evaluated. The W_m value is used for the achievement of the total surface area (S_A) which is in the form:

$$S_A = \frac{W_m \cdot N_A \cdot A_{cs}}{M} \tag{9.8}$$

where N_A is the Avogadro's constant, and A_{cs} is the molecular cross-sectional area of the adsorbate molecule. As reported above, the adsorbate molecule is generally the molecular nitrogen (N_2), showing an A_{cs} value at 77 K of 16.2×10^{-20} m^2. From the total surface area, it is possible to obtain the specific surface area (S_{sp}) as follows:

$$S_{sp} = \frac{S_A}{w} \tag{9.9}$$

here, w represents the sample weight.

9.5.4 Decoding Pore Structures: Theories and Analytical Methods

Studying pore structures is vital in material characterization, particularly when evaluating surface areas, pore volumes, and distributions. Various theoretical and empirical methods provide insights into these attributes using isotherms derived from gas adsorption experiments. These approaches range from macropore analysis to advanced micropore evaluations, each tailored to specific material geometries and adsorption behaviors.

9.5.4.1 Isotherms and Their Classifications

Isotherms provide a graphical representation of the relationship between the amount of gas adsorbed by a material and the relative pressure at a constant temperature. This relationship is fundamental for understanding adsorption phenomena in porous materials. The International Union of Pure and Applied Chemistry (IUPAC) classified isotherms into six distinct types, each reflecting specific adsorption mechanisms and pore structures. This classification is essential for determining the surface area, porosity, and adsorption properties of materials (Figure 9.8)[5].

Type I isotherms are characteristic of microporous materials such as activated carbons and zeolites. These isotherms display a steep initial increase in adsorption at low relative pressures, representing rapid micropore filling. This behavior results from the strong adsorption potential within the narrow pores, where the adsorbed molecules interact closely with the pore walls. Once the micropores are filled, the isotherm plateaus indicate a limit to adsorption capacity governed by the accessible pore volume rather than surface area.

Type II isotherms occur in non-porous or macroporous materials where unrestricted monolayer and multilayer adsorption occurs. The distinctive "knee" in the isotherm represents the completion of monolayer adsorption and the onset of multilayer adsorption. This type of isotherm is common for materials with extensive external surfaces.

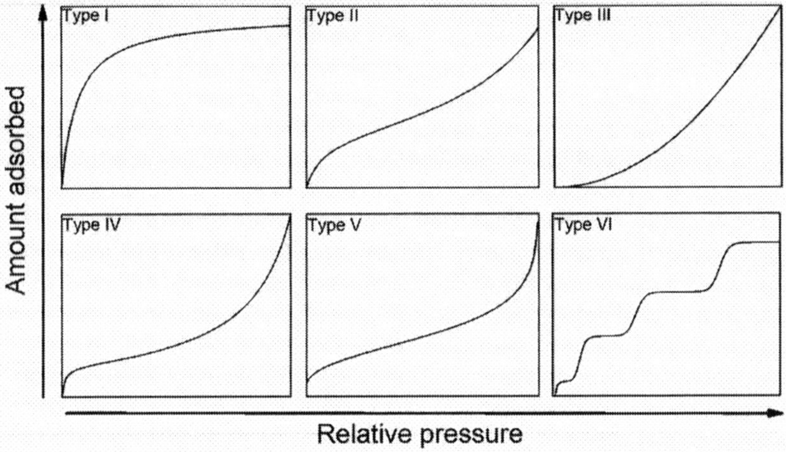

Figure 9.8 Classification of adsorption isotherms according to the IUPAC system. The six isotherm types represent different adsorption behaviors based on pore structure and adsorbate-adsorbent interactions. Type I is typical of microporous materials, Type II and IV correspond to mesoporous and non-porous materials with multilayer adsorption, Type III and V indicate weak adsorbent-adsorbate interactions, and Type VI represents stepwise multilayer adsorption on uniform surfaces.

---

[5] The Isotherm classification was taken from the following papers: Khalfaoui, M., S. Knani, A.B. Lamine and P. Bouaziz. 2003. New theoretical expressions for the five adsorption type isotherms classified by BET based on statistical physics treatment. Journal of Colloid and Interface Science. 263: 350–356.

– Rashid, S.N.A., et al. 2019. Surface properties and pore structure analysis of activated carbons prepared from agricultural wastes using CO_2 activation technique. Adsorption. 25: 757–769.

– Thommes, M., et al. 2015. Physisorption of gases, with special reference to the evaluation. Pure and Applied Chemistry. 87: 1051–1069.

– Yahia, B., et al. 2013. Models for type VI adsorption isotherms from a statistical mechanical formulation. Adsorption Science and Technology. 31: 139–151.

Type III isotherms are less common and exhibit convex curves throughout the adsorption process. They signify weak adsorbate-adsorbent interactions, where adsorbate-adsorbate interactions dominate. Examples include water adsorption on hydrophobic surfaces or nitrogen on polymeric materials.

Type IV isotherms are typical of mesoporous materials and display a pronounced hysteresis loop due to capillary condensation. The initial segment reflects monolayer and multilayer adsorption similar to Type II, while the hysteresis loop arises from the pore-filling and emptying processes. These isotherms are critical for studying mesopore size distributions.

Type V isotherms are associated with weak adsorbate-adsorbent interactions, akin to Type III. However, they also exhibit hysteresis loops indicative of pore condensation, which is characteristic of mesoporous materials. This type is often observed in systems where the adsorbent has a low affinity for the adsorbate.

Type VI isotherms represent stepwise multilayer adsorption on a uniform, non-porous surface. Each step corresponds to the completion of a molecular layer. These isotherms are rare and typically observed with highly structured materials and non-polar adsorptives like argon.

9.5.4.2 Hysteresis in Adsorption Isotherms

Hysteresis is a fundamental phenomenon in adsorption isotherms, particularly in mesoporous and macroporous materials, and is characterized by a difference between the adsorption and desorption branches of the curve. This effect is closely related to capillary condensation, which occurs when the adsorbed gas forms a liquid-like phase within the pores of the solid. The presence of hysteresis provides valuable insights into the structure of the adsorbent, including pore size distribution, connectivity, and adsorption mechanisms.

Figure 9.9 presents the classification of hysteresis loops according to the International Union of Pure and Applied Chemistry (IUPAC). The shape of these loops reflects the nature of the pores and the interactions between the adsorbate and the adsorbent. One of the most distinctive forms is the Type H1 hysteresis loop, which is commonly associated with materials having well-defined cylindrical pores or aggregates of uniform spherical particles. The nearly vertical and parallel adsorption and desorption branches suggest a relatively narrow pore size distribution and strong capillary condensation effects.

In contrast, Type H2 loops are often found in materials with disordered mesoporous networks, where the adsorption and desorption branches are irregular. The steep desorption step observed in these cases indicates pore-blocking or cavitation phenomena, which complicate the interpretation of the pore size distribution. The complexity of these loops reflects the structural heterogeneity of the material.

Another distinctive pattern is the Type H3 hysteresis loop, which is typically observed in materials with slit-like pores. Unlike the previous cases, this type of loop lacks a well-defined adsorption plateau, indicating a progressive and unrestricted multilayer adsorption process rather than capillary condensation. A similar behavior is found in Type H4 loops, although these are more characteristic of materials with both microporous and mesoporous structures. The desorption branch in these cases exhibits a pronounced stepwise behavior, which indicates the simultaneous presence of narrow slit-shaped mesopores and micropores.

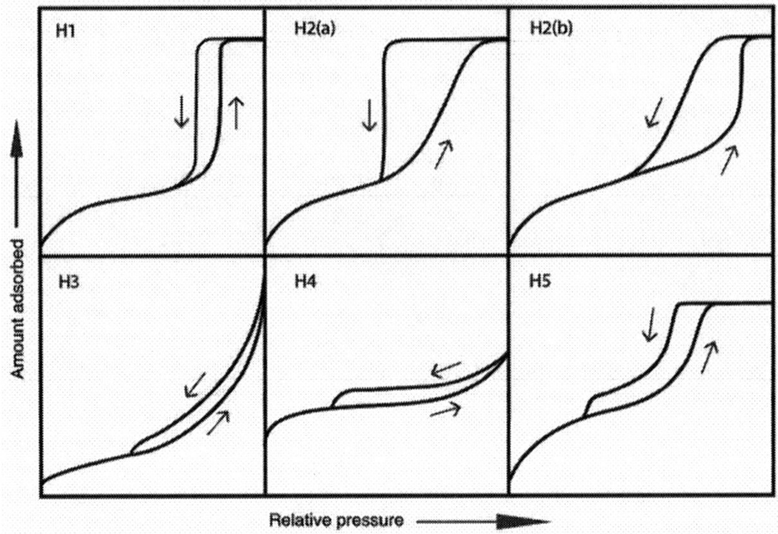

Figure 9.9 Classification of adsorption hysteresis loops according to IUPAC. The six types of hysteresis loops (H1, H2(a), H2(b), H3, H4, and H5) represent different pore structures and adsorption-desorption mechanisms. The arrows indicate the adsorption (up) and desorption (down) branches. The H1 loop is associated with uniform cylindrical pores, while H2(a) and H2(b) reflect complex pore networks with varying connectivity. H3 and H4 loops correspond to slit-shaped pores and microporous materials, respectively, and H5 is linked to partially blocked mesopores. Adapted from Rouquerol et al. (2015). Pure and Applied Chemistry, 87(9–10): 1051–1069. © IUPAC, published by De Gruyter[6].

Studying hysteresis loops is crucial in adsorption science, as it allows for a more detailed characterization of porous materials. By analyzing the shape and extent of the hysteresis, researchers can infer the connectivity of the pore network, the accessibility of adsorption sites, and the specific mechanisms governing adsorption and desorption. This information is essential for optimizing materials used in applications such as catalysis, gas separation, and energy storage, where the control of porosity and surface interactions plays a key role.

[6] Figure 9.9 is reproduced from the IUPAC Technical Report: "Physisorption of gases, with special reference to the evaluation of surface area and pore size distribution," originally published in Pure and Applied Chemistry by Rouquerol, et al. 2015. This reproduction is made in accordance with the copyright policy of IUPAC and De Gruyter, which allows republication and dissemination without formal permission, provided proper acknowledgment is given. The original work is © 2015 IUPAC, published by De Gruyter. All rights reserved.
Full Reference: Rouquerol, J. et al. 2015. Physisorption of gases, with special reference to the evaluation of surface area and pore size distribution (IUPAC Technical Report). Pure and Applied Chemistry. 87(9–10): 1051–1069. © IUPAC, published by De Gruyter.

9.5.4.3 The Isotherms and the Hysteresis Loops to Obtain Porosity, Pore Volume, and Pore Size Distribution

The total pore volume can be estimated from the volume of gas adsorbed at high relative pressures ($P/P_0 \approx 0.99$) assuming that the pores are completely filled with the condensed adsorbate. This is mathematically expressed as:

$$V_p = \frac{V_a \cdot \varrho_{gas}}{\varrho_{liq}} \tag{9.10}$$

where V_a is the adsorbed volume, ϱ_{gas} is the density of the adsorbed gas (that is 0.001256 g cm^{-3} for nitrogen gas in standard conditions), and ϱ_{liq} is the density of the adsorbate in its liquid state (e.g., for nitrogen at 77 K, ϱ_{liq} = 0.808 g cm^{-3})[7]. The porosity is then given by:

$$\Phi = \frac{V_p}{V_t} = \frac{V_a \cdot \varrho_{gas}}{V_t \cdot \varrho_{liq}} \tag{9.11}$$

where V_t represents the total volume of the porous system.

To determine pore size distribution, the main approach is the Barrett-Joyner-Halenda (BJH) method, which applies the Kelvin equation to the desorption branch of the hysteresis loop:

$$r_K = -\frac{2\gamma V_m}{RTln(P/P_0)} \tag{9.12}$$

where r_K is the pore radius in nm, γ is the surface tension of liquid nitrogen (8.85 mN m^{-1}), V_m represents the molar volume of the liquid adsorbate (for molecular nitrogen, it is 34.7 cm^3 mol^{-1}), R is the universal gas constant (i.e., 8.314 J mol^{-1} K^{-1}), T is the temperature in Kelvin (that is, 77 K for liquid nitrogen), and P/P_0 is the relative pressure at which desorption occurs. Operatively speaking, the first step is to determine the r_K value from the Kelvin equation, then a correction for the pore wall thickness (δ) using the t-plot method (see below) must be accounted for:

$$r_P = r_K + \delta \tag{9.13}$$

Here, δ is obtained from the experimental adsorption data. Finally, the differential pore volume (dV_p/dr_p) must be calculated by using the BJH algorithm:

$$dV_p = V_p(P_i) - V_p(P_{i+1}) \tag{9.14}$$

Here, P_i, and P_{i+1} are successive pressures in the desorption isotherm.

The shape of the hysteresis loop provides further insights into pore geometry—H1 loops are associated with uniform cylindrical pores, H2 loops with ink-bottle pores, and H3–H4 loops with slit-like or microporous structures.

[7] When measuring pore volume using gas adsorption techniques, we typically measure the adsorbed gas volume (V_a) at standard conditions (STP: 0°C, 1 atm). However, the total pore volume (V_p) is related to how much of the adsorbate would occupy the pores in its liquid phase, since the assumption is that the gas condenses into the pores. Thus, we need to convert the gas-phase volume into its liquid-equivalent volume. This is done by using the $\varrho_{gas}/\varrho_{liq}$ correction factor.

Figure 9.10 reports the generated pore size distribution graph illustrating how the differential pore volume varies across different pore radii, thereby revealing the characteristic size distribution of the porous network. This figure shows a simulation of a bimodal pore size distribution, reporting two distinct peaks: one around 5 nm and another around 50 nm, which is typical for materials with a combination of mesopores and macropores according to IUPAC pore size classification. The simulated separation reported in Figure 9.10 is rarely obtained for porous systems having a very complex and smoother pore size distribution. Indeed, an overlay among different peaks is usually retrieved.

9.5.4.4 How the T-Plot Method Works

The t-plot method assumes that the thickness (t) of the adsorbed layer follows a predictable function of the relative pressure (P/P_0), which is obtained from reference data for non-porous materials with a similar surface chemistry. This allows the estimation of the micropore volume by identifying deviations from the linearity in the plot, the external surface area by extrapolating the linear portion of the plot, and the pore wall thickness correction, which is added to the Kelvin radius to obtain the total pore radius.

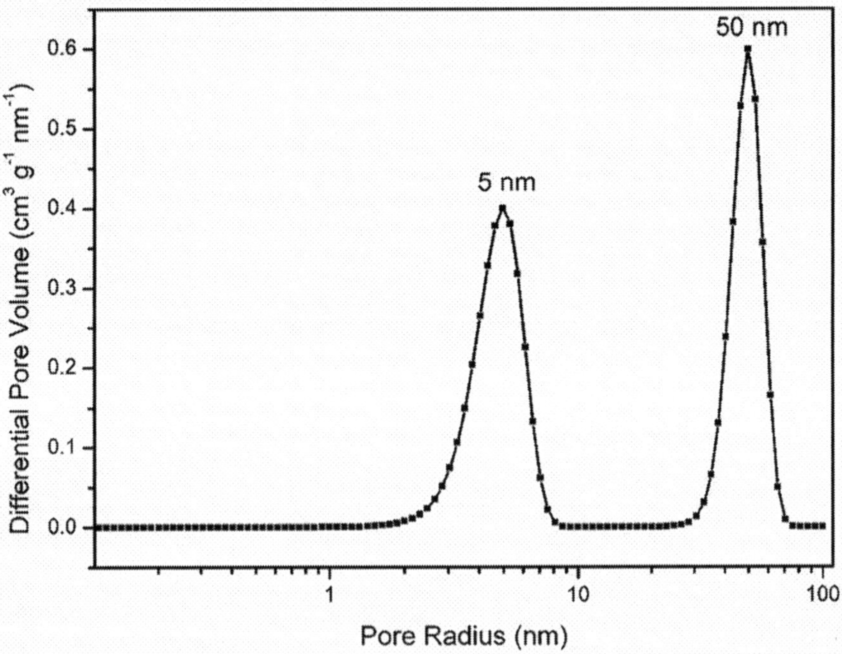

Figure 9.10 Simulated bimodal pore size distribution obtained using the BJH method. The graph illustrates how the differential pore volume varies across different pore radii, revealing the characteristic size distribution of the porous network. According to the IUPAC classification, two distinct peaks can be observed: one centered around 5 nm and another around 50 nm, which is characteristic of a bimodal pore structure that combines mesopores and macropores. Such well-separated peaks are uncommon in real porous materials, as complex and overlapping distributions are typically observed.

The statistical thickness equation used in the t-plot method is typically:

$$t = f(P/P_0) \tag{9.15}$$

A commonly used model for nitrogen adsorption at 77 K is the Halsey equation:

$$t = 0.354 \cdot \left[\ln(P_0/P)\right]^{-0.5} \tag{9.16}$$

Another model is the Harkins-Jura equation as it follows:

$$t = 0.1 \cdot \sqrt{\frac{13.99}{0.034 + \log_{10}\left(\frac{P_0}{P}\right)}} \tag{9.17}$$

Once t (in nm) is determined at different pressures, a t-plot is achieved where the adsorbed volume (V_a) is reported versus the statistical thickness (t).

The relationship between the adsorbed volume and the thickness depends on which type of adsorption model one uses. As examples, for zeolites and activated carbons, the Langmuir expression given below can be applied:

$$V_a = V_m \cdot \frac{b \cdot \frac{P}{P_0}}{1 + \frac{P}{P_0}} \tag{9.18}$$

Here, b is the Langmuir adsorption affinity constant, while V_m has been defined above (i.e., for N_2 at 77 K, it is 34.7 $cm^3\ mol^{-1}$). The next step is to report V_a versus t. Therefore, the linear part in the latter graph must be identified. It must be in the form:

$$V_a = a + bt \tag{9.19}$$

Hence, by identifying the intercept at $V_a = 0$, it comes that:

$$t = \delta = -\frac{a}{b} \tag{9.20}$$

This is the correction factor that must be added to the Kelvin radius previously defined.

Figure 9.11 reports a simulated t-plot graph retrieved using the equations just considered. It accounts for a generic zeolite made mainly by micropores (according to the IUPAC definition). The linear part was identified in the range $0.3 \leq t \leq 0.6$. After interpolation, the red line was obtained. It is described by the equation: $V_a = 34.9 \cdot t - 9.6$. Therefore, by forcing $V_a = 0$, the correction factor, δ, of 0.27 nm can be achieved.

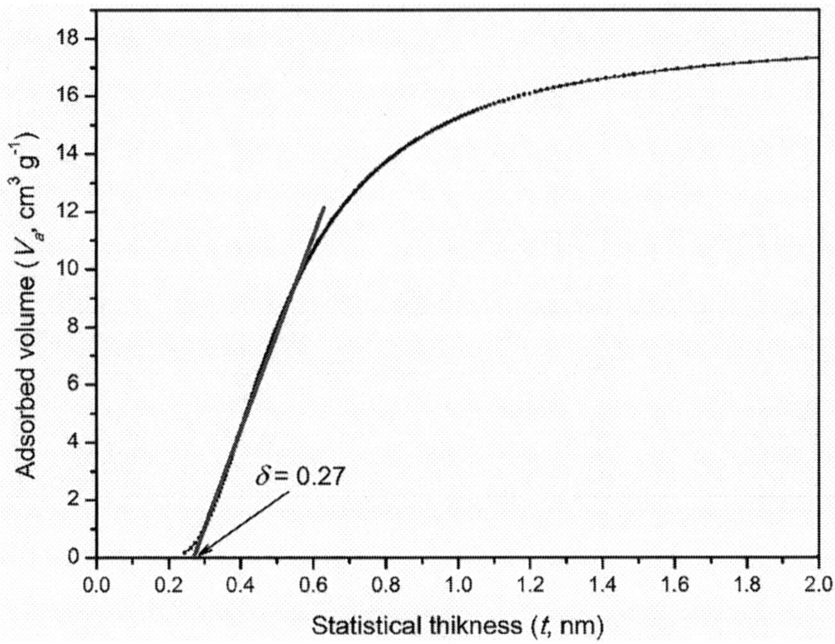

Figure 9.11 Simulated t-plot graph for a microporous zeolite. The plot demonstrates the relationship between the adsorbed volume (V) and the statistical thickness (t) of the adsorbed nitrogen layer at 77 K. The linear region between 0.3 and 0.6 nm indicates the external surface area contribution. The interpolation line follows the equation $V = 34.9 \cdot t - 9.6$, from which a correction factor (δ) of 0.27 nm was derived by extrapolating to $V = 0$.

9.6 SOIL STRUCTURE AND COMPACTION

Soil structure refers to the natural arrangement of soil particles, including sand, silt, clay, organic matter, and other materials, into aggregates that influence critical properties such as water movement, aeration, and root penetration (Figure 9.12). The way these particles cluster together, forming different structural types like granular, blocky, prismatic, or even massive (Figure 9.12), directly affects the soil's ability to retain moisture and nutrients while allowing for proper drainage and root development.

Granular structures, often found in topsoil[8] rich in organic matter, promote easy water infiltration and root growth, whereas blocky and prismatic forms, typical of subsoils, can restrict lateral water movement and root expansion. On the other hand, a massive structure lacks visible aggregation and tends to be compact, impeding water infiltration and limiting plant development due to poor aeration. This leads naturally to the concept of soil compaction, which occurs when soil particles are pressed tightly together, reducing the volume of pore spaces and consequently affecting water infiltration, drainage, and root growth. Compaction is often the result

[8] The term "topsoil" refers to the surficial horizon of a soil.

of heavy machinery, repeated foot traffic, or overgrazing, and its effects are more pronounced in clay soils due to their fine particles, which compact more easily compared to sandy soils with larger pore spaces. The consequences of compaction include reduced water infiltration, leading to surface runoff and erosion, limited root penetration that restricts access to essential nutrients and water, and decreased oxygen availability, which hampers microbial activity and plant respiration. Addressing these issues involves strategies to maintain good soil structure and minimize compaction, such as reducing traffic on wet soils, incorporating organic matter to improve aggregation, and using deep tillage only when necessary to break up compacted layers. Ultimately, understanding the close relationship between soil structure and compaction is key to managing soil health effectively, ensuring sustainable crop production, and maintaining the natural functions of soil in various environmental contexts.

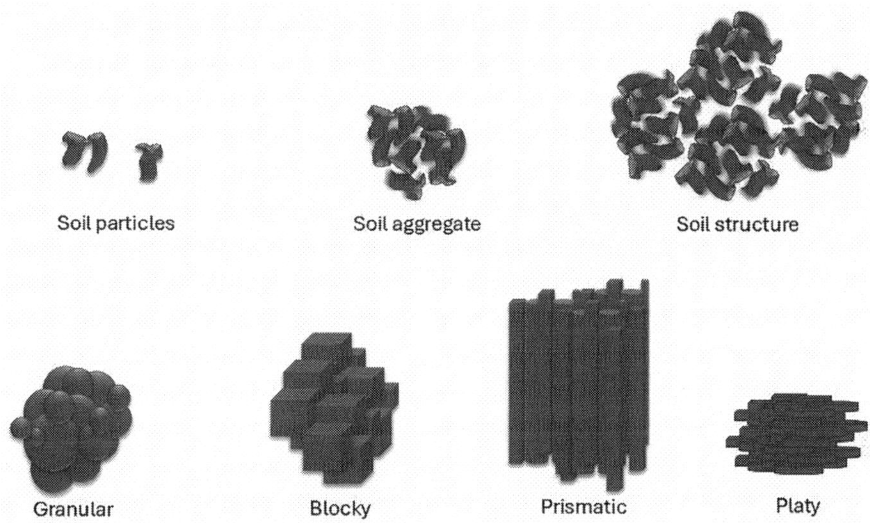

Figure 9.12 Illustration of soil particles, aggregates, and structure with different types of soil aggregate arrangements. The upper part of the figure illustrates the hierarchical organization of soil components, starting from individual soil particles, progressing to the formation of soil aggregates, and culminating in the development of soil structure, where aggregates are organized with visible pores and cracks. The lower part highlights four basic types of soil structures: (A) Granular structure, composed of small, rounded aggregates that promote good aeration and water infiltration; (B) Blocky structure, characterized by angular, block-shaped aggregates that moderate water movement and root growth; (C) Prismatic structure, defined by vertically elongated aggregates that influence water distribution primarily in the vertical direction; and (D) Platy structure, consisting of thin, flat aggregates stacked horizontally, which can restrict water infiltration and root penetration.

9.6.1 The Mechanisms of Soil Aggregate Formation

Soil aggregates are formed through complex chemical mechanisms driven by the interactions between organic and inorganic soil components, with water playing a pivotal role in mediating these processes. The formation of aggregates begins with associating clay minerals, organic matter, and other inorganic particles through a range of physicochemical forces. Organic matter, including polysaccharides, humic substances, and microbial exudates, acts as a binding agent, creating a matrix stabilizing soil particles. Clay minerals, due to their high surface area and charge properties, interact strongly with both organic compounds and water molecules, facilitating aggregate formation. Water serves as a dynamic medium in this system, hydrating soil particles and functioning as a molecular bridge that enhances bonding between organic and inorganic components.

The types of interactions involved in soil aggregation include electrostatic attractions, hydrogen bonding, cation bridging, and van der Waals forces. Electrostatic interactions occur primarily between negatively charged clay surfaces and positively charged sites on organic molecules or metal cations. Cation bridging is a key mechanism where multivalent cations such as Ca^{2+}, Mg^{2+}, and Al^{3+} act as connectors between negatively charged clay particles and organic matter, reducing repulsion and promoting flocculation. Water facilitates this process by solvating these cations, allowing them to maintain their charge while effectively linking different soil components.

Hydrogen bonding is crucial, particularly in connecting polar functional groups within organic matter to clay minerals or other hydrated surfaces. The presence of water molecules, which can form hydrogen bonds with both organic hydroxyl, carboxyl, and amine groups, as well as with oxygen atoms on clay surfaces, strengthens the aggregate structure. Additionally, though weaker, van der Waals forces contribute to particle adhesion at very short distances, especially in dry conditions when water films are minimal. Moreover, water influences the swelling and shrinking behavior of clay minerals, which impacts the physical rearrangement of particles and the formation of micro- and macro-aggregates over time. In wet conditions, thin films of water around particles create capillary forces that draw particles together, while in drying phases, water loss increases direct contact between soil constituents, enhancing physical bonding. Thus, water is not merely a passive solvent but an active participant in soil aggregation, mediating chemical interactions, facilitating cation exchange, and modulating the physical proximity of soil particles, which are fundamental to the development and stability of soil structure.

9.6.2 Soil Aggregate Stability: Methods of Evaluation

Soil aggregate stability is a key indicator of soil health, influencing water infiltration, erosion resistance, and nutrient cycling. Different methods can be applied to evaluate the stability of soil aggregates[9]. For example, the wet-sieving method is one of the most commonly used. It measures the resistance of soil aggregates to disintegration when exposed to water, simulating natural rainfall or irrigation. Aggregates are placed on a sieve and oscillated vertically in water, with

[9] Almajmaie, A., et al. 2017. Evaluation of methods for determining soil aggregate stability. Soil and Tillage Research. 167: 39–45.

the proportion of stable aggregates remaining after a set time indicating stability. This method is sensitive to soil moisture conditions and requires careful sample preparation to ensure consistent results. Another widely used method is the slaking test, which assesses the breakdown of aggregates when rapidly immersed in water. This method highlights the role of air entrapment and rapid wetting, which can cause aggregates to collapse, particularly in soils with low organic matter. The slaking test is quick and simple, but may not fully represent field conditions where wetting occurs more gradually. The turbidity method offers a more indirect approach by measuring the cloudiness of water after aggregates disintegrate. Higher turbidity indicates greater disintegration and, therefore, lower aggregate stability. This method is advantageous for rapid assessments and requires minimal equipment, but it may overlook structural changes within the aggregates that do not result in full disintegration. The rainfall simulation test mimics the impact of raindrops on soil surfaces. This method evaluates the detachment of particles and the sealing of the soil surface, providing insights into erosion risks. While realistic, rainfall simulators can be complex and costly, limiting their use in routine assessments. Finally, the dry sieving focuses on the mechanical stability of aggregates in dry conditions. Although useful for understanding soil resistance to physical disturbance, it does not account for the critical role of water in aggregate stability, making it less relevant for assessing erosion risks. The choice of any of the methods depicted above should depend on the specific objectives of the study, soil type, and environmental conditions. Combining methods can provide a more comprehensive understanding of soil aggregate stability, capturing mechanical and hydrological influences.

9.7 SOIL WATER RETENTION AND AVAILABILITY

Soil water retention and availability are fundamental properties that determine how much water a soil can retain and how much water is accessible for plant use. These characteristics play a crucial role in agriculture, plant health, and the overall functioning of ecosystems. The ability of soil to hold and supply water depends on several interrelated factors, including texture, structure, organic matter content, and porosity.

Soil texture significantly influences water retention. Soils composed primarily of clay have tiny particles with a large surface area, allowing them to hold more water compared to sandy soils, which consist of larger particles with larger pores. These larger pores in sandy soils cause water to drain quickly, leaving less moisture available for plant roots. In contrast, clay soils retain water for longer periods, though they may sometimes hold it so tightly that plants struggle to access it.

Soil structure also affects water retention. Soil particles naturally group to form aggregates, and the spaces between these aggregates determine how water moves and is stored. Well-structured soils with stable aggregates allow water to infiltrate effectively while retaining enough moisture for plants. Conversely, compacted soils have reduced porosity, limiting both water infiltration and retention, which can stress plants during dry periods.

Organic matter is another key factor in enhancing soil's ability to retain water. It improves soil structure by promoting the formation of aggregates and increases soil porosity. Additionally, organic materials like decomposed plant residues can absorb and hold water themselves, slowly releasing it as plants need it. This is one reason why soils rich in organic matter tend to perform better under drought conditions.

As defined earlier, porosity and the size distribution of the pores also determine how much water the soil can hold. In soil science, micropores are also referred to as residual pores (see Section 9.4.3) – retain water tightly through capillary action, making it available for plant roots to absorb. Larger pores or macropores, also known in soil science as transmission pores (see Section 9.4.3), facilitate water movement through the soil, aiding drainage but not contributing much to water retention for plants.

When discussing soil water, it is helpful to distinguish between different key concepts related to soil water retention and availability. These are water holding capacity (WHC), field capacity (FC), and the wilting point (WP). They are summarized as reported in Table 9.1. In particular, water holding capacity refers to the total amount of water a soil can retain against the force of gravity. This includes both water that is available to plants and water that is held too tightly for plant uptake. WHC depends on soil texture, structure, and organic matter, encompassing water stored in all pore sizes. Field capacity represents the amount of water retained in the soil after excess gravitational water drains away, typically 2–3 days after heavy rainfall or irrigation. This is the upper limit of water readily available to plants, with water held primarily in mesopores (i.e., storage pores. see Section 9.4.3). Macropores are filled with air at field capacity, while micropores

Table 9.1 Key soil water concepts

Concept	Definition	Plant Water Availability	Key Pore Types Involved
Water holding capacity (WHC)	The total amount of water a soil can retain against gravity, including both available and unavailable water.	Includes both available and unavailable water.	Macropores, that is transmission pores. Mesopores, i.e., storage pores. Micropores, that is residual pores.
Field capacity (FC)	The amount of water retained in the soil after gravitational water has drained, representing the upper limit of readily available water for plants.	Mostly available to plants.	Primarily mesopores (storage pores).
Wilting point (WP)	The critical soil moisture threshold is where plants can no longer extract water, leading to irreversible wilting unless the soil is rehydrated.	Water is largely unavailable to plants.	Micropores (residual pores).

retain water accessible to plant roots. The wilting point, or permanent wilting point, is the critical soil moisture threshold where plants can no longer extract sufficient water to maintain turgor pressure, causing them to wilt irreversibly unless the soil is rehydrated. At this stage, water remains in the soil but is held so tightly within micropores (that in soil science are referred to as residual pores; see Section 9.4.3) hat plant roots cannot access it. This condition marks a physiological limit for plant survival, as the inability to absorb water disrupts essential metabolic processes, ultimately leading to plant death if prolonged.

When further analyzing soil water, it is useful to distinguish between specific types, each playing a unique role in the soil-water-plant relationship. Therefore, it is possible to differentiate between gravitational water (GW), capillary water (CW), residual water (RW), and hygroscopic water (HW). Table 9.2 simply summarizes the different water types. Namely, gravitational water is the water that drains quickly through the soil due to the force of gravity. It moves freely through large pores, that is transmission pores (also known in soil science as macropores; see Section 9.4.3), especially after heavy rainfall or irrigation. Because it drains so rapidly, it is not retained in the root zone long enough to be utilized by plants. While it is not directly available for plant uptake, gravitational water is crucial for replenishing groundwater reserves and maintaining deep soil moisture levels, which can be important for plants with deep root systems during prolonged dry periods. Capillary water is the primary source of water for plants. It resides in the storage pores (see Section 9.4.3)—also known as mesopores in soil science—where it is held by the cohesive and adhesive forces between water molecules and soil particles. Unlike gravitational water, capillary water does not drain away quickly. Instead, it remains available within the root zone, making it accessible to plant roots. Plants absorb this water through osmosis, a process driven by

Table 9.2 Types of soil water

Water Type	Definition	Availability to Plants	Location in Soil
Gravitational water (GW)	Water that drains quickly through macropores due to gravity is not retained in the root zone	Unavailable	Macropores (i.e., transmission pores)
Capillary water (CW)	Water held in mesopores by capillary forces is mostly available for plant uptake	Mostly available	Mesopores (i.e., storage pores)
Residual water (RW)	Water retained in micropores, held too tightly to be accessible to plants, is associated with the wilting point	Unavailable	Micropores (i.e., residual pores)
Hygroscopic water (HW)	A thin film of water tightly bound to soil particles in bonding pores, completely unavailable to plants	Completely unavailable	Surface of soil particles (i.e., bonding pores)

the difference in water potential (see below) between the soil and the plant root cells. Capillary water supports essential physiological processes such as nutrient transport, photosynthesis, and cellular turgor, maintaining plant structural integrity. Residual water is found in the smallest soil pores, known as residual pores or micropores (see Section 9.4.3), where it is held with such strong adhesive forces that it becomes largely unavailable to plants. As previously described, this water is associated with the wilting point, representing the critical soil moisture threshold at which plants can no longer extract sufficient water to maintain turgor pressure. Although some water remains in the soil at this stage, it is retained so tightly within the residual pores that plant roots cannot access it. Lastly, hygroscopic water forms a thin film tightly bound to the surface of soil particles inside the so-called bonding pores. This water is held with such strong adhesive forces that plant roots cannot exert enough pressure to extract it. It exists even in arid soils, giving the soil a slight dampness to the touch despite the absence of usable moisture for plants. Although hygroscopic water is not available to plants, it plays a minor role in influencing soil temperature and microbial activity. The thin moisture films can help support certain drought-tolerant microorganisms, contributing to soil health even under arid conditions.

9.7.1 Water Potential

Water potential is a key concept in soil science and plant physiology, describing the potential energy of water within a system compared to pure water at a reference state, usually at atmospheric pressure and room temperature. It dictates the direction in which water moves, as water always flows from regions of higher water potential to regions of lower water potential. This movement is crucial for processes such as water uptake by plant roots, nutrient transport, and soil moisture dynamics. Water potential (Ψ) is measured in pressure units (typically megapascals, MPa) and can be influenced by several components, each reflecting different forces acting on water in the soil. These components include matric potential, osmotic potential, and gravitational potential.

The matric potential (Ψ_m) is associated with the adhesive and cohesive forces between water molecules and soil particles. It represents the tension or suction exerted by the soil matrix on water, which holds water in the pores against gravitational pull. Matric potential is always negative, indicating that energy must be expended to remove water from the soil matrix. The matric potential can also be described as the pressure the plant must exert to extract water from the soil. It becomes more negative as soil dries out because water is held more tightly in smaller pores, requiring greater force for extraction.

The osmotic potential (Ψ_o) arises from the presence of dissolved solutes in the soil water. This component affects water movement across semi-permeable membranes, such as plant root cells, where water moves from areas of lower solute concentration (higher water potential) to higher solute concentration (lower water potential). In saline soils, osmotic potential significantly impacts plant water uptake, as the high solute concentration outside the roots reduces the water potential gradient, making it harder for plants to absorb water. It is important to note that osmotic potential reflects water dynamics already discussed in Section 7.7, where the concepts of low-density water (LDW) and high-density water (HDW) domains were introduced. In this sense, the concept of water potential represents the other side of the same coin discussed in Chapter 7, further illustrating how the physical and chemical properties of water govern its movement and availability in soils.

The gravitational potential (Ψ_g) relates to the position of water in a gravitational field. It plays a significant role in water movement in saturated soils or during infiltration events when gravity pulls water downward through the soil profile. This potential is positive when water moves downward and negative when it moves upward against gravity, such as during capillary rise.

The total water potential in the soil is the sum of these components: $\Psi = \Psi_m + \Psi_o + \Psi_g$. In unsaturated soils, matric potential is the dominant factor controlling water retention and availability to plants. As soils dry, water becomes increasingly bound to soil particles, lowering the matric potential and making water less accessible to plant roots. Conversely, gravitational potential predominates in saturated conditions, driving water movement downward.

9.7.2 How to Measure Water Retention?

The Pressure Plate Apparatus is a laboratory device used to determine how much water a soil can retain at different levels of tension, helping us understand soil's ability to hold and release water under varying conditions. It essentially simulates what happens in nature when water infiltrates the ground, drains due to gravity, and becomes harder for plants to extract as it moves deeper into the soil matrix.

The concept of soil water potential is at the core of the method, specifically the matric potential, which reflects how tightly water is held in the soil pores. When the soil is fully saturated, water fills all the pores and drains freely under gravity. However, as the soil dries, water becomes increasingly bound to soil particles, requiring more effort (or energy) to be extracted. The pressure plate apparatus artificially creates these conditions by applying external air pressure to force water out of the soil, mimicking how plants or evaporation would naturally draw it.

Figure 9.13 Sealed chambers (A) and ceramic plates (B) need to be assembled to measure the water retention curve using the Pressure Plate Apparatus. The photos were taken in the hydrology lab of the Department of Agricultural, Food and Forest Sciences of the University of Palermo.

The apparatus comprises a sealed chamber (Figure 9.13A) where soil samples are placed on a porous ceramic plate (Figure 9.13B). This plate is the key: it allows water to pass through but resists air entry up to a specific threshold, known as the air entry value. The chamber is pressurized with air, and this pressure acts on the water within the soil, pushing it through the ceramic plate and out of the chamber.

The process begins with saturating the soil samples to ensure all pores are filled with water. The samples are then placed in the pressure chamber on the ceramic plate, and the chamber is sealed. Air pressure is gradually applied in controlled steps. The applied pressure corresponds to specific matric potentials (for example, –0.33 bar to simulate field capacity or –15 bar to simulate permanent wilting point). Water is slowly forced out of the soil at each pressure level through the ceramic plate and collected outside the chamber.

The critical part of the process is waiting until equilibrium is reached at each pressure step, meaning no more water is being expelled from the soil. This indicates that the water remaining in the soil is held too tightly to be moved by the current level of applied pressure. The soil samples are then weighed to determine how much water remains. Finally, the samples are oven-dried to establish their dry weight, allowing the calculation of the soil water content at each applied pressure[10].

The mathematical foundation of the pressure plate method lies in the relationship between soil water content (θ) and soil water potential (Ψ), often represented as the soil water retention curve. This curve illustrates how water content decreases as the soil becomes drier (i.e., as the matric potential becomes more negative).

The volumetric water content (θ) is calculated using the difference between the wet weight and dry weight of the soil:

$$\theta = \frac{W_{wet} - W_{dry}}{\rho_w \cdot V} \tag{9.21}$$

Here, W_{wet} is the weight of the soil at the equilibrium (that is after the application of the pressure), W_{dry} is the weight of the soil after the oven-drying, ρ_w is the density of pure water, and V is the volume of the soil sample. Relationship (9.21) gives the volumetric water content, which represents the volume of water per unit volume of soil.

The applied air pressure is directly related to the matric potential expressed in pressure units. The capillary theory governs the relationship between the applied pressure and the energy needed to extract water, where smaller pores require higher tension to release water. The resulting data are plotted with the volumetric water content on the x-axis and the corresponding matric potential on the y-axis, often on a logarithmic scale (Figure 9.14). This produces the characteristic S-shaped curve, showing how easily the soil releases water at various moisture levels.

[10] Many other experimental approaches are available to measure soil water retention depending on the applied pressure range. In this book, only the Pressure Plate Apparatus is accounted for. For more details about different procedures, the readers are addressed to more specific textbooks (see Further Reading Section).

Mathematically, the relationship is often described by using empirical models such as the van Genuchten equation:

$$\theta(\Psi) = \theta_r + \frac{\theta_s - \theta_r}{\left[1 + \left(\alpha \cdot |\Psi|\right)^n\right]^m} \tag{9.22}$$

Here, $\theta(\Psi)$ is the water content at a given matric potential, θ_r is the residual water content (i.e., the water that cannot be extracted even at high tensions), θ_s is the saturated water content, $|\Psi|$ is the absolute value of the matric potential, and α, n, m are fitting parameters related to soil properties.

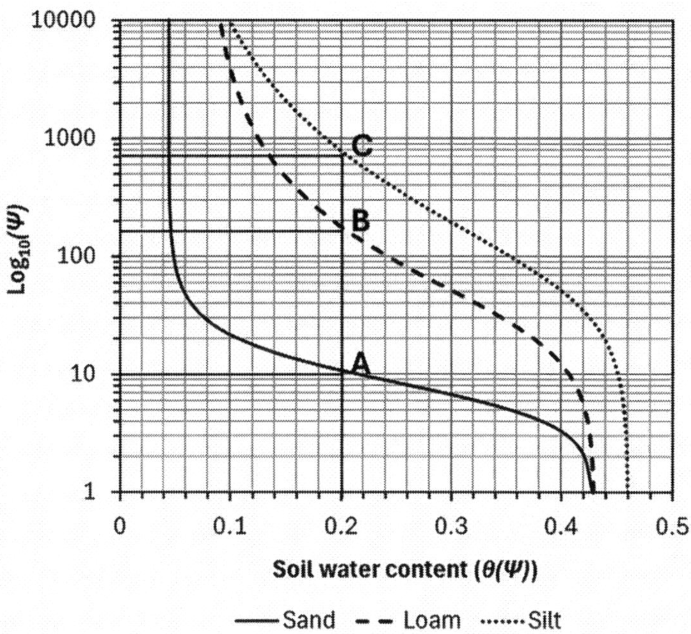

Figure 9.14 Soil water retention curves were simulated using the van Genuchten model for sandy, loamy, and silty soils. The y-axis represents the logarithm of the matric potential ($Log_{10}(\Psi)$), indicating the suction required to extract water. The x-axis shows the volumetric water content ($\theta(\Psi)$). The curves demonstrate that finer soils (silt) retain more water at a given suction compared to coarser soils (sand). Points A, B, and C highlight the increasing retention force required to remove water as the texture becomes finer[11].

Figure 9.14 presents the simulation of the van Genuchten model for sandy, loamy, and silty soils. The graph clearly illustrates that as the soil texture becomes finer—and the surface area increases (i.e., sand < loam < silt)—a greater suction (that is higher $|\Psi|$) is required to remove a given amount of water from the sample

[11] The simulation was performed by using the parameters reported in R.F. Carsel and R.S. Parrish. 1988. Developing joint probability distributions of soil water retention characteristics. Water Resources Research. 24: 755–769.

(e.g., points A, B, and C in Figure 9.14). This result aligns with the earlier findings, considering the chemical and physical mechanisms governing water retention in differently sized soil particles.

9.7.3 SIMPLE EXPERIMENTAL DETERMINATION OF WHC, FC, AND WP

Water holding capacity (WHC), field capacity (FC), and wilting point (WP) are critical soil properties that influence plant water availability.

Equation 9.21, previously outlined, can be applied to estimate the water holding capacity. However, the filter paper method provides a simple and cost-effective approach for determining the soil WHC. This technique is particularly useful for teaching, as it offers a direct and tangible way to observe how soil retains moisture. Unlike more complex methods that require specialized equipment, the filter paper method can be easily performed in a laboratory with minimal resources, making it an ideal tool for introductory soil science experiments.

The procedure begins with the preparation of the soil sample. A representative soil sample is collected and, if necessary, air-dried to remove excess moisture. The sample is then sieved through a 2 mm mesh to eliminate large particles, plant roots, and debris that could interfere with water retention measurements.

To determine WHC, the soil must first be brought to saturation. A measured quantity of soil (typically 20–50 g) is placed in a perforated container or Petri dish, ensuring that it has enough space for water absorption. Distilled water is gradually added until the soil is fully saturated—this is evident when no further water is absorbed and free water starts appearing on the surface. The sample is then left to equilibrate for approximately 24 h, allowing moisture to distribute evenly within the soil matrix. Once saturation is achieved, pre-weighed filter papers (such as Whatman No. 42 or other high-absorption papers) are used to measure the retained water. A dry filter paper is carefully placed in contact with the saturated soil surface, ensuring it absorbs moisture only from the soil and not free-standing water. The setup is then covered to prevent evaporation and left to equilibrate for an additional 24 h at room temperature. Following the absorption period, the filter paper is carefully removed and immediately weighed to determine how much water it absorbed from the soil. The difference between the wet and dry filter paper weight represents the mass of water retained by the soil at saturation. To quantify the water-holding capacity, the following formula is used:

$$WHC(\%) = \frac{\text{Weight of filter paper} - \text{Weight of dry filter paper}}{\text{Oven dried soil weight}} \times 100 \qquad (9.23)$$

The filter paper method can also be adapted to approximate field capacity under controlled conditions. Since field capacity represents the amount of water retained in the soil after gravitational water has drained, this method provides an indirect yet practical approach to estimating it, especially in laboratory settings where more advanced equipment may not be available.

The soil sample must first be saturated to determine field capacity using filter paper to ensure all pores are filled with water. This is done by gradually adding distilled water until no further absorption occurs, indicating complete saturation. Once the soil reaches this state, it is left to drain under gravity in a perforated container or a setup that allows free water to escape, typically for 24 to 48 h. This drainage period replicates natural field conditions, allowing excess water to leave the macropores while finer pores retain moisture through capillary forces. At the end of this process, the soil is assumed to have reached field capacity, holding only the water available to plants before permanent drying begins.

At this stage, pre-weighed filter paper is carefully placed in contact with the soil, ensuring it absorbs moisture directly from the sample without coming into contact with any remaining free water. The setup is covered to minimize evaporation, and the filter paper can equilibrate with the soil moisture over 24 h. Once the equilibrium is reached, the filter paper is removed and immediately weighed to determine the amount of water it has absorbed. An estimate of the soil's water retention at field capacity can be obtained by comparing the difference between the wet and dry filter paperweights.

The wilting point (WP) is determined by growing a test plant (such as a sunflower or bean) in a controlled environment and monitoring soil moisture as the plant undergoes severe water stress. When the plant wilts permanently and cannot recover even after rehydration, the corresponding soil moisture is measured using the gravimetric method or a pressure plate apparatus. The pressure plate method applies controlled suction to soil samples in a chamber, typically at –1,500 kPa, to simulate the wilting condition and extract water, allowing the determination of residual moisture content.

By measuring water-holding capacity, field capacity, and wilting point, the available water capacity (AWC) for plants can be calculated as the difference between field capacity and wilting point, which is essential for irrigation planning and soil water management.

9.8 SOIL BULK AND PARTICLE DENSITY

Bulk density is a fundamental physical property of soil. It is defined as the mass of dry soil per unit volume, including solids and pore spaces. It is expressed in grams per cubic centimeter (g cm^{-3}) and provides critical insights into the soil structure, porosity, compaction, and water-holding capacity.

The formula for calculating dry bulk density (ρ_b) is:

$$\rho_b = \frac{M_s}{V} \tag{9.24}$$

where M_s is the mass of dry soil (in g), while V is the total soil volume (in cm^3), including solids and pore spaces.

Bulk density varies significantly with soil texture and structure. For instance, sandy soils typically exhibit bulk densities in the 1.3 to 1.6 g cm^{-3} range due to larger particle sizes and less pore space. Conversely, clayey soils, with finer particles and more pore spaces tend to have bulk densities within the range of 1.0 to 1.1 g cm^{-3}.

Several factors influence bulk density, including soil compaction, organic matter content, and moisture levels. Compacted soils have higher bulk densities due to reduced pore space, which can impede root growth and water infiltration. In contrast, soils rich in organic matter generally exhibit lower bulk densities, enhancing aeration and water retention.

Bulk density is a critical parameter for agricultural practices, affecting root penetration, seedling emergence, and overall plant health. It also helps estimate the soil water storage capacity and influences irrigation practices, making it essential for soil management and conservation strategies.

Particle density refers to the mass of soil solids per unit volume of solids, excluding pore spaces. It reflects the density of the mineral particles and organic matter constituting the soil. As for the bulk density, particle density is also expressed in grams per cubic centimeter.

The formula for particle density (ρ_s) is:

$$\rho_s = \frac{M_s}{V_s} \tag{9.25}$$

where M_s is the mass in grams of the soil solids and V_s is the volume in cubic centimeters of the soil solids.

Typical particle density values for mineral soils range between 2.6 and 2.7 g cm^{-3}, primarily determined by the composition of minerals such as quartz, feldspar, and other silicates. Soils with higher organic content have slightly lower particle densities due to the lower density of organic materials than mineral particles.

Unlike bulk density, particle density remains relatively constant unless there are significant changes in soil composition, such as through organic matter addition or mineral alteration. It does not account for pore spaces and thus provides a measure of the density of the solid phase of the soil alone.

Understanding particle density is crucial for calculating soil porosity, the fraction of the soil volume occupied by pore spaces. Porosity (Φ), bulk, and particle density are related to each other by the relationship:

$$\Phi = 1 - \frac{\rho_b}{\rho_s} \tag{9.26}$$

which can be easily derived from relationship 6.

9.8.1 Experimental Procedure for Measuring Bulk and Particle Density

The core method is commonly used to determine bulk density. A metal cylinder of known volume (Figure 9.15) is carefully driven into the soil to extract an undisturbed sample. The soil sample is then dried in an oven at 105°C for 24 h to remove all moisture. The mass of the dry soil (M_s) is recorded, and bulk density

(ρ_b) is calculated using Equation 9.24 by dividing the dry mass by the total volume of the cylinder. Alternatively, the clod method, which involves coating an irregular soil clod with paraffin and measuring its volume via water displacement, can be used for cohesive soils.

Figure 9.15 Differently sized cylinders used to measure the bulk density. The picture was taken in the lab of hydrology of the Department of Agricultural, Food and Forest Sciences of the University of Palermo.

For particle density, the pycnometer method is widely employed. A known mass of dry, finely ground soil is placed in a pycnometer (Figure 9.16), and its volume is determined by displacement using a liquid, typically water or ethanol. The particle density (ρ_s) is then computed using Equation 25, dividing soil dry mass by its solid volume. This method ensures that pore spaces do not contribute to the volume measurement, distinguishing it from bulk density. By obtaining both bulk and particle density values, soil porosity can be derived using Equation 9.26, thereby providing insights into soil air and water-holding capacity.

Figure 9.16. A scientific glass pycnometer, a laboratory instrument used for density measurements. The pycnometer consists of a transparent glass bottle with a narrow neck and a precisely fitting stopper, which features a small capillary hole at the top. The flat bottom ensures stability, while the smooth, curved edges allow for precise volume measurements. This image was generated with the assistance of ChatGPT.

EXERCISES

1. A 30 g soil sample is fully saturated and allowed to drain before a pre-weighed filter paper is placed in contact with it. The dry filter paper weighs 1.2 g, and after absorbing moisture from the soil, it weighs 2.5 g. The oven-dried weight of the soil is 28 g.
   (a) Calculate the water-holding capacity (WHC) using Equation (9.23).
   (b) Discuss the implications of WHC in different soil textures.
2. A researcher saturates a 50 g soil sample and allows it to drain for 48 h. After drainage, the remaining wet weight is 65 g, and the oven-dried weight is 50 g.
   (a) Calculate the field capacity (FC) as the percentage of retained water in the soil.
   (b) How would FC differ between sandy soil and clay soil? Explain why.
3. A cylindrical soil core with a volume of 250 cm^3 is collected from a field. After drying at 105°C for 24 h, the mass of the dry soil is 320 g.
   (a) Calculate the bulk density (ρ_b) using Equation (24).
   (b) Compare your result with typical values for sandy, silty, and clayey soils.
   (c) How would soil compaction affect bulk density?
4. A soil sample has a bulk density (ρ_b) of 1.35 g cm^{-3} and a particle density (ρ_s) of 2.65 g cm^{-3}.
   (a) Calculate the porosity (Φ) using Equation (9.26).
   (b) Interpret the result: What does this value indicate about the soil's ability to store air and water?
   (c) How would organic matter addition affect soil porosity?
5. Using the van Genuchten equation (9.22), a soil scientist plots the water retention curve for three soil types: sandy, loamy, and clayey. At a matric potential of –0.33 bar, the water contents (θ) are recorded as follows:
   - Sand: 10%
   - Loam: 28%
   - Clay: 42%

   At –15 bar, the water contents are:
   - Sand: 2%
   - Loam: 10%
   - Clay: 25%

   (a) Calculate the available water capacity (AWC) for each soil type.
   (b) Explain why finer-textured soils retain more water than coarse-textured soils.
   (c) How does soil texture influence irrigation practices?

FURTHER READING

Angulo-Jaramillo, R., V. Bagarello, M. Iovino and L. Lassabatere. 2016. Infiltration Measurements for Soil Hydraulic Characterization, 1st Ed. Springer International Publishing Switzerland.

Conklin, A.R. 2005. Introduction to Soil Chemistry: Analysis and Instrumentation, 1st Ed. Wiley-Interscience.

Radcliffe, D.E. and Jiři Šimůnek. 2010. Soil Physics with Hydrus. Modelling and Applications, 1st Ed. CRC Press.

Sparks, D.L. 2003. Environmental Soil Chemistry, 2nd Ed. Academic Press.

Sposito, G. 2008. The Chemistry of Soils, 2nd Ed. Oxford University Press.

Chapter **10**

The Chemistry of Soil Colloids: The Clay Minerals

10.1 INTRODUCTION TO CLAY MINERALS

Clay minerals are fine-grained, naturally occurring hydrous aluminosilicates that play a pivotal role in environmental and soil chemistry. Formed primarily through the weathering and alteration of primary rock-forming minerals (see Chapter 3), they contribute extensively to the structural and functional integrity of soils. Their unique crystalline structures, such as the 1:1 and 2:1 layer silicates (see below in this Chapter), dictate their swelling behavior, cation exchange capacity, and interactions with water, ions, and organic matter. These properties result in a high surface area and reactivity, making clay minerals fundamental to soil processes such as nutrient retention and cation exchange—both essential for plant growth and ecosystem sustainability.

Beyond their role in soil fertility, clay minerals also influence environmental processes such as water regulation and pollutant mitigation. Their water holding capacity (see Chapter 9), particularly in expansive structures like smectite, helps maintain soil moisture levels, directly affecting plant hydration and resilience to drought. Additionally, their surface charge properties, which vary with *pH* and mineral type, enable the adsorption and immobilization of pollutants, such as heavy metals and organic contaminants. These adsorption mechanisms make clay minerals crucial components in natural soil remediation processes and in reducing environmental contamination.

The interactions between clay minerals and soil microorganisms further highlight their importance in ecosystem dynamics. By providing surfaces for microbial colonization and influencing nutrient cycling—particularly in processes like nitrogen fixation and organic matter decomposition—clay minerals support soil fertility over time. Additionally, their role in soil aggregation enhances structure, reduces erosion, and improves overall land stability, thereby promoting

DOI: 10.1201/9781003780212-10

sustainable land management and protecting water quality in surrounding ecosystems.

In conclusion, clay minerals are integral to soil and environmental chemistry due to their diverse physicochemical properties. Their ability to influence ion exchange, water retention, microbial interactions, and pollutant adsorption underscores their significance in maintaining soil health, supporting plant growth, and mitigating environmental contamination. The following sections will delve deeper into the structural variations of clay minerals, their ion exchange mechanisms, and their broader implications for soil behavior and sustainability.

10.2 STRUCTURAL AND CHEMICAL ORGANIZATION OF SILICON AND ALUMINUM IN CLAY MINERALS

Clay minerals, primarily composed of silicon (Si) and aluminum (Al), derive their fundamental properties from the electronic configurations of these elements, which govern their bonding behavior, coordination environments, and interactions within soil matrices. These minerals belong to the phyllosilicate group, characterized by layered crystal structures influencing their reactivity, cation exchange capacity, and role in soil fertility. The ability of silicon to form tetrahedral coordination units and aluminum to adopt octahedral coordination units is central to the architecture of clay minerals, as these structural units determine the stability, charge distribution, and adsorption properties of different clay types.

The arrangement of tetrahedral and octahedral layers forms the basis for the 1:1 and 2:1 layered silicates, which define the physical and chemical behavior of clays. In 1:1 clays like kaolinite, each unit layer consists of a single tetrahedral sheet bonded to an octahedral sheet, resulting in a relatively stable, non-expanding structure. In contrast, 2:1 clays such as montmorillonite and illite have two tetrahedral sheets sandwiching an octahedral sheet, creating interlayer spaces that influence swelling, water retention, and cation exchange properties.

Beyond their role in soil chemistry, the electronic properties of silicon and aluminum are crucial in determining how clays interact with nutrients, pollutants, and organic matter. The partial substitution of aluminum for silicon in tetrahedral sheets generates permanent negative charges, which are balanced by exchangeable cations such as potassium (K^+), calcium (Ca^{2+}), and magnesium (Mg^{2+}). These charges influence nutrient availability in soils and contribute to the adsorption and immobilization of heavy metals[1] and organic contaminants, making clay minerals important for environmental remediation and soil stability.

---

[1] The term "heavy" when referring to certain metals in the periodic table is based on their density (d). A metal is generally considered "heavy" if $d > 2$ g cm^{-3}. By this definition, aluminum ($d = 2.7$ g cm^{-3}) is classified as "heavy," along with arsenic ($d = 5.8$ g cm^{-3}), chromium ($d = 7.2$ g cm^{-3}), cadmium ($d = 8.7$ g cm^{-3}), copper ($d = 9.0$ g cm^{-3}), lead ($d = 11.3$ g cm^{-3}), and mercury ($d = 13.5$ g cm^{-3}). However, there is scientific debate regarding this classification. Some researchers propose raising the density threshold from 2 g cm^{-3} to 5 g cm^{-3} to define a metal as "heavy," which would exclude aluminum from this category.

Understanding the tetrahedral and octahedral coordination of these elements provides key insights into the chemical flexibility and adaptability of clay minerals in different soil environments. In the following sections, we will explore the electronic structure of silicon and aluminum, their hybridization states, and how these fundamental properties shape the molecular architecture and environmental functions of clay minerals.

10.2.1 Tetrahedral Organization of Silicon

Silicon (Si), the second most abundant element in the Earth's crust after oxygen, has the atomic number 14 and the electron configuration:

$$\text{Si: } 1s^2\ 2s^2\ 2p^6\ 3s^2\ 3p^2$$

The valence shell of silicon consists of four valence electrons in the 3*s* and 3*p* orbitals, which enables it to form four covalent bonds. In silicate minerals, silicon primarily exists in tetrahedral coordination (SiO_4^{4-}), where one silicon atom is surrounded by four oxygen atoms at the corners of a tetrahedron.

To explain the tetrahedral coordination of silicon, the concept of hybridization—already conceptualized in Chapter 5–is employed. The 3*s* and three 3*p* orbitals of silicon hybridize to form four sp^3 hybrid orbitals, each forming a σ-bond with an oxygen atom. The resulting tetrahedral unit (Figure 10.1) is the fundamental building block of silicate minerals, including clays.

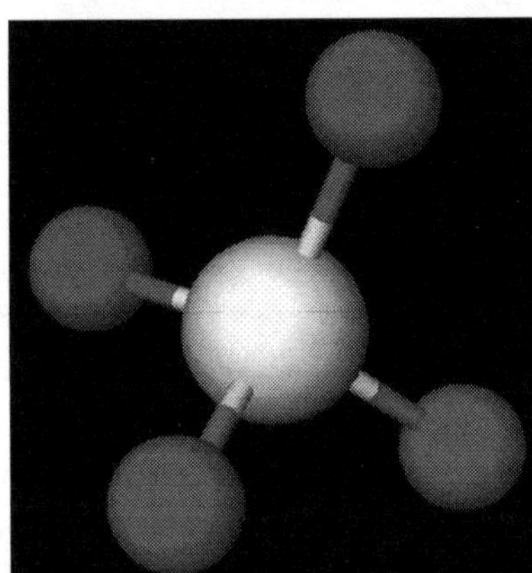

Figure 10.1 Tetrahedral configuration of the SiO_4^{4-} unit. The dark balls are the oxygen atoms which are negatively charged, while the white ball is the silicon.

These tetrahedra can exist as isolated units (as in olivine), or they may polymerize to form chains (Figure 10.2A), sheets (Figure 10.2B), and frameworks in different types of silicate structures. In clay minerals, these tetrahedra are linked at the corners to form tetrahedral sheets, which combine with octahedral layers to create the characteristic phyllosilicate structures (see below).

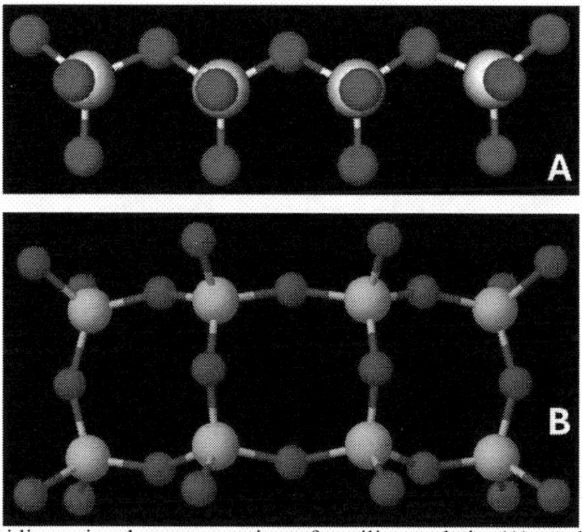

Figure 10.2 Tridimensional representation of a silicate chain (**A**), and sheet (**B**) with molecular formulas $Si_4O_{13}^{10-}$ and $Si_8O_{22}^{12-}$, respectively.

10.2.2 Electronic Structure of Aluminum and the Octahedral Coordination

With an atomic number of 13, aluminum shows the following electronic configuration:

$$\text{Al: } 1s^2\ 2s^2\ 2p^6\ 3s^2\ 3p^1$$

Since aluminum possesses three valence electrons ($3s^23p^1$), it is classified as trivalent, meaning it can form three covalent bonds. However, unlike silicon, which prefers tetrahedral coordination, aluminum in clay minerals is predominantly found in an octahedral arrangement (Figure 10.3).

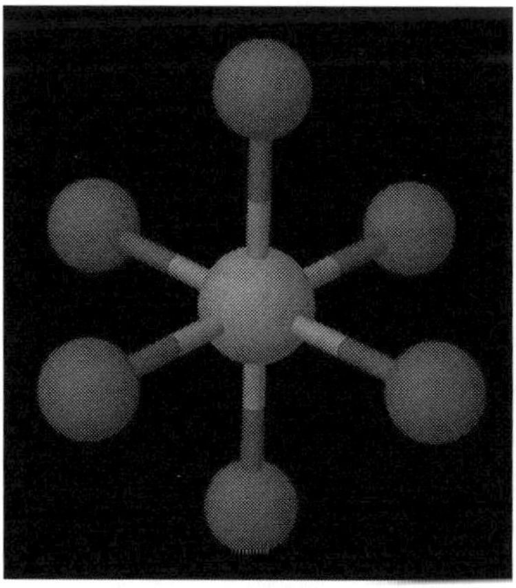

Figure 10.3 Octahedral configuration of AlO_6^{9+} unit. The dark balls are the oxygen atoms which are negatively charged, while the grey ball is the aluminum.

This preference for octahedral coordination arises from the ability of aluminum to undergo sp^3d^2 hybridization. Figure 10.4 illustrates the geometrical shapes of the five d orbitals present in the valence shell of aluminum.

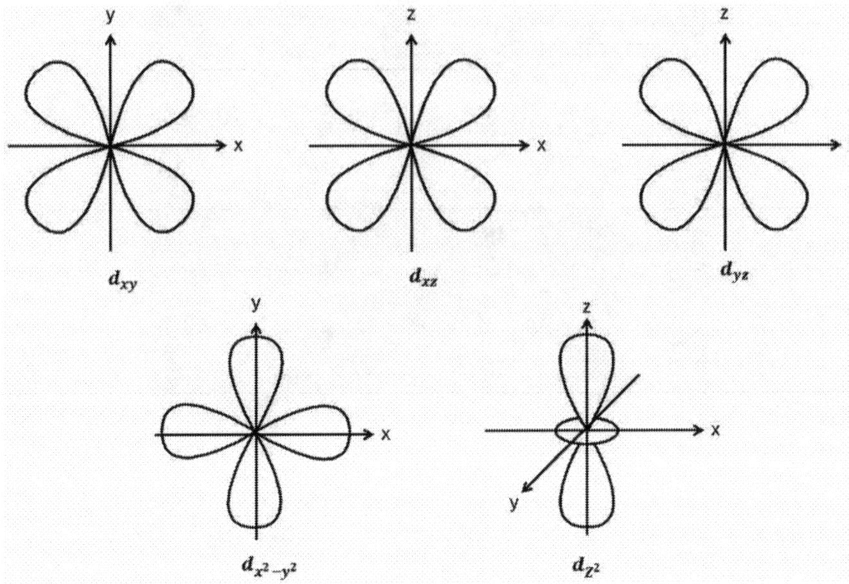

Figure 104 Geometrical shape of the d orbitals in the valence shell of the aluminum element.

Due to the contribution of two of these five d orbitals to the sp^3d^2 hybridization, the linear combination of the six atomic orbitals (one s, three p, and two d) results in six hybrid molecular orbitals. These latter are oriented along the six Cartesian coordinate axes (x, y, and z directions), defining the directions along which ligands may approach aluminum. As a result, the characteristic octahedral geometry emerges. In clay minerals, the ligands are typically oxygen atoms, which may act as bridging units between adjacent octahedra (Figure 10.5A) or connect to silicon-containing tetrahedra (Figure 10.5B), contributing to the formation of the layered phyllosilicate structure.

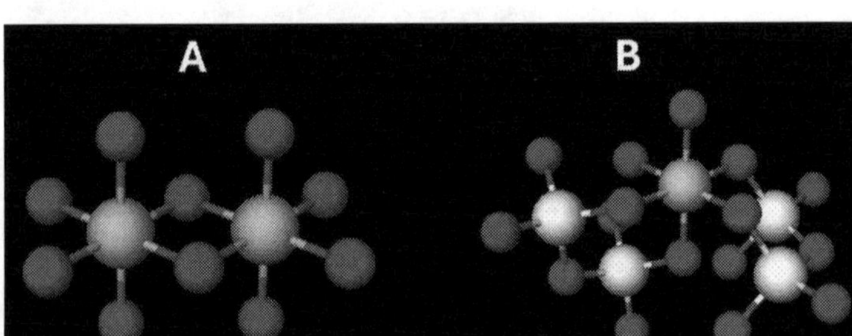

Figure 10.5 Representation of two adjacent octahedra bridged by oxygen atoms (**A**) and a three-dimensional representation of silicon tetrahedra linked to aluminum octahedron via oxygen bridges (**B**).

10.2.3 Isomorphic Substitution and Charge Imbalance in Clay Minerals

One of the most fundamental structural characteristics influencing the charge and reactivity of clay minerals is isomorphic substitution—a phenomenon in which one ion occupies the crystallographic site of another without significantly altering the overall mineral structure. This substitution occurs in both the tetrahedral and octahedral sheets of clay minerals and is responsible for the permanent negative charge that gives rise to the cation exchange capacity (CEC) of these materials (see Chapter 7). Although isomorphic substitution is often described in terms of ionic size compatibility, an equally important factor is the chemical similarity of substituting cations, which is largely dictated by their proximity in the periodic table. Elements that are contiguous in the periodic table tend to share comparable electronic structures and bonding behavior, allowing for smooth replacement within the lattice. For example, Al^{3+} can replace Si^{4+} in tetrahedral sheets because aluminum and silicon belong to neighboring groups (13 and 14, respectively) and form structurally equivalent bonds with oxygen. Similarly, in the octahedral sheet, Mg^{2+}, Fe^{2+}, and Mn^{2+} can replace Al^{3+}, as they exhibit similar valence electron configurations and bonding tendencies, ensuring minimal disruption to the crystal structure.

When a cation with a lower charge replaces one with a higher charge in the crystal structure of a clay mineral, a charge imbalance occurs, leading to the formation of permanent negative charges within the mineral lattice. This is a crucial characteristic of clay minerals, as these excess negative charges cannot be neutralized by the crystal itself and must instead be compensated by external cations from the surrounding environment. These charge-compensating cations (such as K^+, Na^+, Ca^{2+}, and Mg^{2+}) are loosely held in the interlayer spaces or along the surface of clay minerals, making them exchangeable. This phenomenon, known as cation exchange, plays a critical role in the nutrient retention capacity of soils, as these interlayer cations can be readily replaced by other cations from soil solutions.

Furthermore, the magnitude of charge imbalance depends on the specific isomorphic substitution event. For example, when Al^{3+} replaces Si^{4+} in the tetrahedral sheet, the charge imbalance is relatively small (a decrease of +1 per substitution). However, when divalent cations such as Mg^{2+} or Fe^{2+} replace trivalent Al^{3+} in the octahedral sheet, a greater charge imbalance occurs (a decrease of +1 per substitution per replaced cation), potentially leading to higher cation exchange capacities and increased swelling behavior in certain clay minerals.

10.2.4 Crystallographic Vacancies and their Contribution to Charge Imbalance

In addition to isomorphic substitution, another structural feature that influences the electrical properties of clay minerals is the presence of crystallographic vacancies in the tetrahedral and octahedral sheets. These vacancies occur when

Al^{3+}or Si^{4+} are absent from their expected lattice positions during the mineral formation, leaving a void in the crystal structure. The absence of these cations intensifies the charge imbalance, as the missing positively charged ions are not available to counterbalance the negative charge of the oxygen atoms in the structure.

Crystallographic vacancies are particularly common in the octahedral sheets of dioctahedral clays, such as kaolinite and montmorillonite. In dioctahedral structures, only two-thirds of the available octahedral sites are occupied, with the remaining one-third vacant due to charge stabilization and structural constraints. These vacancies increase the net negative charge of the clay mineral, making it even more reliant on interlayer cations for charge compensation.

Similarly, in some cases, vacancies can also occur in tetrahedral sheets, though this is less common. A missing Si^{4+} in the tetrahedral layer can further destabilize the charge balance, requiring additional interlayer cations or influencing the overall reactivity and swelling behavior of the clay.

The presence of both isomorphic substitution and crystallographic vacancies means that the actual charge imbalance in clay minerals can be greater than what would be predicted based on the substitution alone. This explains why certain clays exhibit higher-than-expected cation exchange capacities and enhanced reactivity in soil environments.

10.2.5 Why Silicon cannot Replace Aluminum in Octahedral Coordination

While aluminum can substitute for silicon in tetrahedral sheets, the reverse does not occur—silicon cannot replace aluminum in octahedral coordination within clay minerals. The fundamental reason lies not simply in ionic size differences but more importantly in bonding preferences, electrostatic stability, and energetic constraints. In clay minerals, aluminum is commonly found in octahedral coordination, where six oxygen atoms symmetrically surround the aluminum ion, forming a stable octahedral geometry. This stability arises because Al^{3+}, with an ionic radius of 0.054 nm in octahedral coordination, allows for effective electrostatic interactions with surrounding oxygen ligands, creating bonds that have a balance of covalent and ionic character. In contrast, Si^{4+}, despite having a smaller octahedral ionic radius of 0.040 nm, does not readily adopt octahedral coordination due to its stronger preference for highly directional covalent bonding. Silicon naturally forms sp^3 hybridized orbitals, optimizing its bonding within a tetrahedral arrangement, whereas an octahedral setting would require sp^3d^2 hybridization, leading to unfavorable electronic distribution. Since silicon strong covalent bonding relies on orbital overlap and directional stability, forcing it into an octahedral geometry results in bond distortion, increased lattice strain, and energetic instability, thereby making octahedral coordination highly unfavorable. Unlike aluminum, whose bonds with oxygen have a more ionic character that allows flexible coordination numbers, silicon strong orbital-dependent bonding characteristics restrict it to tetrahedral geometry.

Beyond hybridization and bonding constraints, Pauling's Rule 3[2] further explains why Si^{4+} is incompatible with octahedral coordination in clay minerals. According to this rule, the electrostatic bond strength (s) of a cation in a coordinated structure is determined by its charge (Z) divided by its coordination number (CN). In tetrahedral coordination, Si^{4+} has a bond strength of $s = 4/4 = 1.0$, while Al^{3+} in octahedral coordination has $s = 3/6 = 0.5$. If Si^{4+} were to replace Al^{3+} in an octahedral site, the resulting bond strength would be $s = 4/6 = 0.67$, an intermediate value that disrupts the charge balance within the octahedral sheet. This mismatch in electrostatic bond strength destabilizes the mineral structure by altering the expected bond network, thus making the substitution unfavorable.

Additionally, octahedral sheets in clay minerals involve edge-sharing octahedra, which means that cations in adjacent sites are brought into close proximity. High-valency cations like Si^{4+}, when placed in an octahedral arrangement, would experience strong Coulombic repulsions with neighboring cations, further destabilizing the lattice. This added electrostatic strain reinforces why silicon remains confined to tetrahedral coordination, where its bonding tendencies and charge distribution align with the structural requirements of the mineral.

Experimental and geological evidence further support this restriction—no naturally occurring phyllosilicate minerals contain silicon in octahedral sites, whereas aluminum is frequently observed in both tetrahedral and octahedral positions[3]. However, under extreme high-pressure conditions, such as those found in subduction zones or impact structures, silicon can be forced into octahedral coordination due to lattice compression that shortens Si–O bond lengths and alters electronic interactions, as observed in stishovite, a high-pressure polymorph of SiO_2.

---

[2] Pauling's rules are reported in his original paper from 1929: Pauling, L. 1929. The principles determining the structure of complex ionic crystals. Journal of the American Chemical Society. 51(4): 1010–1026. The five rules can be summarized as follows:

1. Coordination Rule: The coordination number of a cation is determined by the radius ratio between the cation and the surrounding anions. The cation-anion distances approximate the sum of their ionic radii, defining whether coordination is tetrahedral (CN = 4), octahedral (CN = 6), or cubic (CN = 8).
2. Electrostatic Valency Rule: The electrostatic bond strength (s) of a cation is calculated as $s = Z/CN$ (where Z is the cation's charge and CN its coordination number). In a stable structure, the total bond strength received by an anion should approximately equal its charge.
3. Polyhedral Sharing Rule: The stability of a structure decreases when polyhedra share edges or faces, particularly for high-charge, low-coordination cations. Corner-sharing is the most stable, while face-sharing is the least stable due to increased cation-cation repulsions.
4. High-Valency Cation Separation Rule: Cations with high charges and low coordination numbers tend to be spaced apart in a crystal structure to reduce electrostatic repulsion, which helps maintain overall structural stability.
5. Rule of Structural Simplicity (Parsimony): The number of different cation and anion environments in a stable crystal tends to be minimal. That is, crystals prefer simpler, repetitive structural units rather than highly complex arrangements.

[3] Pinkas, J. 2005. Chemistry of silicates and aluminosilicates. Ceramics—Silikáty. 49(4): 287–298.

In such cases, external pressure stabilizes Si^{4+} in octahedral coordination by modifying its bond distances and forcing it into an energetically viable structure. However, these conditions are far removed from those under which clay minerals form, reinforcing that under normal environmental conditions, silicon remains tetrahedrally coordinated. The inability of silicon to occupy octahedral sites in clay minerals is therefore not primarily a matter of ionic size, but rather a consequence of bonding preferences, electrostatic balance, and charge distribution constraints. Its strong preference for directional covalent bonds in tetrahedral geometry prevents it from adopting the less covalently constrained octahedral coordination observed in aluminum-bearing octahedral sheets. The rare cases in which Si^{4+} adopts octahedral coordination further confirm that external pressure, rather than natural crystal chemistry, is required for such a substitution to occur.

10.3 STRUCTURAL COMPOSITION AND CLASSIFICATION OF CLAY MINERALS

The structural composition of clay minerals is defined by their layered arrangement of tetrahedral and octahedral sheets.

The so-called 1:1 clay minerals are characterized by the pairing of one tetrahedral sheet with one octahedral sheet, creating a simple and stable structure (Figure 10.6).

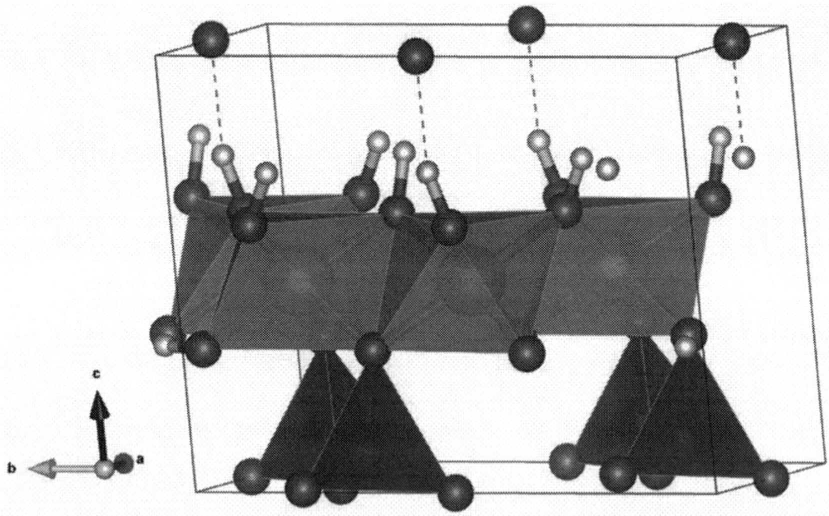

Figure 10.6 Crystal unit of a typical 1:1 clay mineral. This is a representation of kaolinite with the following crystallographic parameters: a = 0.5155 nm, b = 0.8945 nm, c = 0.7404 nm; α = 91.7°, β = 104.9°, γ = 89.8°. The hydrogen bonds between the different crystal units are represented by dashed lines. This figure was generated using the VESTA3 software[4].

[4] Momma, K. and F. Izumi. 2011. VESTA 3 for three-dimensional visualization of crystal, volumetric and morphology data. Journal of Applied Crystallography. 44: 1272–1276.

The sheets are held together by strong hydrogen bonds, which limit the expansion of the interlayer space and restrict water and ion intercalation. This results in a low cation exchange capacity and minimal swelling behavior. Kaolinite, a prime example, is often found in highly weathered soils in tropical and subtropical regions. Due to its structural stability, kaolinite plays an essential role in soil environments where maintaining structure under varying moisture conditions is crucial. Its low CEC makes it less active in nutrient retention, but its stability aids in soil aggregation and erosion resistance.

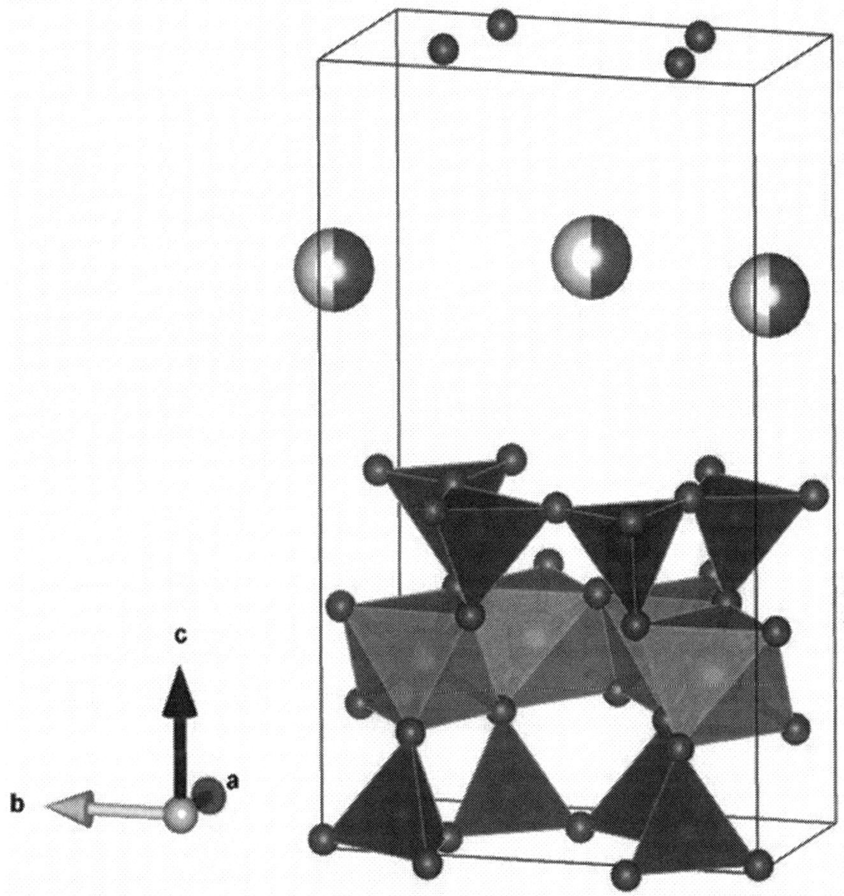

Figure 10.7 Crystal unit of a 2:1 clay mineral. This is a representation of the montmorillonite structure, where different crystal units are interlayered with calcium ions (represented by the large bicolor spheres in the figure). It should be noted that Ca^{2+} is surrounded by water molecules. The crystallographic parameters for montmorillonite are: a = 0.523 nm, b = 0.906 nm, and c—also referred to as the inter-basal distance—ranges between 0.1 and 1.2 nm, depending on the amount of water in the interlayer space. The water content in the inter-basal space is also influenced by the nature of the surrounding cations. The unit cell angles are $\alpha = 90°$, $\beta = 99°$, and $\gamma = 90°$. This figure was generated using the VESTA3 software[4].

The 2:1 clay minerals have a more complex structure, composed of two tetrahedral sheets enclosing an octahedral sheet (Figure 10.7). This configuration generates a larger interlayer space that can accommodate water molecules and various cations, leading to pronounced swelling behavior. Montmorillonite, a prominent member of the smectite group, exhibits high CEC, which is vital for nutrient retention and exchange in soils. The swelling and shrinking properties of 2:1 clays are essential for water retention and influence the physical stability of soils. These characteristics allow smectites to act as natural buffers for soil moisture, playing a critical role in arid and semi-arid environments where water availability is sporadic. The interlayer space capacity to host cations and water molecules also makes smectites important in mitigating the leaching of nutrients and contaminants, thus enhancing soil fertility and environmental protection.

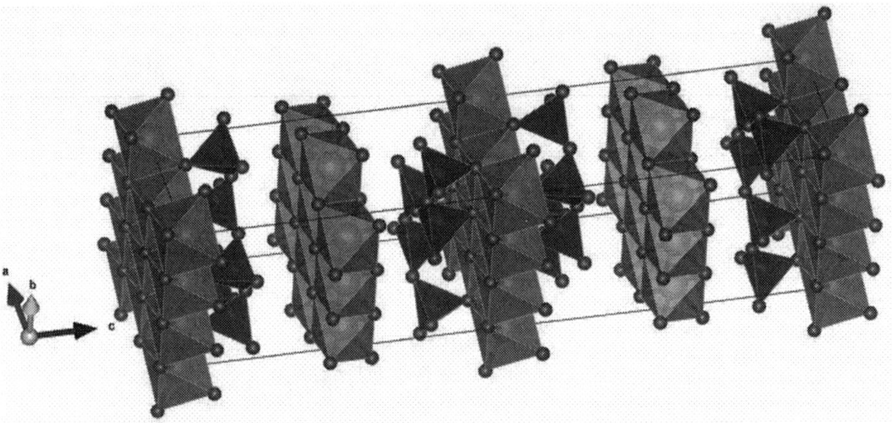

Figure 10.8 Typical crystal unit of a 2:1:1 clay mineral. This is the tridimensional representation of a magnesium chlorite having minimal formula $Al_2H_8Mg_5Si_3O_{18}$. The central part of the crystal reveals the 2:1:1 units affecting the general name of this class of clay minerals. The crystal parameters for this chlorite are: $a = 0.531$ nm, $b = 0.919$ nm, and $c = 2.85$. The latter is the interlayer space. The unit cell angles are $\alpha = 90°$, $\beta = 97.15°$, and $\gamma = 90°$. This figure was generated using the VESTA3 software[4].

The 2:1:1 clay minerals are distinguished by an additional hydroxide sheet sandwiched between successive 2:1 layers (Figure 10.8). This unique structure imparts properties such as moderate CEC and increased thermal stability compared to purely 2:1 minerals. Chlorite, a representative mineral, contains interlayer cations that can interact with surrounding water and ions, but the additional hydroxide layer limits its swelling behavior. This hydroxide sheet makes chlorite more resistant to chemical weathering and more stable under a broader range of environmental conditions. This stability is particularly relevant in soils subject to significant temperature fluctuations or chemical inputs, as chlorite structure helps maintain soil integrity. Its moderate CEC allows for reasonable nutrient retention and interaction with organic and inorganic compounds in the soil, thereby contributing to the complex dynamics of soil chemistry and nutrient cycling.

10.4 OTHER CLAY MINERALS

In addition to the well-known clay minerals such as kaolinite, montmorillonite, and chlorite, there are other important clay and related minerals that play significant roles in soil and environmental chemistry. These minerals include vermiculite, illite, allophane, imogolite, zeolites, as well as various oxides and hydroxides of iron, titanium, and manganese.

10.4.1 Vermiculite

Vermiculite is a 2:1 layer silicate mineral similar to smectite but with a higher degree of interlayer hydration and cation exchange capacity. It is often characterized by its ability to expand when heated, a property that makes it useful in applications like soil conditioning and insulation. Vermiculite's structure comprises two tetrahedral sheets and one octahedral sheet, with water molecules and exchangeable cations (e.g., Mg^{2+}, K^{+}) occupying the interlayer spaces. This allows vermiculite to retain nutrients and moisture, making it valuable for plant growth.

10.4.2 Muscovite Mica

Muscovite is a common phyllosilicate mineral belonging to the mica group, characterized by its perfect basal cleavage, which allows it to be easily split into thin, flexible sheets. It has a 2:1 layer structure composed of two tetrahedral sheets sandwiching an octahedral sheet, with potassium ions (K^{+}) occupying interlayer spaces. Muscovite mica is chemically stable and often found in metamorphic and igneous rocks, where it plays a role in weathering processes leading to the formation of clay minerals such as illite.

10.4.3 Illite

Illite is a 2:1 layer silicate mineral with a structure similar to that of muscovite mica. It contains potassium ions (K^{+}) in its interlayer spaces, which help stabilize the structure and prevent significant swelling. Illite has a moderate CEC and plays a crucial role in soil fertility by retaining essential nutrients such as potassium. It is often found in temperate climates and forms from the weathering of feldspars and micas.

10.4.4 Allophane and Imogolite

Allophane and imogolite are non-crystalline (amorphous) aluminosilicate minerals that are often found in volcanic soils. These minerals are characterized by their high surface area and reactivity. Allophane consists of hollow, spherical structures (Figure 10.9A), while imogolite has a tubular morphology (Figure 10.9B). Both minerals have a high affinity for adsorbing nutrients, water, and pollutants, making them important for soil aggregation, nutrient cycling, and environmental remediation processes.

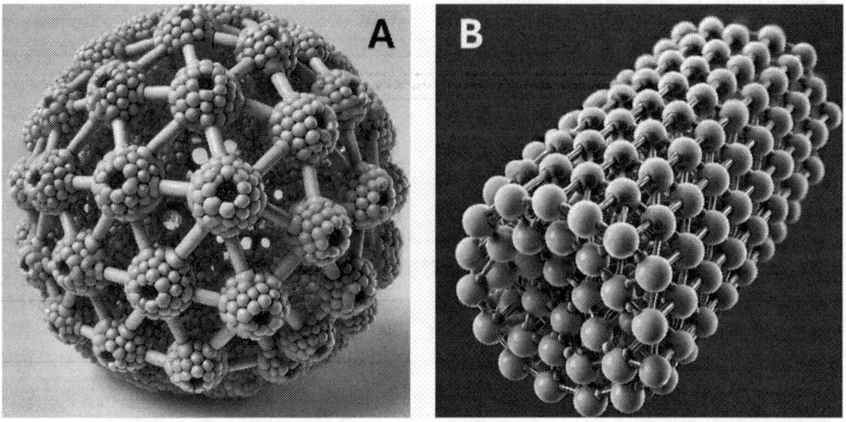

Figure 10.9 Schematic representation of the hollow, spherical structure of allophane (**A**) and the tubular structure of imogolite (**B**). This figure was generated with the assistance of ChatGPT.

10.4.5 Zeolites

Zeolites are crystalline aluminosilicates with a three-dimensional framework of interconnected channels and cavities. These minerals are known for their high cation exchange capacity, selective adsorption properties, and ability to trap and exchange cations such as ammonium (NH_4^+) and potassium (K^+). Zeolites are widely used in water purification, soil conditioning, and as a medium for nutrient retention in agricultural applications.

10.4.6 Oxides and Hydroxides

Iron, manganese oxides, and hydroxides are important soil constituents that influence soil color, redox potential, and nutrient cycling. Common examples include goethite (FeO(OH), Figure 10.10A), hematite (Fe_2O_3, Figure 10.10B), and birnessite (MnO_2, Figure 10.10C). These minerals play key roles in the adsorption and immobilization of heavy metals and other pollutants and in facilitating electron transfer processes that influence soil redox reactions. Iron oxides are known for their high reactivity and strong ability to bind phosphate ions, affecting soil phosphorus availability.

10.5 ENVIRONMENTAL IMPLICATIONS OF THE CLAY MINERAL STRUCTURE

Understanding the structural differences among clay minerals is crucial for assessing their environmental impact. 1:1 clays like kaolinite provide stability in tropical soils, reducing erosion and supporting sustainable land use, while 2:1 clays like smectites enhance nutrient retention and water availability in semi-arid environments. Chlorite, with its 2:1:1 structure, contributes to soil stability in areas subjected to thermal and chemical fluctuations.

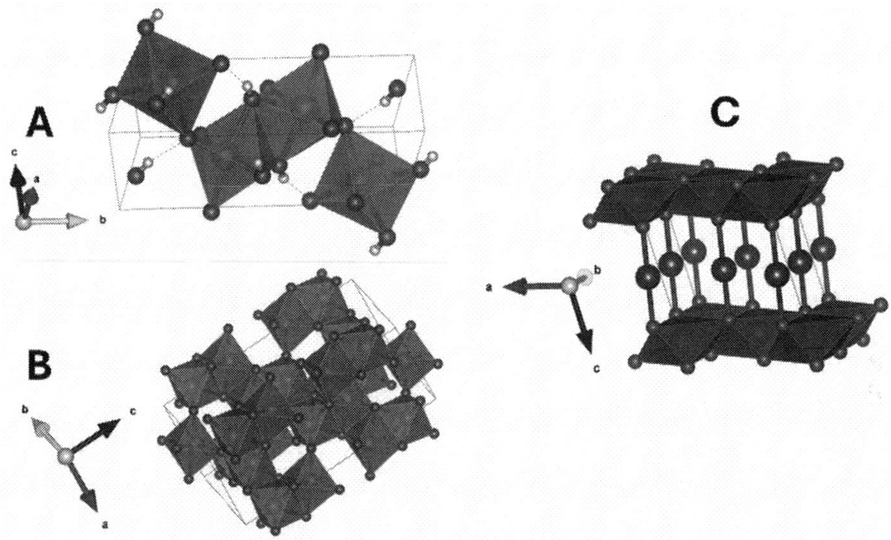

Figure 10.10 Crystal unit of goethite with crystal parameters: a = 0.4598 nm; b = 0.9951 nm; c = 0.3018 nm; $\alpha = \beta = \gamma = 90°$ (**A**); crystal unit of hematite having crystal parameters $a = b$ = 0.4986 nm; c = 1.3501 nm; $\alpha = \beta = 90°$, $\gamma = 120°$ (**B**); crystal unit of birnessite where the interlayer space is occupied by copper ions (**C**). Birnessite's interlayer space can also be filled with water molecules and other different types of cations. The crystal parameters of the birnessite reported here are: a = 0.5530 nm; b = 0.2884 nm; c = 0.5898 nm; $\alpha = \gamma = 90°$, $\beta = 104.6°$. This figure was generated using the VESTA3 software[4].

Beyond their role in soil fertility, clay minerals play a key role in pollution mitigation. Their variable charge properties and interlayer spaces allow them to adsorb heavy metals, pesticides, and organic contaminants, preventing groundwater contamination and aiding natural remediation. Smectites, in particular, have a high cation exchange capacity (CEC) and expandable layers, making them effective in pollutant sequestration.

The interaction between clay minerals and soil microbial communities further influences nutrient cycling and organic matter stabilization (discussed in Chapter 11). These properties make clay minerals essential in sustainable soil management and environmental conservation[5].

[5] To deepen the information contained in this paragraph, the reader is referred to the Further Reading section at the end of this chapter.

10.6 FORMATION AND WEATHERING PROCESSES

Chapter 3 described the processes of lithogenesis and pedogenesis, which include the formation of clay minerals. This intricate process involves the chemical and physical weathering of silicate rocks. As previously discussed, weathering is influenced by environmental factors such as *pH*, temperature, mineral composition of the parent rock, and water availability. These conditions collectively determine the type of clay mineral that will develop, as different minerals require specific environmental conditions for their formation.

Kaolinite, a 1:1 clay mineral, predominantly forms in tropical and subtropical climates where intense weathering occurs. These environments are typically characterized by acidic *pH* and substantial rainfall, which leach base cations such as calcium, magnesium, potassium, and sodium. This process removes mobile elements, leaving behind aluminum and silicon, which combine to form kaolinite. The mineral strong hydrogen bonding between layers results in a highly stable structure with low cation exchange capacity and minimal swelling. Kaolinite is often found in well-drained soils, where continuous water flow removes soluble cations, leading to the accumulation of aluminum oxides and silica. Its presence indicates advanced weathering and nutrient depletion, which influences soil fertility and vegetation.

Smectite minerals, including montmorillonite, form under conditions of moderate weathering, typically in semi-arid to arid regions with periodic water saturation and drainage. Unlike kaolinite, smectites develop in environments with a neutral to slightly alkaline *pH*, where base cations and silica remain available. Volcanic ash or parent materials rich in basic cations further promote smectite formation. These minerals have a high cation exchange capacity and significant swelling potential due to water retention in their interlayer spaces, making them essential for soil moisture retention. However, this swelling behavior can also lead to soil instability and cracking in dry conditions.

Chlorite, a 2:1:1 clay mineral, forms in low-temperature diagenetic environments or during the early stages of metamorphism. It typically develops from ferromagnesian minerals such as biotite and amphiboles, which release magnesium and iron during weathering. Chlorite's additional hydroxide sheet imparts thermal stability and moderate cation exchange capacity. Unlike smectites, chlorite exhibits minimal swelling, allowing it to persist in soils derived from metamorphic or igneous parent materials, contributing to the physical integrity of the soil matrix.

The presence of different clay minerals in soil has significant environmental implications. Kaolinite-dominated soils, common in tropical regions, often exhibit low fertility due to their limited CEC and require management practices to enhance productivity. Smectite-rich soils contribute to nutrient retention and water management, although their expansive nature can pose structural challenges. Chlorite, due to its stability, plays a role in long-term nutrient retention in specific geological environments.

Additionally, the mineralogical composition of soils affects their capacity to mitigate environmental pollutants. Smectites, with their large interlayer spaces, can trap heavy metals and contaminants, preventing their leaching into groundwater. This property makes them particularly valuable in pollution-prone areas, where they act as natural barriers to toxin mobility. While kaolinite has limited cation exchange capacity, it can still contribute to the adsorption of organic molecules, assisting in natural soil remediation. Chlorite, due to its stable structure and moderate cation exchange properties, supports the long-term sequestration of metals, playing a role in sustained plant growth and environmental protection.

10.7 PHYSICAL AND CHEMICAL PROPERTIES OF CLAY MINERALS

The physicochemical properties of clay minerals define their behavior in soil and environmental systems. Their cation exchange capacity, specific surface area, and structural flexibility influence their ability to retain nutrients, regulate water, and adsorb contaminants.

Cation exchange capacity (CEC) arises from isomorphic substitution, where lower-valence ions replace higher-valence ions, creating negative charges that attract essential cations like K^+, Ca^{2+}, and Mg^{2+}. Smectites, with their high CEC, efficiently retain nutrients, while kaolinite, with its lower CEC, contributes more to soil stability than fertility.

The swelling and shrinkage behavior of clay minerals significantly impacts soil structure. Smectites exhibit high swelling, which enhances moisture retention but can lead to cracking in dry conditions. In contrast, kaolinite remains dimensionally stable, making it more suited for erosion-resistant soils.

Clay minerals also regulate soil *pH*, buffer nutrient availability, and influence contaminant adsorption, playing a crucial role in environmental sustainability.

10.8 ROLE OF CLAY MINERALS IN SOIL CHEMISTRY AND NUTRIENT CYCLING

Clay minerals are fundamental in soil chemistry and nutrient cycling due to their reactive surfaces and cation exchange capacity. Their ability to adsorb and release essential nutrients, such as potassium, calcium, and magnesium, ensures a steady supply for plant growth and long-term soil fertility. 2:1 clay minerals like smectites exhibit high CEC, enhancing nutrient retention, while 1:1 minerals like kaolinite, with lower CEC, play a more limited role.

Clay minerals also interact strongly with soil organic matter (SOM; this is deeply described in Chapter 11), forming stable complexes that influence nutrient availability and soil structure. These interactions occur through cation bridging, hydrogen bonding, and van der Waals forces, protecting organic matter from rapid microbial decomposition. This protection enhances humus formation, improves soil aggregation, and increases porosity, contributing to better aeration, water retention, and root penetration. Additionally, these mineral-organic associations support microbial communities, which are essential for nutrient decomposition and recycling.

Another crucial function of clay minerals is their buffering capacity, which helps regulate soil *pH*. By adsorbing and releasing hydrogen (H^+) and aluminum (Al^{3+}) ions, clay minerals mitigate rapid shifts in acidity or alkalinity. This property is particularly significant in acidic soils, where high aluminum solubility can lead to toxicity. Through ion exchange processes, clay minerals stabilize *pH*, fostering optimal conditions for plant growth and microbial activity.

Clay minerals facilitate nutrient cycling by holding cations on exchange sites and releasing them as needed. This mechanism prevents nutrient leaching, maintains soil fertility over time, and protects groundwater quality, even during periods of low external input, such as reduced fertilization. Clay minerals sustain nutrient availability, ensuring continued plant growth.

In addition to nutrient dynamics, clay minerals influence soil moisture retention. Their layered structure, particularly in 2:1 minerals like smectites, allows for water absorption in interlayer spaces, enhancing water-holding capacity. This property is vital in agricultural and natural ecosystems, where moisture availability impacts plant health. By moderating water retention and movement, clay minerals help regulate temperature fluctuations and reduce drought stress in plants.

Beyond their role in nutrient management, clay minerals contribute to environmental protection by adsorbing and immobilizing pollutants. Their high surface area and negative charge enable them to bind with heavy metals, pesticides, and organic contaminants, reducing their mobility and bioavailability. This property is particularly valuable in soil remediation, where clay minerals stabilize contaminated soils and prevent hazardous substances from leaching into groundwater. Smectites and vermiculites, with their high CEC and expandable layers, are especially effective in trapping pollutants, helping maintain soil and water quality.

10.9 WATER RETENTION AND MOVEMENT IN CLAY-RICH SOILS

The water retention and movement properties of clay minerals shape soil hydrology, plant growth, and environmental stability. Their microporous structure enables them to retain significant moisture, making clay-rich soils vital in arid and semi-arid regions.

Capillary action in clay soils ensures that moisture remains available to plant roots, even after surface drying. However, excessive water retention can lead to poor drainage and waterlogging, which affects root aeration and microbial activity.

The swelling behavior of smectites reduces permeability, making soils prone to water retention issues. In contrast, kaolinite, with low swelling capacity, allows better drainage and stability. To optimize soil moisture balance, organic matter incorporation and gypsum amendments can improve aeration and structure in clay-rich soils.

Beyond agriculture, clay soils regulate surface runoff and groundwater recharge, playing a role in flood mitigation and soil erosion prevention.

10.10 CLAY MINERALS AND ENVIRONMENTAL REMEDIATION

Clay minerals are essential in environmental remediation due to their high specific surface area, ion exchange capacity, and ability to form stable complexes with contaminants. These properties make them highly effective in immobilizing pollutants and mitigating environmental hazards, particularly in addressing heavy metals, organic pollutants, and radionuclides. Their applications extend to soil and water treatment, playing a crucial role in maintaining ecosystem health.

One of their most critical functions is the adsorption and immobilization of heavy metals[1] such as lead (Pb^{2+}), cadmium (Cd^{2+}), zinc (Zn^{2+}), and arsenic (As^{3+}). Adsorption occurs through ion exchange and surface complexation, where heavy metal cations replace naturally occurring cations, such as calcium (Ca^{2+}) or sodium (Na^{+}), on the clay surface. The efficiency of this process is *pH*-dependent, as higher *pH* levels increase negative surface charge, thereby enhancing metal adsorption. Smectites like montmorillonite, with their high CEC and expandable interlayers, are particularly effective in heavy metal sequestration.

Clay minerals also contribute to radionuclide containment, preventing the migration of hazardous elements such as cesium-137 (Cs^{+}) and strontium-90 (Sr^{2+}) into groundwater. Illite and vermiculite, with their selective adsorption affinity, effectively trap cesium due to their frayed edge sites, preventing contamination spread in areas near nuclear facilities or industrial waste sites.

In water treatment, clay minerals efficiently adsorb dissolved contaminants, including heavy metals, organic pollutants, and microbial pathogens. Their high surface area and porosity make them valuable for filtration and flocculation, aiding in the removal of suspended particles, dyes, and pharmaceutical residues. Clays such as kaolinite and bentonite are widely used in filtration systems to enhance water quality. Additionally, they serve as sorbents in constructed wetlands and permeable reactive barriers (PRBs), immobilizing contaminants as water percolates through remediation systems.

The ability of clay minerals to interact with organic pollutants, including pesticides, phenols, and polychlorinated biphenyls (PCBs), further expands their role in environmental cleanup. The adsorption efficiency depends on the surface charge, specific surface area, and the nature of organic molecules. Modified clay materials, such as organoclays, improve hydrophobic interactions, enhancing the sequestration of petroleum hydrocarbons and chlorinated solvents in soil and water systems.

Combining clay minerals with other natural amendments, such as activated carbon or biochar[6], has proven effective in enhancing pollutant immobilization. These composite materials provide dual adsorption mechanisms, increasing efficiency in capturing and degrading contaminants. Such approaches are gaining traction as sustainable, cost-effective strategies for managing industrial discharge and agricultural runoff.

---

[6] The chemistry of biochar is discussed in Chapter 13.

Despite their effectiveness, clay-based remediation strategies face challenges such as saturation of adsorption sites, environmental *pH* fluctuations, and potential desorption of bound contaminants under changing conditions. Ongoing research focuses on functionalizing clay minerals and developing hybrid materials to improve stability and long-term efficiency in remediation efforts.

10.11 CLAY MINERAL INTERACTIONS WITH SOIL ORGANIC MATTER (SOM)

The interaction between clay minerals and soil organic matter plays a critical role in soil stability, fertility, and resilience. The formation of clay-organic complexes influences aggregation, water retention, nutrient cycling, and carbon sequestration, enhancing soil health and long-term productivity. These interactions contribute to erosion resistance, compaction mitigation, and sustainable land management.

Clay minerals, due to their reactive surfaces and cation exchange capacity, interact with organic molecules through adsorption, ion exchange, and hydrogen bonding. Electrostatic attraction between negatively charged clay surfaces and functional groups in SOM (e.g., carboxyl and phenolic groups) facilitates the formation of stable clay-organic complexes. Smectites and vermiculites, with their high surface area and CEC, form stronger interactions than kaolinite, which has lower reactivity.

The formation of these complexes contributes to SOM stabilization, reducing microbial decomposition rates by physically protecting organic matter from enzymatic degradation. This stabilization influences carbon sequestration, limiting CO_2 release and supporting long-term soil carbon storage (see Chapter 11 for details on SOM properties).

Clay-organic complexes also act as binding agents, enhancing soil aggregation and structure. Well-aggregated soils improve porosity, infiltration, and water-holding capacity, ensuring moisture availability for plant roots, particularly in dry conditions. The stability provided by these interactions strengthens soil resistance to erosion, preserving topsoil rich in nutrients and organic matter.

The nutrient cycling process is also regulated by clay-SOM interactions, influencing the release and retention of essential elements such as nitrogen, phosphorus, and potassium. These complexes prevent nutrient leaching, ensuring their gradual availability for plant uptake. For example, phosphorus adsorbed onto clay-organic complexes remains bioavailable, preventing fixation in insoluble forms. Similarly, nitrogen bound within these complexes undergoes gradual mineralization, supplying ammonium or nitrate over time.

Clay minerals also contribute to soil *pH* buffering, helping to stabilize acidity fluctuations caused by acid rain, fertilizers, and microbial activity. The adsorption of organic acids onto clay surfaces neutralizes excess hydrogen ions (H^+), maintaining a balanced *pH* that supports nutrient availability and microbial function.

10.12 IMPLICATIONS FOR SOIL MANAGEMENT AND AGRICULTURE

A thorough understanding of clay mineral properties is essential for effective soil management and sustainable agriculture. Their cation exchange capacity, water retention, and swelling behavior directly influence soil fertility, structure, and productivity. By applying targeted management strategies, farmers and land managers can optimize soil conditions, improve crop yields, and enhance long-term sustainability.

The composition and proportion of clay minerals determine soil water-holding capacity and nutrient retention. Smectite-rich soils retain high amounts of water and nutrients but may suffer from poor permeability and waterlogging, affecting root respiration. Effective irrigation management, such as drip or subsurface irrigation, helps regulate moisture balance, preventing excess saturation or prolonged drought stress. Selecting moisture-tolerant crops, such as rice and legumes, can further enhance productivity in clay-heavy soils.

Soil amendments play a crucial role in improving fertility and soil structure. Adding lime to acidic soils neutralizes *pH*, enhancing CEC and nutrient availability, particularly in kaolinite-dominated soils with naturally low CEC. Organic amendments, such as compost and manure, improve soil aggregation, aeration, and permeability, reducing the risk of compaction and erosion. These amendments also introduce additional cation exchange sites, thereby stabilizing nutrient availability over time.

Clay-rich soils, particularly those rich in swelling clays like smectites, are prone to cracking when dry and compaction when wet, leading to structural instability and erosion risks. The incorporation of organic matter helps stabilize clay particles, reducing erosion and runoff. Cover crops further enhance soil cohesion, protect against raindrop impact, and promote aggregate formation through root interactions, contributing to water retention and improved soil structure.

Sustainable soil management practices are key to maintaining soil health and long-term productivity. Rotational cropping systems help balance nutrient demands, with leguminous crops replenishing nitrogen while clay minerals retain nutrients for future growth. Reduced tillage and no-till farming preserve soil aggregates, minimizing erosion and compaction, while enhancing water infiltration and nutrient cycling.

10.13 FUTURE DIRECTIONS AND RESEARCH

The study of clay minerals continues to advance as researchers explore modifications and applications to address soil degradation, pollution, and climate change. Enhancing the intrinsic properties of clay minerals for targeted environmental and agricultural applications is a growing focus, particularly in developing sustainable solutions.

One promising area of research is nano-clays, which offer increased surface area, enhanced reactivity, and improved ion exchange capacity compared to bulk clay minerals. These properties make nano-clays highly effective for adsorbing and immobilizing pollutants, particularly in water treatment and soil remediation. Functionalizing nano-clays through surface treatments or composite materials has shown potential for tailored pollutant removal, improving environmental cleanup strategies.

Another key frontier is the integration of clay minerals with biological agents, such as microbial communities and plant roots, to enhance bioremediation. Clays can immobilize toxic substances, creating a stable environment where microbes metabolize and detoxify pollutants. Phytoremediation, when combined with clay amendments, can further stabilize soils, reduce erosion, and support microbial activity, complementing the pollutant-adsorbing capacity of clay minerals.

With growing concerns about climate change, research into the role of clay minerals in carbon sequestration is gaining importance. Their ability to stabilize soil organic carbon (SOC) by forming protective complexes with organic matter helps mitigate carbon dioxide emissions and regulate the global carbon cycle. Future studies must investigate how climate variables, such as temperature fluctuations and altered precipitation, influence the stability of clay-organic complexes and their role as carbon sinks. Additionally, exploring the impact of land management practices, including tillage and organic amendments, can inform strategies to enhance climate resilience through clay-mediated carbon storage.

Advancements in analytical techniques are essential for a deeper understanding of clay mineral behavior. Improved spectroscopic, microscopic, and computational methods enable precise analysis of surface interactions and structural changes at the nanoscale, providing insights into their role in soil processes, pollutant interactions, and environmental stability.

For these research findings to have real-world impact, collaboration between scientists, policymakers, and land managers is crucial. Policy incentives promoting the use of modified clays for pollution control and soil enhancement can support sustainable land management. Education and outreach efforts will also be key to ensuring that farmers and environmental professionals adopt best practices for utilizing clay minerals in agriculture and remediation.

EXERCISES

1. Explain the differences between 1:1, 2:1, and 2:1:1 clay minerals in terms of structure, swelling behavior, and cation exchange capacity (CEC). Provide examples of each type and discuss their implications for soil fertility.
2. A soil sample contains high proportions of smectite and kaolinite.
   (a) Compare their CEC values and explain their impact on nutrient retention.
   (b) How would adding lime or organic amendments modify the soil's nutrient-holding capacity?

3. Describe how clay minerals contribute to pollutant immobilization in contaminated soils and water.
   (a) What role do montmorillonite and illite play in heavy metal sequestration?
   (b) How could modified clays (e.g., organoclays, nano-clays) improve remediation efficiency?
4. Consider clay-heavy agricultural soil with high water retention but poor drainage.
   (a) What management strategies (e.g., irrigation techniques, and soil amendments) could be applied to improve aeration and root health?
   (b) How does swelling and shrinking behavior affect soil structure stability?
5. Discuss two emerging areas of research in clay mineral applications, such as nano-clays, bio-remediation, or carbon sequestration. How could these developments enhance agricultural sustainability and environmental protection?

FURTHER READING

Bergaya, F., B.K.G. Theng and G. Lagaly (eds). 2006. Handbook of Clay Science. Elsevier.

Dixon, J.B. and D.G. Schulze (eds). 2002. Soil Mineralogy with Environmental Applications. Soil Science Society of America.

Sparks, D.L. 2003. Environmental Soil Chemistry, 2nd Ed. Academic Press.

Weil, R.R. and N.C. Brady. 2017. The Nature and Properties of Soils, 15th Ed. Pearson Prentice-Hall.

Chapter **11**

The Chemistry of Soil Colloids: The Soil Organic Matter

11.1 INTRODUCTION TO SOIL ORGANIC MATTER (SOM)

Soil Organic Matter (SOM) is a fundamental component of soil, crucial in soil fertility, structure, and biological activity. It comprises organic compounds derived from plant and animal residues at different stages of decomposition, along with microbial biomass and its metabolic by-products. These materials undergo continuous biological, chemical, and physical transformations, shaping soil health and productivity.

SOM influences nearly all soil properties, including water retention, nutrient cycling, cation exchange capacity, and microbial dynamics. Stabilizing soil aggregates improves porosity, aeration, and water infiltration while reducing erosion and compaction. Additionally, SOM is a key factor in carbon sequestration, functioning as both a source and a sink for atmospheric carbon dioxide, thus playing a significant role in global climate regulation.

Chemically, SOM is highly diverse, consisting of carbohydrates, proteins, lipids, lignins, nucleic acids, and humic substances. These compounds vary in stability and bioavailability: simple sugars and amino acids rapidly decompose and provide immediate nutrients, whereas humic substances persist for centuries, gradually releasing essential elements. This complexity makes SOM a dynamic and integral part of soil ecosystems.

Despite its recognized importance, the molecular nature of SOM remains an area of active scientific research. Climate, soil texture, land management practices, and microbial activity shape its composition and behavior, making its long-term stability difficult to predict. Advances in analytical techniques have deepened our understanding of SOM dynamics, but key aspects of its formation, stabilization, and degradation remain unresolved.

DOI: 10.1201/9781003780212-11

SOM is commonly estimated through soil organic carbon (SOC), constituting about 58% of its mass. The relationship SOM ≈ 1.724 × SOC[1] provides a useful approximation, although variations occur depending on composition. In agricultural soils, SOM levels range from less than 1% in degraded arid regions to over 10% in well-managed temperate soils, reaching 30–40% in organic-rich Histosols.

Given its central role in soil function and environmental sustainability, understanding and managing SOM is essential for maintaining agricultural productivity, restoring degraded lands, and mitigating climate change. This chapter examines SOM composition, formation, and functions, highlighting its role in soil ecosystems and its importance for sustainable land management.

11.2 SOM CLASSIFICATION AND COMPOSITION

Soil organic matter plays a vital role in terrestrial ecosystems, encompassing a diverse range of organic materials at different stages of decomposition. It originates from plant litter, animal residues, microbial tissues, and their biochemical by-products, forming a heterogeneous mixture that directly influences soil fertility, structure, and nutrient cycling. Through continuous interactions with physical, chemical, and biological processes, SOM regulates key soil functions, including carbon sequestration and microbial activity, contributing to long-term soil stability.

To better understand the complexity of SOM, it is commonly divided into three major fractions, each representing a different stage in the decomposition process: fresh organic inputs, humified organic matter, and microbial biomass. These fractions interact dynamically, undergoing transformations that influence soil health and ecosystem function (Figure 11.1).

---

[1] The constant 1.724 in the relationship SOM ≈ 1.724 × SOC originates from an assumed carbon content of soil organic matter (SOM) based on historical studies. This conversion factor is derived from the assumption that organic matter in soils contains approximately 58% carbon by weight. The equation is formulated as: $SOM = SOC/0.58 = 1.724 \cdot SOC$. This 58% value was originally established from elemental analysis of soil organic matter in various soil types. It was proposed as an average approximation, recognizing that the actual carbon content of SOM varies depending on its composition, age, and decomposition stage. Different organic materials (e.g., fresh plant residues, humified organic matter, microbial biomass) contain different percentages of carbon, generally ranging between 45 and 60%. The 1.724 conversion factor is not universal and may not apply to all soils. Some organic-rich soils (e.g., peat and Histosols) may have a lower carbon content (~50%), leading to a different conversion factor. In some modern studies, region-specific or material-specific conversion factors are used instead of assuming a fixed 1.724 value. Some researchers use site-specific carbon-to-SOM ratios, determined through direct analysis. In peatlands and organic soils, a lower factor (~1.9–2.0) is sometimes used. In arid or mineral-associated soils, a lower carbon content (~52%) may suggest a higher factor (~1.9–2.1). The 1.724 factor is an empirical approximation, widely used in soil science for quick estimations of SOM from measured SOC values. However, it is not a fixed natural law and may require adjustment based on soil type, land use, and regional conditions.

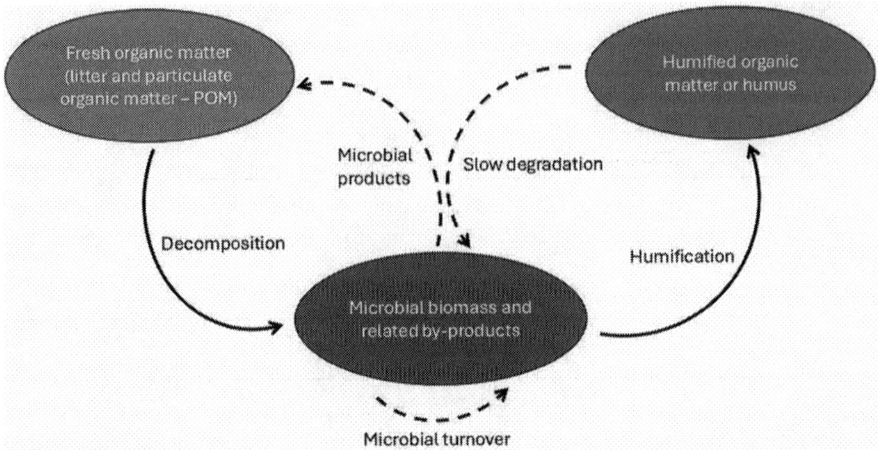

Figure 11.1 This illustration reports on the dynamic interrelationships between the three main soil organic matter pools and how they interact with each other. Fresh Organic Matter (top left-hand side) is litter and particulate organic matter that provides readily available energy and nutrients. The humified organic matter (top right-hand side) shows the stable humic substances that enhance soil structure and nutrient retention capacity. The microbial biomass (bottom-side of the figure) represents the living component that mediates transformations between the other pools while contributing its biomass to the soil. The arrows indicate the key processes. The solid arrows represent primary transformation pathways, i.e., decomposition (microbes break down the fresh organic matter) and humification (microbial processing transforms organic materials into stable humic compounds). The dashed arrows show secondary processes: 1. The slow degradation of humic substances; 2. The return of microbial products to the fresh organic matter pool; 3. The microbial turnover (self-cycling).

Fresh organic matter, depicted in Figure 11.1, consists primarily of recently incorporated plant and animal residues, including litter and particulate organic matter (POM). This fraction is an immediate source of energy and nutrients for soil microorganisms, fueling biological activity and the early stages of decomposition. The breakdown of fresh organic matter releases simple organic compounds, such as sugars and amino acids, which are readily utilized by microbial communities. As decomposition progresses, some organic materials undergo humification, a process that leads to the formation of stable compounds collectively known as humified organic matter or humus. Humus consists of complex organic substances that persist in the soil for extended periods, enhancing soil structure and long-term carbon storage. Humic substances, mainly made by humic acids, fulvic acids, and humin, interact with soil minerals, improving aggregate stability, cation exchange capacity, and water retention. These properties make humus a critical component in maintaining soil fertility, as it helps to prevent nutrient leaching and supports the slow release of essential elements.

The third fraction of SOM, microbial biomass, plays a dual role as both an active decomposer of fresh organic matter and a contributor to the formation of stable humified compounds through microbial by-products and enzyme-driven reactions. Microorganisms break down organic matter into simpler molecules, facilitating nutrient cycling and contributing to the continuous renewal of SOM. Additionally, microbial residues become incorporated into the soil matrix, further influencing soil chemistry and structure.

Both primary and secondary transformation processes govern the interplay between these three fractions. As illustrated in Figure 11.1, decomposition (represented by solid arrows) is the primary pathway through which fresh organic matter is broken down by microbial activity. Humification, another primary process, transforms decomposed materials into stable humic compounds. Secondary processes (dashed arrows) include the gradual degradation of humic substances, the recycling of microbial by-products into fresh organic matter, and microbial turnover, where microorganisms decompose and reform within the soil ecosystem.

From a functional and analytical perspective, SOM can also be classified based on its turnover rate and bioavailability, which reflect how quickly different organic matter pools cycle through the soil system. These pools—active, slow, and passive—differ in their decomposition rates and contributions to soil function, as conceptually illustrated in Figure 11.2.

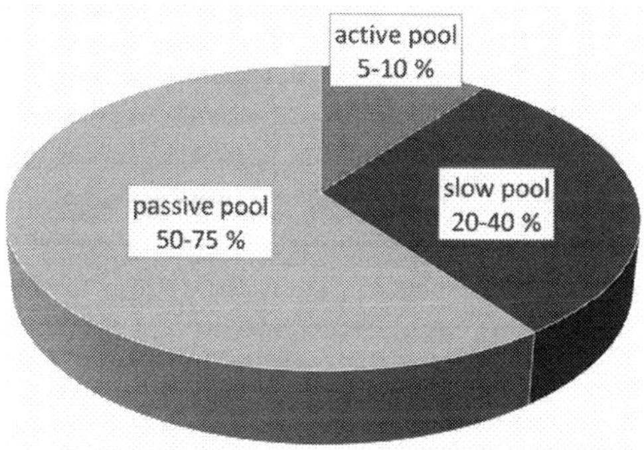

Figure 11.2 Conceptual representation of soil organic matter (SOM) pools based on turnover rates and stability. The active pool (5–10%) consists of rapidly decomposing organic materials that provide immediate nutrients. The slow pool (20–40%) includes partially stabilized organic matter with intermediate resistance to decomposition. The passive pool (50–75%) comprises highly recalcitrant materials that persist for decades, contributing to long-term carbon storage and soil structure stabilization. The percentages presented here are only indicative, as the distribution of SOM pools varies depending on soil type, climate, and land management practices.

The active pool, constitutes about 5–10% of SOM and comprises rapidly decomposing organic materials, such as simple carbohydrates and proteins. These compounds break down within days to a few years, supplying readily available nutrients to plants and microorganisms. Although small in proportion, the active pool is crucial in maintaining nutrient cycling and microbial activity.

The slow or intermediate pool, accounts for approximately 20–40% of SOM and includes partially stabilized organic matter. This fraction consists of more resistant compounds that persist in the soil for years to decades, often protected within soil aggregates or bound to minerals. It contributes significantly to soil structure, water retention, and the gradual release of nutrients.

The passive or recalcitrant pool, comprises 50–75% of SOM and consists of chemically stable organic compounds that resist decomposition. These materials, including humified organic matter bound to soil minerals, have turnover times that range from decades to millennia. This fraction is particularly important for long-term carbon sequestration, as it prevents the rapid release of organic carbon back into the atmosphere, contributing to soil resilience and stability.

While these three pools provide a useful framework for understanding SOM dynamics, their distribution varies depending on soil type, climate, and land management practices. For example, organic-rich soils tend to have a higher proportion of passive SOM, while degraded soils may lack sufficient active organic matter to sustain biological activity. The balance between these pools significantly influences soil carbon storage, nutrient availability, and structural integrity.

Environmental conditions play a critical role in regulating SOM transformation and stability. Temperature directly affects microbial metabolism, with decomposition rates typically doubling for every 10°C increase within biologically active ranges. Moisture availability determines oxygen levels in the soil, with intermediate moisture conditions optimizing decomposition rates by balancing hydration and aeration. Soil *pH* also influences SOM breakdown, as microbial activity and enzyme function are more efficient in neutral to slightly alkaline soils. These environmental factors interact in complex ways, leading to seasonal and spatial variations in SOM composition across different landscapes.

Management practices significantly impact SOM stability and decomposition rates. Conventional tillage, for instance, disrupts soil structure, accelerating organic matter decomposition by increasing aeration. Conversely, conservation practices, such as reduced tillage, cover cropping, and organic amendments, help retain SOM by increasing organic inputs and slowing decomposition rates. By adopting sustainable soil management strategies, it is possible to enhance SOM stability, improve soil fertility, and increase carbon sequestration, making these practices essential for long-term agricultural productivity and climate change mitigation.

As research advances, improving our ability to analyze, manage, and optimize SOM composition will ensure soil health and resilience in different ecosystems. With the development of high-resolution analytical techniques, scientists are gaining deeper insights into SOM chemistry, allowing for better prediction and management of soil carbon dynamics. Understanding these intricate relationships will enable the development of targeted strategies to maintain SOM levels, improve soil fertility, and promote sustainable land use practices.

11.2.1 Litter and Particulate Organic Matter (POM)

Litter and particulate organic matter represent the least decomposed fraction of soil organic matter, consisting of fresh plant residues, roots, and detritus that retain much of their original structure. As the primary input of organic material into the soil, POM serves as a readily available energy source for soil microorganisms, fueling biological activity and initiating the early stages of decomposition. This fraction is rich in cellulose, hemicellulose, and lignin, which vary in their susceptibility to microbial breakdown. While simple carbohydrates degrade quickly, more complex structures, such as lignin, persist longer, influencing the overall turnover rate of organic matter.

The decomposition of POM plays a critical role in soil structure development and aggregation. As microbial activity breaks down organic residues, it releases sticky polysaccharides and microbial by-products that bind soil particles together, forming stable aggregates. This process improves soil porosity, aeration, and water infiltration, particularly in sandy soils where POM enhances moisture retention. By increasing soil cohesion, POM also reduces erosion risks and contributes to overall soil stability.

Beyond its structural influence, POM is a temporary reservoir of essential nutrients, including carbon, nitrogen, phosphorus, and sulfur. As microorganisms metabolize POM, these nutrients are gradually released into the soil, fueling microbial growth and supporting plant uptake. This rapid nutrient cycling is particularly important in agricultural systems, where maintaining soil fertility depends on continuously renewing organic inputs. In particular, POM plays a key role in nitrogen mineralization, as microbial decomposition transforms organic nitrogen into plant-available forms, directly impacting crop productivity.

Over time, POM undergoes progressive decomposition and transformation into more stable organic matter. Some carbon is lost as CO_2 through microbial respiration, while a portion of the degraded material interacts with clay minerals and metal oxides, forming organo-mineral complexes that protect organic carbon from further breakdown. This stabilization process is fundamental to long-term carbon sequestration, as it facilitates the transition from labile organic matter to the more persistent humified fraction of SOM.

The decomposition rate of POM is controlled by environmental conditions such as temperature, moisture availability, and microbial activity. Warmer and wetter environments promote faster breakdown, whereas drier or colder conditions slow microbial metabolism, prolonging the persistence of POM in the soil. The balance between rapid decomposition and stabilization mechanisms determines POM contribution to soil carbon storage, fertility, and organic matter accumulation.

POM is fundamental to soil function and ecosystem sustainability as a transitional stage between fresh organic inputs and humified matter. Its influence on microbial activity, soil aggregation, and nutrient cycling highlights its importance in sustainable land management. Implementing organic amendments, conservation tillage, and residue retention strategies can enhance the role of POM in maintaining soil health, improving carbon sequestration, and ensuring long-term agricultural productivity.

11.2.2 Humified Organic Matter (Humus)

Humified organic matter, commonly referred to as humus, represents the most stable fraction of soil organic matter. It consists of highly decomposed organic residues and newly synthesized compounds undergoing extensive microbial processing, forming complex colloidal structures that influence soil chemistry and physical stability. Unlike the more transient particulate organic matter, humus persists in the soil for decades or even centuries, playing a crucial role in maintaining soil fertility and carbon sequestration.

At the molecular level, humus is composed primarily of humic substances, a diverse group of organic compounds that interact with soil minerals and influence nutrient availability. Among these, humic acids are large, pseudo-high-molecular-weight molecules that improve soil aggregation and cation exchange capacity, enhancing the retention of essential nutrients such as calcium, magnesium, and potassium. Fulvic acids, in contrast, are smaller and more chemically reactive, facilitating nutrient chelation and transport within the soil, making them more readily available for plant uptake. The most recalcitrant fraction, humin, consists of highly stable organic compounds that resist microbial decomposition and persist in the soil for the longest time, contributing significantly to long-term carbon storage.

The presence of humus in the soil enhances multiple soil properties, particularly nutrient dynamics, structure, and water regulation. By increasing CEC, humus improves nutrient retention and availability, preventing the leaching of key elements and supporting plant nutrition over extended periods. Its ability to promote soil aggregation helps reduce erosion, while its impact on soil porosity enhances aeration and water movement. The humus also contributes to water retention, making soils more resilient to drought and excessive moisture by balancing water availability. Another critical function of humus is its role as a buffer against *pH* fluctuations, stabilizing soil chemical reactions and preventing extreme acidity or alkalinity from disrupting microbial and plant processes.

The stability of humus is largely attributed to its chemical interactions with soil minerals. It forms stable complexes with clay particles, iron and aluminum oxides, and polyvalent cations, which shield it from microbial attack and further decomposition. This mineral-associated organic matter is crucial in maintaining long-term soil fertility and carbon sequestration, as it prevents the rapid turnover of organic carbon and contributes to soil resilience against environmental changes.

As the final stage in the decomposition of organic matter, humus plays an indispensable role in soil sustainability and agricultural productivity. Its ability to store nutrients, enhance soil structure, and regulate water dynamics underscores the importance of maintaining adequate humus levels through sustainable land management practices. Incorporating organic amendments, minimizing excessive soil disturbance, and promoting conservation tillage can enhance humus formation, ensuring soil remains fertile and productive for future generations.

11.2.2.1 Historical Perspective on Humus and Its Chemical Composition

Historically, humus has always been divided into three distinct fractions, differentiated based on their solubility in aqueous solutions at different *pH* levels. Specifically, humic acids have been distinguished by their solubility in alkaline conditions, fulvic acids by their solubility in all *pH* conditions, and humin by their insolubility under any *pH* condition (Figure 11.3).

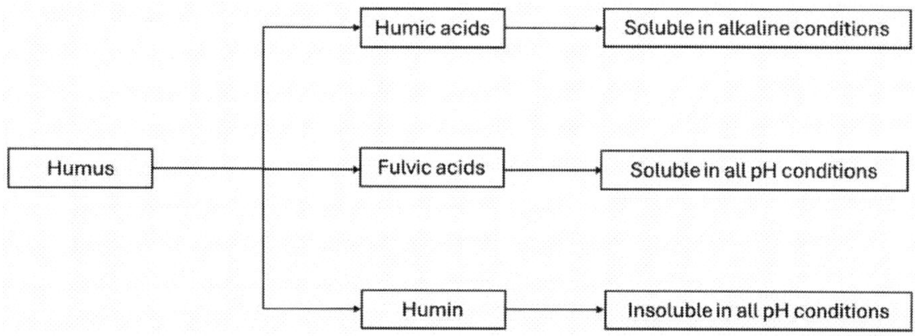

Figure 11.3 Operational definitions of the three fractions of humus.

From a chemical standpoint, however, these three fractions are quite similar, as they all contain saccharidic and polysaccharidic components, proteins and oligopeptides, and more or less complex aromatic systems. The key difference among them lies in the relative proportion of these components. In this regard, humic acids are characterized by a higher content of aromatic systems compared to fulvic acids and humin; fulvic acids contain a higher proportion of hydrophilic functional groups (such as carbohydrates and peptidic systems) compared to humic acids and humin; finally, humin exhibits a greater presence of alkyl systems than the other two fractions.

At this point, it is essential to understand why their solubility appears to vary under different *pH* conditions.

When chemistry was less advanced than today, researchers had minimal tools for studying the organic components of soils. One such tool was *pH* measurement. When a known quantity of soil was suspended in a caustic soda solution (i.e., an alkaline *pH*), the acidic functional groups in the soil organic matter—called humus—underwent deprotonation. This deprotonation resulted in the formation of negative charges, leading to electrostatic repulsion and subsequent solubilization, following mechanisms already analyzed in Chapter 5 of this book.

Filtering the brown-colored solution (which is a colloidal dispersion) from the residual solid soil component allowed the isolation of a mixture that, after treatment with strong acids—such as concentrated hydrochloric acid—led to the precipitation of a solid organic system identified as humic acid and a liquid system with a yellowish hue, identified as fulvic acid. The precipitation process—better known as flocculation since these are colloidal systems—was due to protonation reactions, which resulted in hydrogen bonds and Van der Waals interactions, leading to the aggregation of molecules. The size of these aggregates counteracted the dispersive forces described in previous chapters and became subject to gravitational forces.

And what about humin? The residual soil obtained from the previously described procedures still contained some organic matter, which remained physically protected from the alkaline solution due to its inclusion in clay minerals (which are discussed in Chapter 10). To extract this organic matter, it is necessary to break down the clay structures using diluted hydrofluoric acid (HF) solutions, which, by forming SiF_6^{2-} complexes, exposes the acidic functional groups of the organic matter to the action of the alkaline solutions used for extracting humic and fulvic acids. Consequently, the same conditions applied to extract humic and fulvic acids could also be used for humin extraction.

11.2.2.2 Implications of Humic Substance Classification

The discussion on humic substances leads to several key conclusions regarding their classification, properties, and analytical treatment.

First, the distinction between humic acids, fulvic acids, and humin is purely operational, based on solubility differences in acidic and basic solutions rather than intrinsic molecular structures. This classification reflects solubilization and dispersion processes extensively studied in Chapter 5 of this book and General and Organic Chemistry courses.

Second, humic substances function as colloidal systems, meaning their physical size and behavior align with colloid chemistry principles. Under acidic conditions, humic acids and humin form unstable colloids, flocculating rapidly upon the addition of strong acids. In contrast, fulvic acids remain in dispersion for significantly longer periods, demonstrating greater colloidal stability.

Finally, the physical properties of humic substances may lead to their misclassification as clay in certain soil analyses. Because organic colloids share size and dispersion characteristics with inorganic clay minerals, soils rich in organic matter may be erroneously identified as clay-rich. Conversely, soil texture analysis may yield vastly different results when organic matter is removed, emphasizing the need to carefully interpret textural and compositional data in soil science.

11.2.2.3 The Myth of Pure Humic and Fulvic Acids

Although in this discussion both singular (humic acid, fulvic acid) and plural (humic acids, fulvic acids) forms have been used interchangeably to refer to two of the three fractions of humus, it is crucial to highlight that humic acid and fulvic acid do not exist as chemically pure entities.

In chemistry, when a compound is identified with a singular name, it typically refers to a chemically pure substance. For example, the term "acetylsalicylic acid" uniquely identifies the molecule illustrated in Figure 11.4, representing a single, well-defined chemical structure.

In contrast, the chemical composition of humus, and therefore of its various components, depends on multiple factors, the most important of which is the biomass (whether animal or plant-derived) present in a given geographical area. In other words, the organic matter in soil reflects the nature of the living organisms in a given location, as their residues contribute to the SOM formation.

Ultimately, humic acid and fulvic acid cannot be isolated as chemically pure systems in the same way as acetylsalicylic acid can. Although exhibiting similar behavior in response to acidic and basic solutions, humic acids and fulvic acids possess variable chemical compositions, which are directly influenced by the types of living organisms present in a specific area.

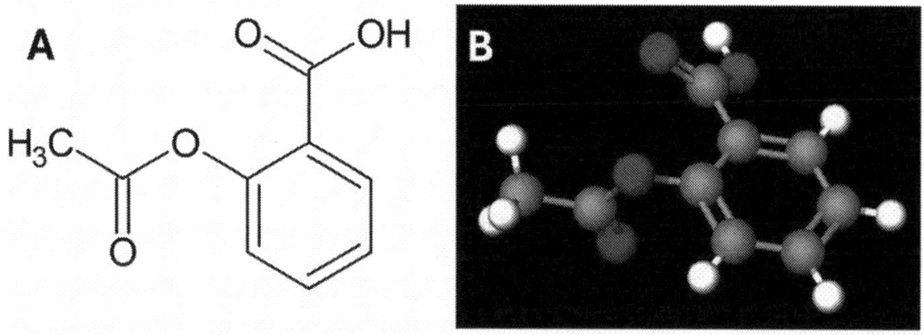

Figure 11.4 Chemical structure of pure acetylsalicylic acid (**A**) and its three-dimensional representation (**B**).

11.2.2.4 Elemental Composition and Structural Variability of Humus Fractions

The chemical composition of humic acids, fulvic acids, and humin is not fixed, as it depends on the biomass from which they originate and the environmental conditions in which they form. However, despite this variability, their elemental composition generally falls within defined ranges: they contain 40–60% carbon (C), 30–50% oxygen (O), 4–5% hydrogen (H), 1–4% nitrogen (N), 1–2% sulfur (S), and 0–0.3% phosphorus (P). While these values provide a general reference, a more effective way to distinguish the three humus fractions is through the analysis of elemental ratios, particularly H/C (hydrogen/carbon), O/C (oxygen/carbon), and N/C (nitrogen/carbon), which offer valuable insights into their molecular structure and reactivity.

One of the key indicators of structural composition is the H/C ratio, which reflects the proportion of hydrogen atoms relative to carbon. This ratio provides important information about the type of molecular structures present in each humus fraction. Humic and fulvic acids generally exhibit H/C ratios of around 1.0, while humin shows a higher value of approximately 1.5. This increased H/C ratio in humin suggests a greater presence of aliphatic (alkyl) structures, which are composed primarily of sp^3-hybridized carbon atoms forming only single bonds. These saturated systems incorporate more hydrogen atoms than aromatic and unsaturated structures, where sp^2 and sp carbon atoms form double or triple bonds, reducing the hydrogen content. As indicated by its elevated H/C ratio, the predominance of alkyl structures in humin contrasts with the higher aromaticity of humic and fulvic acids, which contain more conjugated π-systems.

Another fundamental parameter is the O/C ratio, which reflects the presence of oxygen-containing functional groups and is crucial in determining the solubility, stability, and chemical behavior of humic substances. Humic acids have an average O/C ratio of around 0.5, whereas fulvic acids exhibit a higher O/C ratio, approximately 0.7. This difference indicates a greater concentration of hydrophilic functional groups, such as hydroxyl (–OH), carboxyl (–COOH), and carbonyl (–C=O), in fulvic acids. The presence of these groups increases water solubility and colloidal stability, particularly under acidic conditions, where humic acids tend to flocculate and precipitate. The oxygen-rich nature of fulvic acids explains their ability to remain in colloidal dispersion for extended periods, making them more mobile within the soil matrix than humic acids.

In addition to H/C and O/C, the N/C ratio provides further insight into the molecular complexity of humic substances. This ratio generally follows the trend humic acids > fulvic acids > humin, reflecting differences in nitrogen incorporation. While all three fractions contain peptidic structures, humic acids have a significantly higher aromatic nitrogen content, contributing to their greater structural complexity and chemical reactivity. The presence of nitrogen in aromatic rings and heterocyclic structures enhances the ability of humic acids to interact with soil minerals, influencing nutrient retention and organic matter stability in the soil.

These elemental and structural variations among humic acids, fulvic acids, and humin have significant implications for their behavior in soils and their role in environmental processes. The higher aromaticity of humic acids makes them more resistant to microbial degradation, contributing to long-term carbon sequestration in soils. The oxygen-rich composition of fulvic acids enhances their ability to chelate metal ions and retain essential nutrients, facilitating their mobility within the soil matrix. Meanwhile, the alkyl-dominated nature of humin gives it a more hydrophobic character, influencing its interactions with soil minerals and organic pollutants, making it less biodegradable. The prevalence of saturated alkyl chains in humin reduces the availability of reactive sites for the microbial enzymatic attack, further increasing its persistence in the soil.

11.2.2.5 The Molecular Size and Conformational Behavior of Humic Substances

Humic substances (HS, another term referring to the three humus fractions) are characterized by their intrinsic structural flexibility, which allows them to adjust their molecular conformation dynamically in response to environmental conditions. Unlike rigid macromolecules with well-defined polymeric backbones, humic substances do not exist as single, covalently bonded entities with a fixed molecular weight. Instead, they exhibit supramolecular behavior, meaning they are dynamic associations of smaller molecular components held together by weak interactions such as hydrogen bonding, van der Waals forces, and electrostatic attractions. This self-assembling nature makes humic substances fundamentally different from classical polymers, a distinction often overlooked in the literature. Their apparent molecular weight, therefore, is not an intrinsic property but rather an emergent characteristic of their aggregation state, influenced by environmental conditions and external interactions.

A key feature of humic substances is their variable molecular size, which depends on their chemical composition and conformational state. In solution, humic molecules do not behave as isolated monomers but spontaneously aggregate into supramolecular assemblies that vary in compactness depending on *pH*, ionic strength, and molecular interactions. Under acidic conditions, the protonation of acidic functional groups (such as carboxyl and phenolic groups) reduces electrostatic repulsion, leading to more compact structures where hydrophobic domains cluster. In contrast, at neutral to alkaline *pH*, deprotonation of these functional groups increases electrostatic repulsion, causing molecular expansion and exposing more reactive sites to the surrounding environment.

This relationship between molecular size and conformation is particularly relevant when comparing humic acids and fulvic acids. Due to their higher content of aromatic and hydrophobic components, humic acids tend to aggregate more extensively, forming pseudo-high molecular weight structures that appear large in size but are, in reality, supramolecular associations of smaller units rather than true macromolecules. These aggregates exhibit greater resistance to breakdown and play a major role in organic-mineral interactions in soil. Fulvic acids, on the other hand, are richer in oxygenated functional groups, leading to a lower degree of aggregation and a more hydrophilic nature. As a result, fulvic acids remain more dispersed in solution, interacting readily with water, metal ions, and other solutes, making them key players in nutrient transport and bioavailability.

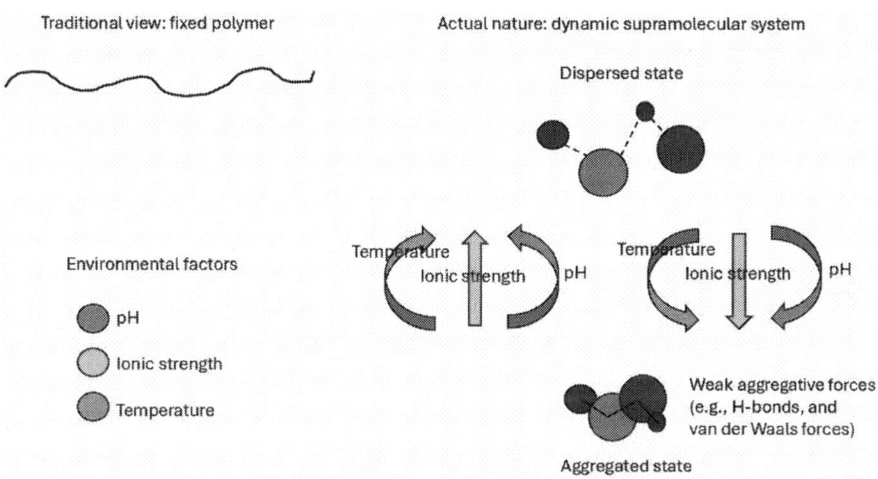

Figure 11.5 The top-left portion of the figure represents the outdated view of humic substances as fixed, high-molecular-weight polymers. The bottom-left section highlights the various environmental factors (e.g., *pH*, ionic strength, and temperature) that influence the structural organization of humic substances. The right-hand side of the figure illustrates the modern, dynamic view, where humic substances continuously transition between dispersed and aggregated states in response to changes in environmental conditions. In the aggregated state, humic substances exhibit an apparent molecular size that resembles that of true polymers, though this is merely an emergent property of their supramolecular organization rather than a reflection of actual covalent polymerization. For this reason, it is more appropriate to describe humic substances in the aggregated state using the term pseudo-molecular size, emphasizing their non-polymeric nature.

The conformational state of humic substances is further influenced by their interactions with external molecules, such as metal ions, organic pollutants, and natural organic matter. For instance, metal cations can induce bridging effects, forming larger aggregates and influencing the mobility of both humic substances and associated elements in soils and water systems. Similarly, hydrophobic organic compounds may integrate into humic structures, altering their solubility, aggregation behavior, and environmental transport mechanisms. These interactions demonstrate how humic substances act as molecular carriers, modulating the fate and bioavailability of diverse chemical species in natural environments.

From a functional perspective, the conformational plasticity of humic substances plays a fundamental role in soil structure stabilization, carbon sequestration, and regulating chemical exchanges in natural systems. Their ability to associate and dissociate reversibly in response to environmental factors enables them to function as buffers for chemical reactions, reservoirs for essential nutrients, and mediators in organic-inorganic interactions. This dynamic behavior ensures that humic substances remain highly adaptable, making them indispensable in soil fertility, water purification, and sustainable agricultural practices.

In summary, humic substances should not be considered high-molecular-weight polymers but rather dynamic supramolecular systems whose apparent molecular size is an emergent property of their aggregation state. Their structural organization is not fixed but continuously changes depending on chemical composition and external environmental factors. Understanding this supramolecular nature is essential to accurately describe their functionality in soil, water, and organic matter cycles. It reinforces their role as key modulators of biogeochemical processes in natural and managed ecosystems (Figure 11.5)[2].

11.2.2.6 The Role of Water in the Structural Dynamics of Humic Substances

Beyond their supramolecular organization and conformational flexibility, humic substances exhibit a remarkable interaction with water, which plays a crucial role in their self-assembly, aggregation behavior, and structural transitions. Recent studies suggest that humic substances possess hydrogel-like[3] properties, allowing them to retain and modulate water within their structure. This characteristic is

[2] Angelico, R., C. Colombo, E. Di Iorio, M. Brtnický, J. Fojt and P. Conte. 2023. Humic Substances: From Supramolecular Aggregation to Fractal Conformation—Is There Time for a New Paradigm? Applied Sciences. 13: 2236. https://doi.org/10.3390/app13042236

[3] A hydrogel-like structure refers to a material that has the ability to absorb and retain large amounts of water while maintaining its structural integrity. Hydrogels are typically composed of hydrophilic (water-attracting) polymer networks, which can expand and contract depending on environmental conditions such as humidity, temperature, and *pH*. In the case of humic substances, their hydrogel-like behavior means that they can trap water molecules within their structure due to the presence of oxygen-rich functional groups (such as carboxyl, –COOH, hydroxyl, –OH, and carbonyl, –C=O, groups). These groups interact with water through hydrogen bonding, creating a semi-flexible network capable of swelling and deswelling in response to environmental factors.

particularly evident in fulvic acids, which exhibit enhanced water retention and swelling capacity due to their higher content of oxygenated functional groups (Figure 11.6). These properties influence the solubility and mobility of humic substances, as well as their ability to bind and transport contaminants in soil and aquatic environments.

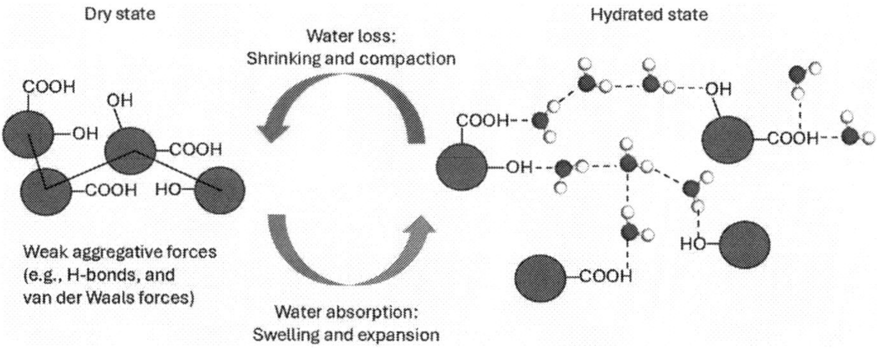

Figure 11.6 The hydrogel-like behavior of humic substances. The left side represents humic molecules in a low-hydration state, where limited water interaction results in a compact structure. The right side shows the hydrated state, where humic substances incorporate water molecules, leading to swelling, increased solubility, and greater molecular mobility. The two curved arrows arrows indicate the reversible nature of hydration and dehydration, highlighting how external conditions, such as moisture availability, govern the equilibrium between these states. This property is particularly evident in fulvic acids, which exhibit enhanced water retention due to their higher content of oxygenated functional groups. The ability of humic substances to modulate water retention plays a critical role in soil moisture dynamics, nutrient availability, and contaminant transport in natural and agricultural systems.

The hydration state of humic substances is not static. Still, it undergoes significant changes with temperature fluctuations, leading to structural hysteresis—a phenomenon where humic molecules exhibit a delayed response in reorganizing their hydrogen bond network upon heating and cooling cycles. This hysteresis effect is attributed to the time-dependent restructuring of water clusters within the humic matrix, which governs their aggregation state and conformational arrangement.

Experimental evidence indicates the presence of two major structural transitions, occurring around 20°C and 40°C, which correspond to the disruption and reformation of hydrogen bonds between humic substances and surrounding water molecules. These transitions highlight the dynamic and thermos-responsive nature of humic aggregates, drawing parallels to synthetic polymeric gels and surfactant micelles that undergo similar structural rearrangements in response to external stimuli.

Additionally, humic substances exhibit micelle-like behavior at low concentrations, with hydrophobic domains clustering to minimize interaction with water. This self-assembly tendency significantly enhances their ability to solubilize hydrophobic organic compounds (HOCs), facilitating the transport of otherwise poorly soluble pollutants in natural systems. The pre-micellar aggregation of humic molecules suggests that their apparent solubility is not solely a function of their chemical composition but is also dictated by the extent of intermolecular interactions and hydration dynamics.

These findings reinforce that humic substances do not behave as traditional macromolecules but instead function as adaptive supramolecular structures whose apparent molecular weight, conformational state, and environmental reactivity are strongly influenced by water interactions. Understanding these hydration-driven mechanisms is essential for predicting the role of humic substances in soil moisture regulation, pollutant mobility, and nutrient transport, further demonstrating their crucial function in environmental and agricultural systems[4].

11.2.2.7 The Kinetics of Structural Transitions in Humic Substances

One of the most fascinating properties of humic substances is their ability to change conformation dynamically depending on environmental conditions. However, these structural rearrangements do not occur instantaneously. Instead, humic molecules exhibit time-dependent responses, meaning that their reorganization follows specific kinetic patterns rather than being an immediate consequence of external changes. For example, when *pH* increases, deprotonation of functional groups causes humic molecules to transition from a compact to an expanded conformation, but this process is not instantaneous. Instead, the system requires time to establish a new equilibrium, during which intermolecular hydrogen bonds and van der Waals interactions progressively adjust to the new conditions.

This kinetic aspect is particularly important in environmental and soil chemistry because it determines how quickly humic substances respond to seasonal fluctuations, agricultural amendments, or contamination events. Similarly, the structural hysteresis observed in humic substances suggests that their aggregation and solubilization processes involve delayed reorganization of water networks and weak intermolecular forces. Such behavior is reminiscent of polymer relaxation dynamics, reinforcing the concept that humic substances function as adaptive supramolecular assemblies rather than rigid macromolecules.

11.2.2.8 Interactions Between Humic Substances and Other Organic Compounds

Humic substances are often described as metal complexing agents, but their interactions with organic molecules are just as significant. Due to their amphiphilic nature, humic substances can stabilize, solubilize, or inhibit the activity of a wide range of naturally occurring organic compounds, playing a fundamental role in nutrient bioavailability, microbial metabolism, and organic matter cycling.

For example, lipids and fatty acids, which are inherently hydrophobic, can associate with the non-polar regions of humic molecules, forming organized

[4] Conte, P. and J. Kucerik. 2016. Water Dynamics and Its Role in Structural Hysteresis of Dissolved Organic Matter. Environmental Science and Technology. 50: 2210–2216. DOI: 10.1021/acs.est.5b04639

supramolecular domains that enhance their stability in aqueous environments. This ability to interact with lipophilic compounds explains why humic substances play a key role in transporting hydrophobic organic pollutants, including pesticides and hydrocarbons. Likewise, proteins and enzymes often undergo structural modifications in the presence of humic matter, as the surrounding humic matrix influences their hydrogen bonding networks and tertiary conformations. This interaction is particularly relevant in soil microbiology, where humic substances can modulate enzymatic activity by protecting proteins from degradation or altering their accessibility to substrates.

Another important interaction occurs with polyphenols and tannins (Figure 11.7), plant-derived secondary metabolites contributing to soil organic matter formation. The condensation of polyphenols with humic substances influences humification processes, gradually forming aromatic-enriched humic acids. These organic interactions underscore the role of humic substances as chemical mediators, capable of altering the fate of natural and anthropogenic organic molecules in the environment.

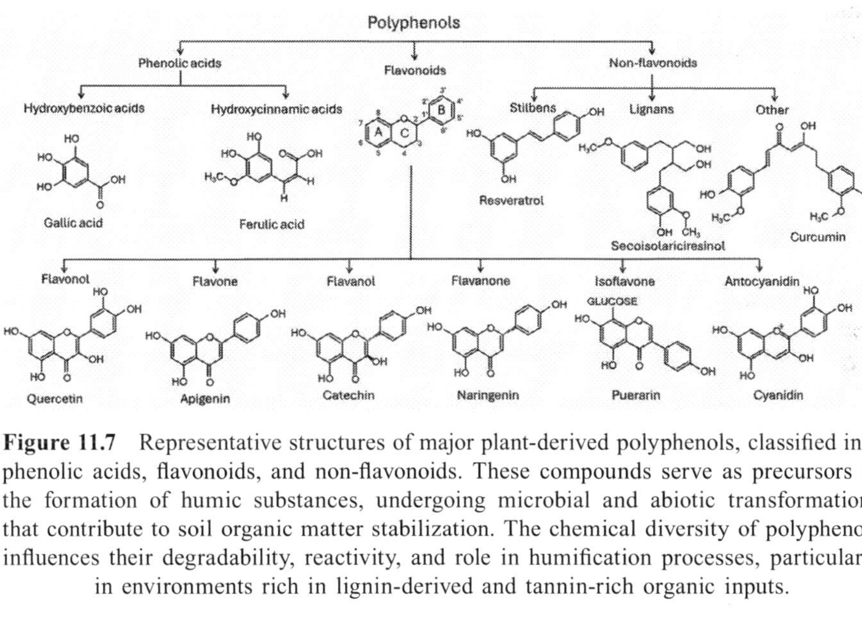

Figure 11.7 Representative structures of major plant-derived polyphenols, classified into phenolic acids, flavonoids, and non-flavonoids. These compounds serve as precursors in the formation of humic substances, undergoing microbial and abiotic transformations that contribute to soil organic matter stabilization. The chemical diversity of polyphenols influences their degradability, reactivity, and role in humification processes, particularly in environments rich in lignin-derived and tannin-rich organic inputs.

11.2.2.9 Humic Substances in Carbon Sequestration and Climate Change Mitigation

Given their persistence in soils, humic substances are often regarded as key players in long-term carbon sequestration. Their role in stabilizing organic carbon is crucial in climate change mitigation, as they influence the balance between carbon storage and carbon release from terrestrial ecosystems. The presence of humic substances reduces carbon mineralization rates, preventing organic carbon from being rapidly decomposed into CO_2 by microbial activity.

The stability of humic substances as carbon sinks depends on both intrinsic and extrinsic factors. Their aromatic content contributes to biochemical recalcitrance, meaning that microbial enzymes struggle to break down these complex structures. Meanwhile, interactions with soil minerals, such as iron and aluminum oxides, create organo-mineral complexes that physically protect humic matter from microbial degradation. However, with increasing global temperatures, the behavior of humic substances may shift, potentially accelerating their decomposition and releasing previously stabilized carbon into the atmosphere.

These considerations highlight why soil management strategies must account for humic substances in carbon farming approaches. Agricultural practices such as reduced tillage, organic amendments, and cover cropping can enhance humic substance formation and preservation, contributing to climate-resilient soil systems.

11.2.2.10 The Humification Process: From Organic Matter to Humic Substances

Understanding the formation of humic substances is just as important as analyzing their properties. Humification is a complex and gradual transformation of organic matter, primarily driven by biological, chemical, and environmental factors. The process begins with the microbial decomposition of plant and animal residues, leading to the breakdown of macromolecules such as polysaccharides, proteins, and lipids into smaller, chemically diverse molecular fragments. These degradation products serve as the building blocks for humic substances (Figure 11.8).

Unlike older models that proposed humic substances as large macromolecules formed via polymerization, modern understanding suggests that humics are dynamic supramolecular associations of relatively small molecules held together by weak interactions such as hydrogen bonding, van der Waals forces, and electrostatic attractions. While large macromolecules, such as cellulose and proteins, may undergo only partial degradation, the resulting larger molecular fragments are remnants of incomplete decomposition rather than newly polymerized entities. These fragments, along with smaller degradation products, can further self-assemble into humic aggregates, influencing their functional properties in soil and water environments.

Both microbial metabolism and abiotic factors contribute to humification. Microbial catabolism plays a dominant role in the early stages of transformation, breaking down complex biopolymers into lower-molecular-weight compounds. Abiotic reactions, such as oxidative coupling and interactions with minerals, further modulate the structural organization of humic substances. These processes do not lead to covalently bonded polymeric structures but instead favor the formation of stabilized supramolecular associations. Furthermore, interactions with metal ions and clay minerals help stabilize humic aggregates, influencing their persistence and functionality within ecosystems.

The humification process varies significantly across different environments. In oxygen-limited settings, such as wetlands, humic substances tend to accumulate due to reduced microbial degradation, whereas in well-aerated soils, microbial activity accelerates organic matter turnover. These differences highlight the context-dependent nature of humification and underscore the importance of sustainable land management in regulating soil organic matter dynamics.

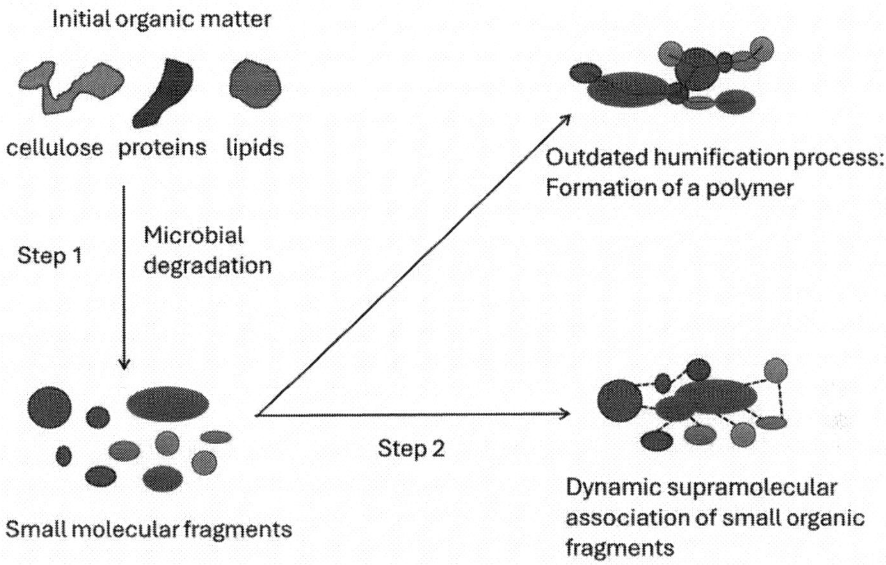

Figure 11.8 This figure illustrates the contemporary understanding of the humification process, emphasizing the transition from organic matter to humic substances as a result of microbial and abiotic transformations. The top-left section represents the initial organic matter, primarily composed of plant and animal residues containing macromolecules such as cellulose, proteins, and lipids. These complex biomolecules undergo microbial degradation, progressively breaking down into smaller molecular fragments and partially degraded larger molecules, as shown in the bottom-left section. The bottom-right section highlights the formation of humic substances from these degradation products, demonstrating how small molecules and larger residual fragments associate rather than polymerize. In contrast to older theories that described humics as high-molecular-weight polymers, the bottom-right section of the figure reflects the modern supramolecular view, where humic substances consist of dynamic aggregates stabilized by weak interactions such as hydrogen bonding, van der Waals forces, and electrostatic attractions. This structural flexibility allows humic substances to interact with minerals, metals, and organic compounds, influencing their role in soil stability, nutrient cycling, and carbon sequestration.

11.2.2.11 Applications of Humic Substances in Agriculture and Industry

Beyond their environmental significance, humic substances have a wide range of practical applications. In agriculture, they are extensively used as soil amendments to improve nutrient retention, soil structure, and plant health. Their ability to enhance cation exchange capacity makes them effective in fertilizer efficiency, reducing nutrient leaching while increasing plant uptake. In addition, humic substances influence root development and microbial activity, contributing to more resilient agricultural systems.

In industry, humic substances have been found to have applications in water treatment, pharmaceuticals, and environmental remediation. Their ability to chelate heavy metals makes them useful for detoxifying polluted water sources, while their antioxidant properties have been explored in biomedical applications. Additionally, humic-based materials are being studied for their potential in sustainable packaging and bio-based industrial processes.

11.3 MICROBIAL BIOMASS: THE LIVING COMPONENT OF SOIL ORGANIC MATTER

Microbial biomass represents the active and dynamic fraction of soil organic matter, encompassing a vast and diverse community of bacteria, fungi, archaea, protozoa, and other microorganisms that continuously drive biochemical transformations within the soil matrix. Despite accounting for only 1–5% of total SOM, this living component plays an outsized role in regulating the decomposition of organic residues, cycling nutrients, and stabilizing soil structure, thereby exerting a profound influence on overall soil functionality.

Unlike the more passive organic matter fractions, microbial biomass actively mediates the flux of carbon, nitrogen, phosphorus, and sulfur between different soil pools, governing both the short-term availability of nutrients and the long-term stability of SOM. Its capacity to store and recycle nutrients makes it an essential buffer between organic inputs and plant-available forms, ensuring that nutrient release and uptake remain synchronized with plant and microbial demands. This dynamic function positions microbial biomass as a catalyst for soil fertility and a key determinant of ecosystem resilience.

Beyond its role in decomposition and nutrient mobilization, microbial biomass is central to soil aggregation and structural stability. Through the secretion of extracellular polymeric substances (EPS) and the entanglement of fungal hyphae with soil particles, microbial communities contribute to aggregate formation, enhancing porosity, water retention, and resistance to erosion. This structural reinforcement not only protects organic matter from rapid degradation but also fosters a stable and habitable environment for microbial life, reinforcing a self-sustaining cycle of organic matter transformation.

The composition and activity of microbial biomass are highly responsive to environmental conditions, fluctuating with soil moisture, temperature, *pH*, substrate availability, and land management practices. In agricultural and natural systems alike, the balance between microbial growth, turnover, and dormancy determines the efficiency of organic matter breakdown and the persistence of carbon in the soil. Practices such as conservation tillage, organic amendments, and crop diversification enhance microbial biomass and diversity, whereas intensive land use and chemical disturbances can suppress microbial communities, leading to reduced SOM stability and impaired nutrient cycling.

By understanding microbial biomass as a fundamental driver of soil health, researchers and land managers can develop strategies to optimize carbon sequestration, improve nutrient use efficiency, and sustain productive ecosystems. This chapter explores the composition, functions, and ecological significance of microbial biomass, illustrating how this vital component of SOM regulates soil fertility, influences biogeochemical cycles, and shapes the long-term sustainability of terrestrial ecosystems.

11.3.1 Microbial Biomass: A Dynamic Driver of Soil Functionality

Though a small fraction of soil organic matter, microbial biomass plays a pivotal role in biochemical cycles due to its dynamic turnover and nutrient-cycling capacity. Composed primarily of carbon, nitrogen, phosphorus, and sulfur, its elemental makeup mirrors that of SOM. The C:N:P:S stoichiometry of microbial biomass remains relatively stable across ecosystems, with a C/N ratio typically ranging from 6 to 10, reflecting microbial metabolic efficiency. A ratio below this range indicates nitrogen-rich, fast-growing microbial communities adept at decomposing organic inputs, while a higher ratio suggests carbon-dominant communities, often fungal-driven, that degrade complex, lignin-rich materials more slowly.

Unlike more recalcitrant SOM fractions, microbial biomass is highly labile. Its turnover rates range from days to months, making it a transient yet essential reservoir of nutrients. Its size and activity fluctuate in response to soil moisture, temperature, organic substrate availability, and disturbance regimes. Under favorable conditions, microbial proliferation enhances decomposition, nutrient mineralization, and soil aggregation, reinforcing overall soil health. Conversely, stressors such as drought, extreme temperatures, salinity, heavy metals, or excessive fertilization reduce microbial biomass, leading to lower enzymatic activity, impaired organic matter decomposition, and diminished soil fertility.

Land management practices significantly shape microbial biomass dynamics. Sustainable approaches such as reduced tillage, cover cropping, and organic amendments have enhanced soil microbial communities and improved overall soil health[5]. Research highlights that organic amendments increase microbial activity by raising soil organic carbon and nitrogen levels[6], while cover crops contribute to microbial diversity, promoting better soil structure and nutrient cycling[7]. Conversely, intensive agricultural practices, including continuous cropping and high synthetic input use, have been linked to declining microbial diversity

[5] Cerecetto, V., et al. 2021. Reduced tillage, cover crops and organic amendments affect soil microbiota and improve soil health in Uruguayan vegetable farming systems. FEMS Microbiology Ecology. 97: fiab023.

[6] Liu, W. et al. 2023. Positive effects of organic amendments on soil microbes and their functionality in agro-ecosystems. Plants. 12: 3790.

[7] Castellano-Hinojosa, A. and S.L. Strauss. 2020. Impact of cover crops on the soil microbiome of tree crops. Microorganisms. 8: 328.

and functionality[8]. Excessive reliance on synthetic fertilizers[9] alters microbial community composition, often leading to reduced biomass and ecosystem resilience[10].

Maintaining an active and diverse microbial community is essential for sustaining soil fertility, enhancing biological resilience, and ensuring long-term ecosystem stability. By adopting sustainable agricultural practices, microbial diversity and function can be preserved, ultimately fostering a more adaptable and productive soil system.

11.3.2 Microbial Biomass and Nutrient Cycling

Microbial biomass serves as a biological engine within the soil, governing the transformation, immobilization, and release of essential nutrients. This nutrient cycling process is fundamental to maintaining soil fertility and ensuring that plants receive a steady supply of bioavailable elements. The decomposition of organic matter by microbial communities not only releases carbon, nitrogen, phosphorus, and sulfur but also regulates the balance between nutrient retention and losses, directly influencing plant growth, soil structure, and environmental sustainability.

11.3.2.1 Carbon Cycling: Decomposition, Sequestration, and Microbial Contributions

The carbon cycle in soil is largely driven by microbial respiration, organic matter decomposition, and carbon stabilization. Heterotrophic microbes degrade plant residues, root exudates, and SOM, breaking complex organic compounds into simpler molecules (Figure 11.9).

During decomposition, microbial respiration releases CO_2 into the atmosphere, while microbial assimilation transforms labile carbon into biomass, contributing to soil carbon storage (Figure 11.9). A critical aspect of microbial involvement in carbon cycling is the formation of microbial necromass, which consists of dead

[8] Niveditha, S.S., et al. 2024. Effects of intensive agricultural management practices on soil biodiversity and implications for ecosystem functioning: A review. International Journal of Research in Agronomy. SP-7(9): 166–169.

[9] A synthetic fertilizer (also called chemical fertilizer or inorganic fertilizer) is a commercially manufactured substance designed to supply essential nutrients to plants. These fertilizers are synthesized through chemical processes rather than being derived from organic or natural sources like compost, manure, or bone meal. Most synthetic fertilizers are produced through industrial chemical reactions. For example, nitrogen fertilizers (like urea or ammonium nitrate) are often made using the Haber-Bosch process, which synthesizes ammonia (NH_3) from atmospheric nitrogen and hydrogen. Phosphorus fertilizers (like superphosphate) are made by chemically treating phosphate rock with sulfuric acid. Potassium fertilizers are typically mined from potash deposits and purified through industrial processes. A deeper acknowledgment about synthetic fertilizers is given in Chapter 13.

[10] Dincă, L.C., et al. 2022. Fertilization and soil microbial community: A review. Applied Sciences. 12(3): 1198.

microbial cells that integrate into SOM (Figure 11.9). This microbial-derived organic matter binds with mineral surfaces, forming stable carbon pools that persist in soil for decades or even centuries (Figure 11.9). The microbial contribution to carbon sequestration is particularly significant in no-tillage systems and organic amendments, where soil structure preserves microbial residues (Figure 11.9).

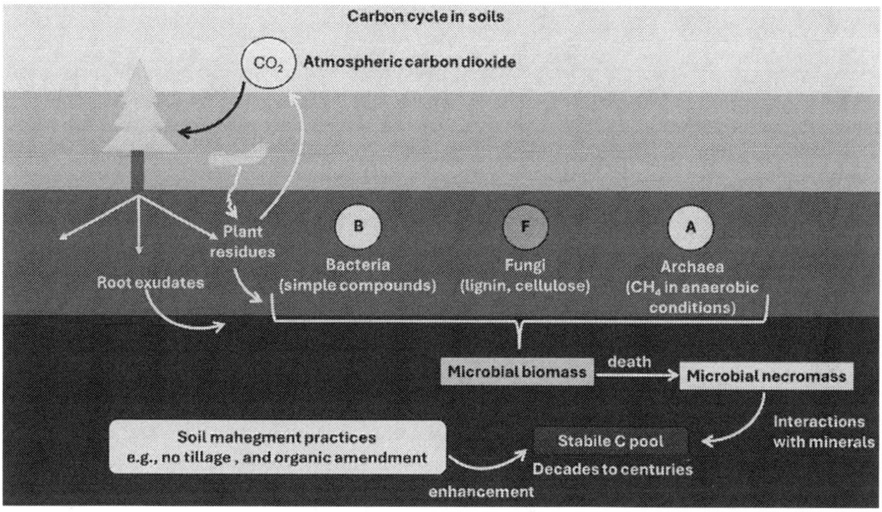

Figure 11.9 Microbial Contributions to the Soil Carbon Cycle. This diagram illustrates the key pathways of carbon cycling in soils. Plants contribute carbon through root exudates and decomposing residues, which are broken down by bacteria (simple compounds), fungi (lignin and cellulose decomposition), and archaea (methanogenesis under anaerobic conditions). Microbial biomass processes these inputs, with dead microbial cells contributing to necromass, which interacts with minerals to form a stable carbon pool. Soil management practices such as no tillage and organic amendments enhance this process, promoting long-term carbon sequestration and soil health.

Different microbial groups play distinct roles:

(a) Bacteria (e.g., *Actinobacteria* and *Proteobacteria*): Rapid decomposition of simple organic compounds.
(b) Fungi (e.g., *Basidiomycetes* and *Ascomycetes*): Breakdown of complex, recalcitrant carbon sources like lignin and cellulose.
(c) Archaea (e.g., methanogens in anaerobic soils): Methane (CH_4) production in waterlogged environments.

11.3.2.2 Nitrogen Cycling: Balancing Retention and Losses

Nitrogen, a critical element for plant growth, is transformed through microbial processes that regulate its availability, fixation, and potential losses. Microbial biomass actively governs ammonification, nitrification, denitrification, and biological nitrogen fixation, ensuring a continuous nitrogen supply while preventing excessive leaching or gaseous losses (Figure 11.10).

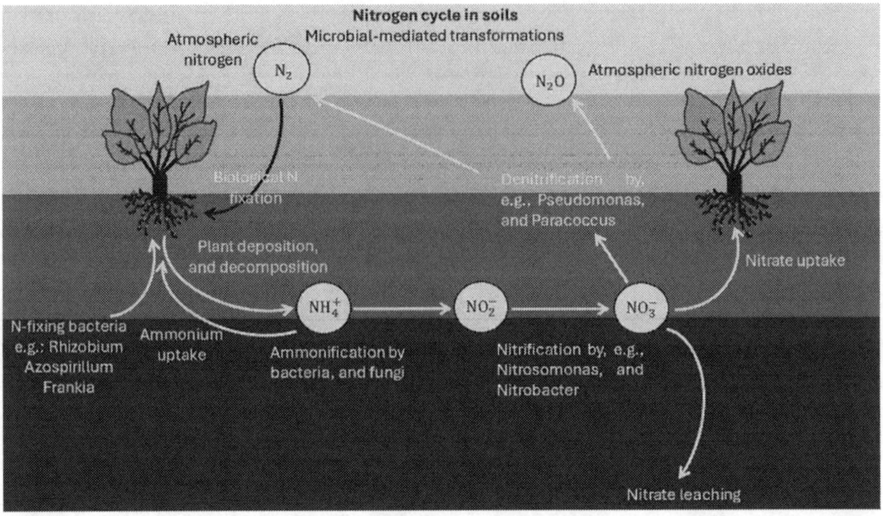

Figure 11.10 Microbial-Mediated Transformations in the Soil Nitrogen Cycle. This diagram illustrates the key pathways of nitrogen cycling in soils, emphasizing microbial roles. Nitrogen-fixing bacteria (Rhizobium, Azospirillum, Frankia) convert atmospheric N_2 into NH_4^+, which is then assimilated by plants or transformed through ammonification by bacteria and fungi. Nitrification, carried out by Nitrosomonas and Nitrobacter, converts NH_4^+ to NO_3^-, making nitrogen available for plant uptake. However, nitrate is also susceptible to leaching into groundwater. Under low-oxygen conditions, denitrifying bacteria (Pseudomonas, Paracoccus) reduce nitrate to N_2 and N_2O, returning nitrogen to the atmosphere. These interconnected microbial processes regulate nitrogen availability, influencing soil fertility and ecosystem sustainability.

(a) **Ammonification:** Organic nitrogen (from decomposed proteins, amino acids) is converted into ammonium (NH_4^+) by decomposer bacteria and fungi (Figure 11.10).

(b) **Nitrification:** Ammonium is oxidized into nitrite (NO_2^-) and then nitrate (NO_3^-) by nitrifying bacteria such as *Nitrosomonas* and *Nitrobacter*. This process increases nitrogen mobility but also makes it vulnerable to leaching (Figure 11.10).

(c) **Denitrification:** Under oxygen-limited conditions, denitrifying bacteria (e.g., *Pseudomonas*, *Paracoccus*) reduce nitrate to gaseous nitrogen (N_2 and N_2O), which can be lost to the atmosphere. This step is a major contributor to nitrous oxide (N_2O) emissions, a potent greenhouse gas (Figure 11.10).

(d) **Biological nitrogen fixation:** Some bacteria (*Rhizobium*, *Azospirillum*, *Frankia*) and cyanobacteria convert atmospheric N_2 into ammonium, making nitrogen available in soils with low fertility (Figure 11.10).

These processes are highly sensitive to oxygen levels, soil *pH*, organic matter inputs, and microbial diversity, making nitrogen cycling one of the most dynamic and management-dependent cycles.

11.3.2.3 Phosphorus Cycling: Unlocking an Essential Nutrient

Phosphorus is often a limiting nutrient in soils due to its strong binding with minerals (e.g., iron, aluminum oxides), making it inaccessible to plants. Microbial biomass enhances phosphorus availability through:

(a) Phosphatase enzyme production (by bacteria and fungi), mineralizing organic phosphorus compounds into bioavailable inorganic forms (Figure 11.11).

(b) Solubilization of mineral-bound phosphorus via organic acid secretion (Pseudomonas, Bacillus) (Figure 11.11).

(c) Mycorrhizal fungi associations (Glomeromycota) extend hyphal networks into the soil, enhancing plant phosphorus uptake.

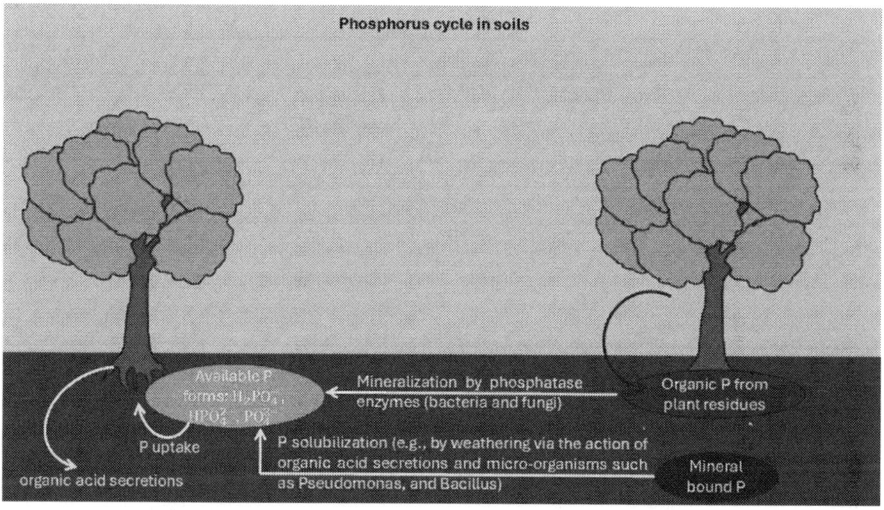

Figure 11.11 Microbial and Biochemical Processes in the soil phosphorus cycle. This diagram illustrates the key transformations of phosphorus in soils, driven by microbial and plant-mediated processes. Organic phosphorus from plant residues undergoes mineralization via phosphatase enzymes released by bacteria and fungi, converting it into bioavailable forms ($H_2PO_4^-$, HPO_4^{2-}, PO_4^{3-}). Additionally, P solubilization occurs through organic acid secretions and the activity of microorganisms such as Pseudomonas and Bacillus, which enhance phosphorus release from mineral-bound forms. Plants absorb soluble phosphorus for growth, while some phosphorus becomes fixed in soil minerals, reducing its immediate availability.

Microbial-driven phosphorus solubilization is particularly important in weathered or degraded soils, where natural phosphorus reserves are tightly bound in mineral complexes.

11.3.2.4 Sulfur Cycling: Microbial Control Over Sulfur Availability

Sulfur, essential for protein synthesis and enzymatic functions, cycles between organic and inorganic forms under microbial mediation. Key microbial processes include:

(a) Sulfate Reduction (*Desulfovibrio* and *Desulfobacter*): Conversion of sulfate (SO_4^{2-}) into hydrogen sulfide (H_2S) under anaerobic conditions (Figure 11.12).

(b) Sulfide Oxidation by, e.g., *Thiobacillus*, *Paracoccus*, and *Acidithiobacillus*: Oxidation of H_2S back into sulfate, making it available for plant uptake (Figure 11.12).

These microbial activities regulate sulfur retention, plant nutrition, and sulfur gas emissions, particularly in flooded environments (e.g., wetlands and rice paddies).

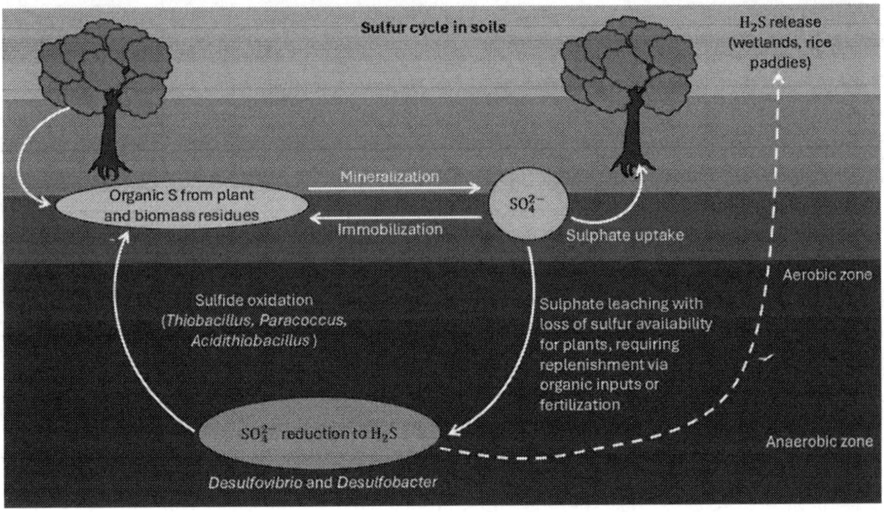

Figure 11.12 This diagram illustrates the key transformations of sulfur in soils, emphasizing microbial oxidation, reduction, and plant uptake. Organic sulfur from plant residues undergoes mineralization, releasing sulfate (SO_4^{2-}) that can be immobilized by microbes, absorbed by plants, or lost via leaching. In anaerobic environments, sulfate-reducing bacteria (Desulfovibrio, Desulfobacter) convert SO_4^{2-} into hydrogen sulfide (H_2S), which can accumulate in waterlogged soils, wetlands, and rice paddies. In contrast, sulfur-oxidizing bacteria (Thiobacillus, Paracoccus, Acidithiobacillus) convert H_2S and elemental sulfur into sulfate in aerobic zones, replenishing available sulfur. Sulfate leaching can result in nutrient loss, requiring replenishment through organic amendments or fertilization.

11.3.3 Microbial Biomass and Soil Structure Stabilization

Beyond its biochemical functions, microbial biomass modifies the physical architecture of soil, playing a key role in aggregation, porosity, and erosion resistance. One of the most significant contributions of microbes to soil structure is through the production of exopolysaccharides (EPS)—gel-like extracellular substances secreted by bacteria and fungi that enhance particle cohesion and promote the formation of stable soil aggregates. These microbial-derived compounds improve water retention, aeration, and root penetration, directly impacting plant growth and soil fertility.

Fungal hyphae, particularly those from mycorrhizal fungi, further reinforce soil structure by weaving an intricate network that physically binds soil particles together. This biological binding agent enhances aggregate stability, reducing susceptibility to compaction and erosion while creating microhabitats that support microbial diversity. Aggregates also serve as protective environments for organic matter, reducing its exposure to decomposition and extending carbon retention within the soil matrix.

Microbial activity creates hotspots of biochemical reactions where enzymes, organic exudates, and nutrient transformations occur at high rates. These microsites, often formed around rhizospheres, decomposing organic residues, and biofilm-covered mineral surfaces, drive soil functional diversity, influencing nutrient availability and microbial interactions. Such localized activity zones buffer soil systems against environmental stressors, ensuring functional redundancy and resilience even under fluctuating conditions. Moreover, microbial-driven soil stabilization mitigates land degradation, helping prevent desertification and soil loss in fragile ecosystems. Agricultural practices that foster microbial biomass—such as cover cropping, organic amendments, and minimal soil disturbance—enhance these structural benefits, ultimately improving soil tilth, fertility, and long-term productivity.

11.3.4 Microbial Biomass and Environmental Sustainability

As a fundamental bioindicator of soil health, microbial biomass is a key determinant of sustainable land management. Its influence extends beyond local soil processes, playing a central role in global biogeochemical cycles, carbon sequestration, and climate resilience. Agricultural strategies that enhance microbial diversity and functionality—including organic amendments, reduced tillage, biochar [Note 8; Biochar is investigated in Chapter 13] application, and diversified cropping systems—contribute to long-term SOM stability while mitigating environmental risks. In contrast, intensive land use practices, excessive chemical fertilization, monoculture cropping, and pesticide overapplication can disrupt microbial communities, leading to declines in microbial biomass, organic matter depletion, and imbalanced nutrient cycling. These disturbances not only degrade soil health but also contribute to increased greenhouse gas emissions, particularly through enhanced CO_2 release from microbial respiration and nitrogen losses via denitrification (N_2O emissions).

Adapting climate-smart soil management approaches requires a deep understanding of microbial biomass dynamics. Promoting soil carbon storage, minimizing nutrient leaching, and enhancing microbial resilience are essential goals for mitigating the adverse effects of climate change. Strategies such as compost application, rotational grazing, and agroforestry leverage microbial functions to maintain soil fertility while reducing environmental footprints.

Fostering a biologically rich soil ecosystem ensures its ability to recover from disturbances, maintain nutrient efficiency, and support productive, sustainable agricultural landscapes. The health of microbial biomass is intrinsically linked to the sustainability of terrestrial ecosystems, reinforcing its critical role in environmental conservation and food security.

EXERCISES

1. Which of the following statements about Soil Organic Matter (SOM) is false?
   (a) SOM improves soil structure by enhancing aggregation and porosity.
   (b) Humic substances are chemically pure compounds with well-defined molecular structures.
   (c) Microbial biomass contributes to SOM turnover and nutrient cycling.
   (d) The passive pool of SOM consists of highly stable organic materials that can persist for millennia.
2. A soil sample is analyzed and found to contain 2.5% Soil Organic Carbon (SOC). Using the commonly used approximation SOM $\approx$ 1.724 $\times$ SOC, estimate the percentage of Soil Organic Matter (SOM) in the sample. Why might this calculation be an approximation rather than an exact value? Discuss at least two factors that can influence the SOC-to-SOM conversion ratio.
3. Soil Organic Matter (SOM) can be divided into active, slow, and passive pools based on turnover rates.
   (a) Define each of these SOM pools and describe their relative stability in the soil.
   (b) Explain how agricultural practices such as tillage and organic amendments influence the balance between these SOM pools.
4. Humic substances play a role in carbon sequestration and climate change mitigation.
   (a) Explain how humic substances contribute to long-term carbon sequestration in soils.
   (b) Discuss how global climate change might impact the stability and decomposition of humic substances in different environments.
   (c) Suggest two sustainable land management practices that can enhance SOM stability and carbon sequestration.

5. Imagine you are designing an experiment to study the effect of microbial biomass on soil fertility.
   (a) What three experimental treatments would you set up to compare microbial biomass activity under different conditions?
   (b) Which soil properties (e.g., pH, moisture, organic carbon) would you measure to assess microbial biomass's impact?
   (c) How would you interpret an increase in microbial biomass activity over time in terms of soil health and nutrient availability?
6. The table below shows the Soil Organic Matter (SOM) content (%) in three different land-use systems:

Land Use Type	SOM Content (%)
Natural forest	8.5
Agricultural land	3.2
Urban soil	1.1

   (a) analyze the trend in SOM content across the three land-use types.
   (b) What are the main factors that could contribute to the differences in SOM content among these systems?
   (c) If a degraded agricultural soil currently has 1.8% SOM, propose two land management practices that could help increase SOM levels and explain their mechanisms.
7. For each statement below, indicate whether it is True (T) or False (F) and provide a brief explanation for false statements.
   (a) Microbial biomass is a passive reservoir of organic matter and does not actively contribute to nutrient cycling.
   (b) Humic substances consist of chemically pure compounds with a fixed molecular structure.
   (c) Soil Organic Matter can enhance water retention and reduce soil erosion.
   (d) The decomposition of fresh organic inputs into SOM is solely a chemical process independent of microbial activity.
   (e) Agricultural practices like cover cropping and reduced tillage can help maintain or increase SOM content over time.

FURTHER READING

Benbi, D.K. and R. Nieder (eds). 2003. Handbook of Processes and Modeling in Soil-Plant System, 1st Ed. Food Products Press® and The Haworth Reference Press. ISBN: 1560229144.

Clapp, C.E., M.H.B. Hayes, N. Senesi, P.R. Bloom and P.M. Jardine (eds). 2001. Humic Substances and Chemical Contaminants, 1st Ed. Soil Science Society of America, Inc. ISBN: 0891188371

Huang, P.M., N. Senesi and J. Buffle (eds). 1998. Structure and Surface Reactions of Soil Particles, 1st Ed. John Wiley and Sons. ISBN: 0471959367.

Violante, A., P.M. Huang, J.M. Bollag and L. Gianfreda (eds). 2002. Soil Mineral-organic Matter-microorganism Interactions and Ecosystem Health. Volume 28A. Elsevier. ISBN: 0444510389

Weil, R.R. and N.C. Brady. 2017. The Nature and Properties of Soils, 15th Ed. Pearson Prentice Hall.

Chapter 12

The Telluric Air

12.1 INTRODUCTION TO THE SOIL GAS PHASE

The soil gas phase refers to the portion of soil not occupied by liquid or solid matter, comprising gases such as oxygen (O_2), carbon dioxide (CO_2), nitrogen (N_2), and other trace gases. This phase is far from passive. Indeed, it is a dynamic component of soil health, influencing key biological and chemical processes essential for plant growth, microbial activity, and overall ecosystem balance.

Soil aeration, or the movement of gases between the soil and the atmosphere, regulates root respiration and microbial metabolism, affecting processes such as nitrogen fixation, organic matter decomposition, and the release of greenhouse gases. Oxygen availability is crucial as it dictates the balance between aerobic and anaerobic microbial processes, with significant implications for nutrient cycling and plant health. Meanwhile, carbon dioxide accumulates as a byproduct of respiration, potentially altering soil *pH* and influencing mineral solubility.

The soil gas phase is also a crucial factor in soil-plant-atmosphere interactions, where gas exchanges not only sustain biological functions but also contribute to broader environmental processes. For instance, the diffusion of nitrous oxide (N_2O) and methane (CH_4) from soils plays a role in atmospheric composition and climate regulation. Moreover, variations in soil texture, porosity, and moisture content dictate the movement and retention of gases, leading to distinct aeration profiles in different soil types.

Understanding the composition and dynamics of the telluric air is essential for agronomic management, land use planning, and environmental conservation. Whether in agricultural systems, natural ecosystems, or urban soils, the balance of gases within the soil directly impacts productivity, carbon sequestration, and soil resilience. This chapter investigates the intricate relationships governing soil aeration, the factors influencing gas exchange, and the broader implications for sustainable land management.

DOI: 10.1201/9781003780212-12

12.2 COMPOSITION OF SOIL ATMOSPHERE

The composition of soil gases is not static, as it fluctuates depending on factors such as depth, soil type, moisture content, and biological activity. While it may seem an invisible and intangible component of the soil environment, the gas phase plays a vital role in sustaining life beneath the surface. The balance of these gases is shaped by a continuous interplay of biological respiration, chemical reactions, and atmospheric diffusion, creating a dynamic underground atmosphere that influences plant health, microbial communities, and broader ecological processes.

Among the key gases present in the soil is oxygen (O_2), which is perhaps the most critical for sustaining life. As the primary driver of aerobic respiration, it supports the metabolic activities of plant roots, fungi, and a vast array of microorganisms. However, oxygen levels in the soil are typically lower than in the open atmosphere due to ongoing consumption by roots and soil organisms. While atmospheric oxygen is stable at around 20%, its concentration in the soil can drop significantly, especially in compacted or waterlogged conditions where diffusion is restricted. In extreme cases, O_2 levels may fall below 5%, creating hypoxic environments that limit root function and favor anaerobic microbial processes. In contrast, carbon dioxide (CO_2) follows an opposite trend. While its concentration in the atmosphere hovers around 0.04% (or 400 parts per million, ppm), in the soil, CO_2 levels are dramatically higher due to continuous production from root respiration and microbial decomposition. Typical CO_2 concentrations in well-aerated soils range between 0.2% and 2% (2,000 to 20,000 ppm), but in poorly ventilated soils, levels can even exceed 10% (that is, 100,000 ppm). This sharp contrast underscores a fundamental characteristic of the soil gas phase: gas exchange with the atmosphere is not free-flowing. If diffusion were unrestricted, CO_2 would not accumulate to such high levels. Instead, various physical factors —soil compaction, high moisture content, or structural barriers—can impede gas movement, leading to CO_2 buildup that affects soil *pH*, nutrient solubility, and overall plant growth (Figure 12.1).

Another major component of the soil atmosphere is nitrogen (N_2). Although it is the most abundant atmospheric gas (~78%), its proportion in soil air can vary due to restricted gas exchange, microbial nitrogen cycling, and shifts in O_2 and CO_2 concentrations. In well-aerated soils, N_2 remains dominant, but in poorly ventilated or compacted soils, its percentage may decrease as CO_2 accumulates and O_2 is depleted. Despite these variations, nitrogen remains biologically crucial as the primary reservoir for nitrogen-fixing bacteria, which convert it into bioavailable forms such as ammonium (NH_4^+) and nitrate (NO_3^-), essential for plant nutrition (see Chapter 11).

Beyond these primary gases, the soil also contains trace gases that, despite their lower concentrations, have significant ecological and atmospheric implications. Methane (CH_4) is produced in anaerobic conditions by methanogenic bacteria, often found in waterlogged or organic-rich soils, and is a potent greenhouse gas. Similarly, nitrous oxide (N_2O) emerges as a byproduct of microbial nitrogen cycling (see Chapter 11), particularly under conditions of fluctuating soil moisture and high nitrogen availability. These gases not only contribute to soil chemistry but also influence broader environmental dynamics, as they can diffuse into the atmosphere and play a role in climate change. Additionally, volatile organic compounds (VOCs) released by plant roots and microbial activity can affect soil microbial communities and even serve as chemical signals in plant communication.

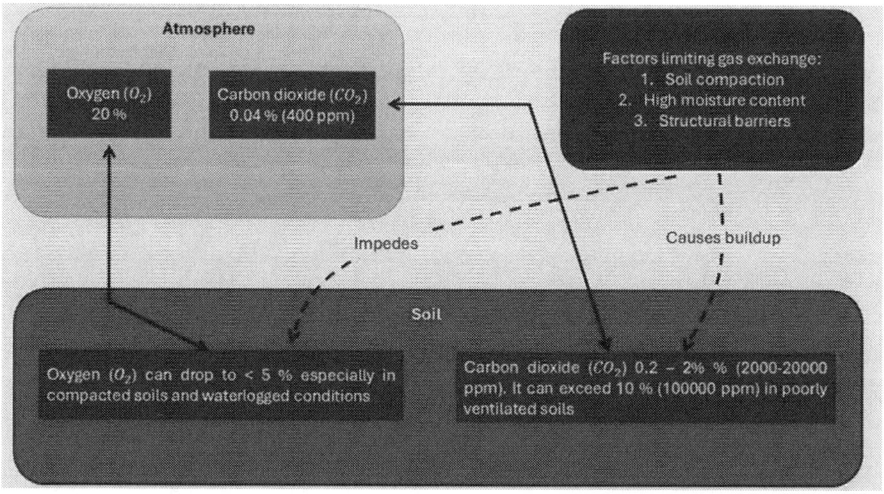

Figure 12.1 Illustration of the exchange of oxygen (O_2) and carbon dioxide (CO_2) between the atmosphere and the soil. Oxygen from the atmosphere diffuses into the soil, where roots and microbial respiration consume it, while CO_2 is produced and released back into the atmosphere. This exchange occurs efficiently in well-aerated soils, but in compacted, waterlogged, or poorly structured soils, gas diffusion is impeded. This leads to decreased oxygen availability (below 5%) and an accumulation of CO_2, which can exceed 10% (100,000 ppm) in poorly ventilated conditions. Factors limiting gas exchange include soil compaction, high moisture content, and structural barriers.

Ultimately, the restricted exchange of gases between soil and atmosphere is a defining feature of the subterranean ecosystem. Unlike the open atmosphere, where gases mix freely, the soil operates as a semi-closed system where physical and biological processes regulate diffusion. When diffusion is hindered, gases like CO_2 can reach concentrations that significantly alter soil chemistry and biological function. Understanding the dynamics of these gases is essential for maintaining soil health, optimizing plant productivity, and addressing broader environmental challenges.

12.3 GAS EXCHANGE MECHANISMS IN SOIL

The movement of gases in the soil is a continuous process influenced by the physical properties of the medium, the biological activity, and the environmental conditions. Unlike the atmosphere, where gases mix freely, their movement in soil is constrained by the porous structure of the medium and the varying composition of gases within it.

Two primary mechanisms govern gas exchange in soil: diffusion and convection, each playing a distinct role in regulating the balance of oxygen, carbon dioxide, and other gases within the subsurface environment (Figure 12.2).

Diffusion is the dominant process driving gas movement in unsaturated soils, where air-filled pores allow molecular migration. Governed by Fick's laws, diffusion occurs as gas molecules move from areas of higher concentration to areas of lower concentration, seeking equilibrium. This passive transport mechanism is responsible for the gradual exchange of gases between the soil and the atmosphere and for the redistribution of gases within the soil profile (Figure 12.2).

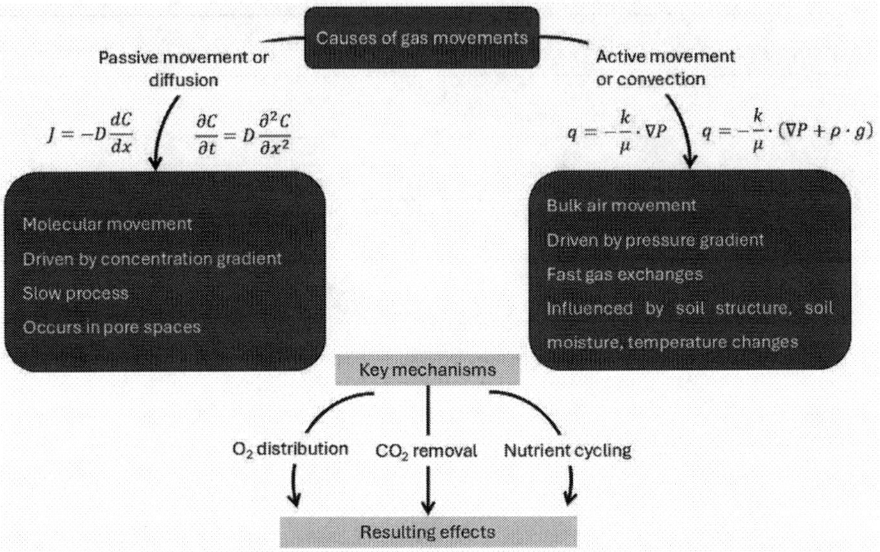

Figure 12.2 This diagram illustrates the primary mechanisms driving gas movement in soils: passive movement (diffusion) and active movement (convection). Fick's Laws of Diffusion describes diffusion as a slow molecular process governed by concentration gradients occurring through pore spaces. The diffusion flux is expressed by Fick's First Law ($J = -D\ dC/dx$; see the main text for symbol definitions). The time-dependent variation of gas concentration follows Fick's Second Law ($\partial C/\partial t = D\ \partial^2 C/\partial x^2$; see the main text for an explanation) which describes how the concentration of a gas changes over time due to diffusion. In contrast, convection (bulk air movement) occurs due to pressure or temperature gradients, allowing gases to move more rapidly through soil pores. This mechanism is governed by Darcy's Law, which models gas flow in porous media ($q = -(k/\mu)\nabla P$; see the main text for details). When temperature-driven buoyancy effects are considered, Darcy-Boussinesq's equation includes the term for gas density (ρ) and gravitational acceleration (g) ($q = -(k/\mu)(\nabla P + \rho \cdot g)$; see the text for further explanations)

Fick's First Law describes the steady-state flux of a gas:

$$J = -D\frac{dC}{dx} \tag{12.1}$$

where J represents the flux of the gas (in mol m^{-2} s^{-1}), D is the diffusion coefficient (in m^2 s^{-1}), C is the concentration of the gas (in mol m^{-3}), and x is the spatial coordinate (in m). This equation expresses how gases move in response to concentration gradients, with the negative sign indicating that the flow is directed from higher to lower concentrations.

Over time, gas concentration in the soil varies dynamically, following Fick's Second Law, which describes how the concentration of a gas changes as a function of both space and time:

$$\frac{\partial C}{\partial t} = D\frac{\partial^2 C}{\partial x^2} \tag{12.2}$$

Here, the first term on the left-hand side represents the rate of the concentration change in a given area, while the second derivative on the right-hand side of the equation describes how the gradient of concentration changes with distance. This equation shows that when gas concentration is not at equilibrium (non-steady-state), it continuously changes over time as gases diffuse through the soil. The rate of this change depends on how easily the gas moves (diffusion coefficient, D) and how steeply concentration varies across space (spatial gradient, $\partial^2 C/\partial x^2$). In other words, diffusion redistributes gases dynamically, responding to differences in concentration across the soil matrix (Figure 12.3).

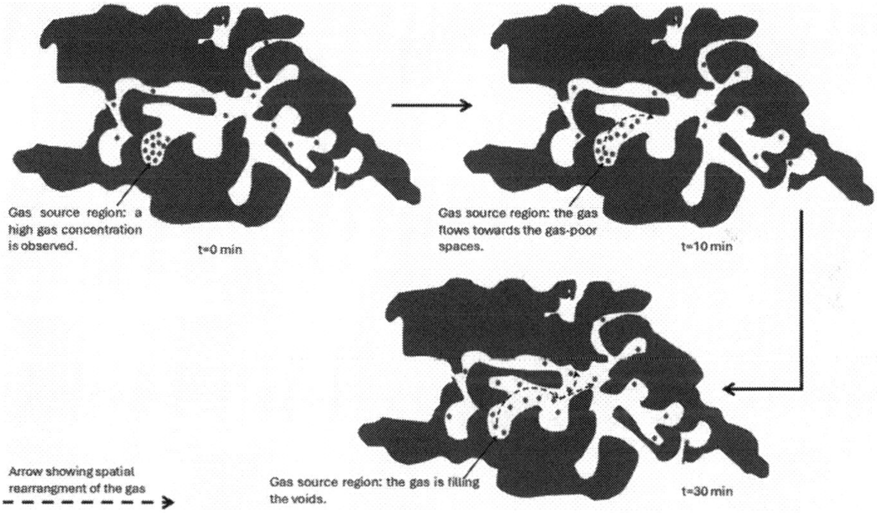

Figure 12.3 Temporal and spatial evolution of gas diffusion in soil pores. Initially (0 min), gas is concentrated in a localized area. Over time (10 min), diffusion moves gas towards lower-concentration regions. Eventually (30 min), gas spreads throughout the available pore spaces, demonstrating how diffusion is governed by concentration gradients and pore connectivity.

The diffusion coefficient is not constant in soils but depends on factors such as soil porosity, moisture content, and compaction. A well-structured, dry soil with a high proportion of interconnected air-filled pores allows for more efficient diffusion, whereas a saturated or compacted soil significantly slows gas exchange, leading to oxygen depletion and CO_2 buildup.

While diffusion governs gas transport in most cases, convection can also play a role, particularly in response to pressure or temperature gradients. Unlike diffusion, which is driven by concentration differences, convection involves bulk air movement through the soil pores, typically occurring in highly permeable soils or when external forces – such as wind or barometric pressure fluctuations—drive gas flow.

The convection dynamics in soils can be described using Darcy's Law for gas flow through porous media, which is analogous to fluid flow in soils. While concentration gradients primarily drive diffusion, convection involves the bulk movement of gases due to pressure or temperature gradients. The governing equation for convective gas flow in porous media is:

$$q = -\frac{k}{\mu} \cdot \nabla P \tag{12.3}$$

where q is the volumetric flux of the gas (m s^{-1}), k is the soil permeability (m^2), μ is the dynamic viscosity of the gas (Pa s), ∇P is the pressure gradient (Pa m^{-1})[1]. This equation indicates that the flow rate of gas in soil is directly proportional to the pressure gradient and the permeability of the soil, while being inversely proportional to the viscosity of the gas. In loose, well-aerated soils, permeability is high, allowing easier convective movement of gases, whereas compacted or fine-textured soils restrict gas flow.

Temperature variations between the soil surface and deeper layers can also create convective currents, causing gases to move in response to density differences. For thermally driven convection, a variation of the Darcy-Boussinesq equation can be used to describe the buoyancy-driven movement of gases due to temperature gradients:

$$q = -\frac{k}{\mu} \cdot (\nabla P + \rho \cdot g) \tag{12.4}$$

where ρ is the gas density (kg m^{-3}), and g is the gravitational acceleration (m s^{-2}).

Although convection is generally a secondary transport mechanism in soils compared to diffusion, it can become significant in certain conditions, such as strong barometric fluctuations, soil heating, or rapid gas exchange with external airflows.

The efficiency of these mechanisms ultimately determines soil's ability to sustain life below the surface. When diffusion is restricted—whether due to excessive moisture, compaction, or poor soil structure—gas exchange becomes limited, leading to an accumulation of CO_2 and depletion of O_2, with consequences for root respiration and microbial activity.

12.4 FACTORS INFLUENCING SOIL GAS DYNAMICS

The movement of gases in soil is influenced by several interrelated factors that govern the rate and efficiency of diffusion and convection. These factors include

[1] The *nabla* operator has already been defined in Chapter 3, Section 3.8 of this book.

the porosity and structure of the soil, its moisture content, temperature, and the tortuosity of gas pathways (Figure 12.4). Each parameter plays a crucial role in determining how oxygen, carbon dioxide, and other gases exchange between the soil and the atmosphere, ultimately affecting soil aeration, microbial activity, and plant root respiration.

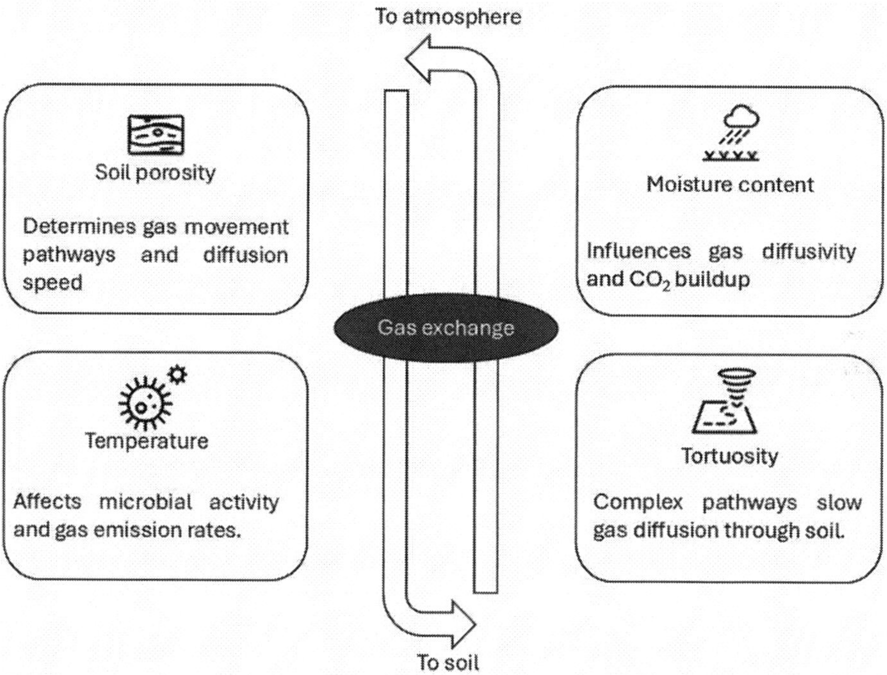

Figure 12.4 This diagram illustrates the key factors regulating gas exchange in soils, including soil porosity, moisture content, temperature, and tortuosity. Soil porosity determines the pathways available for gas movement, with well-structured, coarse-textured soils facilitating faster diffusion than compacted or fine-textured soils. Moisture content influences gas diffusivity by altering air-filled pore space—higher moisture levels reduce oxygen availability and can lead to CO_2 accumulation. Temperature affects microbial activity and gas emission rates, with higher temperatures promoting respiration and decomposition, leading to increased CO_2 production. Tortuosity describes the complexity of gas movement through the soil matrix, where highly convoluted pathways slow diffusion rates. These factors collectively regulate soil aeration and gas movement, influencing plant root respiration, microbial processes, and overall soil health.

One of the most fundamental properties affecting gas dynamics is soil porosity and structure, which dictates the available pathways for gas movement. Coarse-textured soils, such as sandy soils, tend to have larger, well-connected pores, allowing for faster gas exchange. In contrast, fine-textured soils like clays have smaller pore spaces that restrict airflow, often resulting in slower diffusion rates and a higher likelihood of gas entrapment. Soil aggregation and compaction further modify porosity by enhancing connectivity in well-structured soils or obstructing gas exchange in densely packed soils.

Soil moisture content is another critical factor that regulates gas diffusion. As moisture levels increase, air-filled pore spaces shrink, reducing the pathways available for gaseous movement. Water displaces air in the soil matrix, creating a barrier to diffusion since gases diffuse much more slowly in water than in air. In highly saturated soils, where nearly all pore spaces are filled with water, gas exchange is severely restricted, leading to hypoxic (low oxygen) or even anoxic (oxygen-deprived) conditions that favor anaerobic microbial processes such as methane (CH_4) production. The relationship between moisture and gas diffusion highlights the delicate balance required to maintain optimal soil aeration for biological processes (Figure 12.4).

12.4.1 Temperature and Pressure Effects on Soil Gas Behavior

Temperature plays a significant role in soil gas dynamics. It influences the rate of microbial activity, which in turn affects the consumption and production of gases such as O_2, CO_2, and NO_2. Warmer temperatures generally accelerate microbial respiration and organic matter decomposition, leading to higher CO_2 concentrations in the soil. Additionally, temperature affects gas solubility and viscosity, indirectly influencing diffusion rates. In frozen or extremely cold soils, gas exchange is impeded due to reduced microbial activity and the physical blocking of pore spaces by ice.

Beyond biological effects, temperature and pressure variations also impact the physical behavior of gases in soil pores. The ideal gas law provides a fundamental model for understanding how gas volume and pressure change with temperature fluctuations:

$$PV = nRT \tag{12.5}$$

Here, P is the pressure in Pa, V is the volume of the gas in m^3, n is the number of mole of the gas in mol, R is the universal gas constant, and T is the temperature of the system in Kelvin. However, gases in soil do not behave as ideal gases due to the presence of water films, soil particles, and high-pressure microsites, especially in compact or saturated soils. In such cases, the van der Waals equation provides a more realistic model by accounting for intermolecular interactions and finite molecular volume:

$$\left(P+\frac{a}{V^2}\right)\cdot\left(V-b\right)=nRT \tag{12.6}$$

Here, P, V, n, R, and T have the same meaning as above, while a is the van der Waals constant, which accounts for attractive forces between gas molecules, and b is the van der Waals constant, which accounts for the finite volume occupied by gas molecules (Figure 12.5).

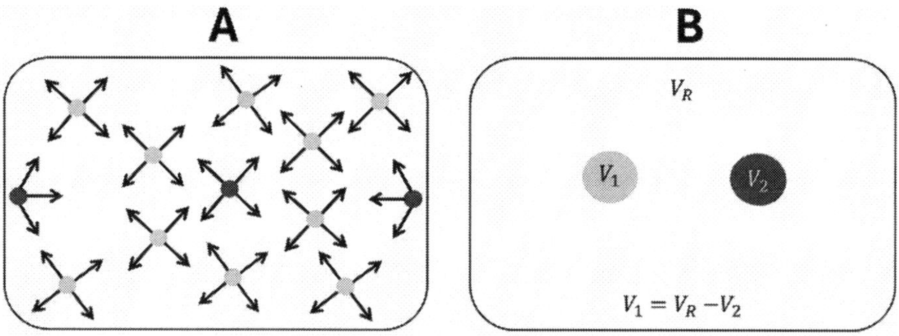

Figure 12.5 Part **A** illustrates gas molecules (dark and light) with arrows indicating intermolecular forces. The light molecules interact with their neighbors, demonstrating attractive forces, which are accounted for by the term an in the van der Waals equation. The dark molecules near the boundary experience fewer neighboring interactions, contributing to deviations from ideal gas behavior. Part **B** represents the excluded volume due to the finite size of gas molecules. The light gray sphere (V_1) denotes the available volume for free movement after subtracting the space occupied by the dark gray molecule (V_2). The relationship $V_1 = V_R - V_2$ corresponds to the van der Waals correction factor b, which accounts for the volume occupied by gas molecules, reducing the space available for movement.

Equation (12.6) improves upon the ideal gas law by incorporating corrections for gas behavior under non-ideal conditions, which are particularly relevant in soils where gases interact with water, organic matter, and soil particles. For example, pressure buildup due to restricted gas diffusion can enhance intermolecular forces in waterlogged soils, reducing gas mobility and increasing the likelihood of anaerobic microsites. Similarly, in dry, well-aerated soils, gas molecules behave more like an ideal gas, freely diffusing through large, connected pore spaces.

12.4.2 Gas Solubility in Soil Water: Henry's Law

The movement and availability of gases in the soil are not only influenced by temperature, pressure, and diffusion mechanisms but also by their interaction with soil moisture. Henry's Law describes how gases dissolve in water, playing a crucial role in determining the concentration of oxygen, carbon dioxide, and other gases in soil water. The law is mathematically expressed as:

$$C_g = k_H \cdot P_g \tag{12.7}$$

where C_g is the concentration of the dissolved gas, k_H is the Henry's constant referring to the equilibrium *dissolved gas* $\rightleftarrows$ *gas*, and P_g is the pressure of the soil air phase. Henry's Law illustrates that the higher the partial pressure of a gas in soil air, the greater its solubility in soil water. However, temperature plays a key role, as the solubility of most gases decreases with increasing temperature, influencing biological and geochemical reactions.

Carbon dioxide has a much higher solubility in soil environments than oxygen, which impacts soil respiration and acidification processes. When CO_2 dissolves, it reacts with water to form carbonic acid, which can dissociate, influencing soil *pH* (refer to Chapter 5 for the equilibria of carbonic acid). Additionally, soil moisture content directly affects gas diffusion and aeration. Low oxygen solubility and limited diffusion lead to oxygen depletion in waterlogged soils, favoring anaerobic microbial processes such as denitrification and methanogenesis. Conversely, lower water content in drier soils reduces gas dissolution, making gaseous exchange with the atmosphere more dominant.

12.4.3 Tortuosity and Gas Diffusion Efficiency

Tortuosity is a less visible but equally important factor in gas movement, which describes the complexity of the path that gas molecules must navigate through the soil matrix. Unlike in open space, where gases move freely, in soil they must follow a convoluted and irregular path due to the arrangement of soil particles, aggregates, and moisture content. The effective diffusion coefficient (D_e, expressed in $m^2\ s^{-1}$) is used to quantify how efficiently gases can move through soil under these constraints. It is defined as:

$$D_e = \frac{D \cdot \vartheta_a}{\tau} \tag{12.8}$$

where D is the diffusion coefficient in free air (in $m^2\ s^{-1}$), ϑ_a is the air-filled porosity (dimensionless), and τ is the tortuosity factor (dimensionless).

This equation expresses how the diffusion of gases is reduced in soil compared to free air. The higher the tortuosity factor, the more complex and indirect the gas pathway, leading to slower diffusion rates. Similarly, lower air-filled porosity limits gas movement, reinforcing the role of soil moisture and texture in regulating gas transport.

Together, these factors shape the spatial and temporal variability of soil gas dynamics (Figure 12.4). A well-structured, moderately moist, and aerated soil supports efficient gas exchange, fostering optimal plant growth and microbial activity. In contrast, poorly drained or compacted soils can create oxygen-deprived conditions, altering biogeochemical cycles and influencing greenhouse gas emissions. Understanding these factors is essential for managing soil health in both natural and agricultural systems, ensuring proper aeration, and minimizing negative environmental impacts.

12.5 SOIL GAS EXCHANGE AND ITS ENVIRONMENTAL IMPLICATIONS

The exchange of gases between the soil and the atmosphere plays a crucial role in maintaining soil health, supporting plant growth, and regulating broader environmental processes. The balance between oxygen diffusion, carbon dioxide release, and the emission of other gases directly influences biological activity, nutrient cycling, and atmospheric composition. When this balance is disrupted, it can have significant consequences for both agricultural productivity and global climate dynamics.

One of the most important aspects of soil gas exchange is oxygen availability. Adequate aeration ensures that plant roots and soil microorganisms can perform aerobic respiration, a process essential for energy production and metabolic functions. When oxygen levels are sufficient, roots remain healthy, and beneficial microbial populations thrive. However, in compacted or waterlogged soils, where gas diffusion is restricted, oxygen depletion can occur, leading to anaerobic conditions. In such environments, plant roots may suffer from hypoxia, resulting in reduced growth, impaired nutrient uptake, and increased susceptibility to disease. Additionally, anaerobic conditions favor microbial pathways that produce alternative metabolic byproducts, which can be detrimental to plant health.

Closely linked to oxygen dynamics is the accumulation of carbon dioxide in soils. As a natural byproduct of root respiration and microbial decomposition, CO_2 levels fluctuate depending on soil aeration and biological activity. In well-aerated soils, CO_2 readily diffuses out, maintaining a stable balance. However, in poorly ventilated soils, CO_2 can accumulate to levels exceeding 10% (100,000 ppm), creating a toxic environment for roots and shifting microbial communities toward anaerobic processes. Prolonged CO_2 buildup can alter soil *pH*, affecting nutrient solubility and availability influencing plant nutrition and microbial interactions.

Beyond its effects on plant growth, soil gas exchange plays a pivotal role in nutrient cycling, particularly in the nitrogen cycle. The presence or absence of oxygen determines the dominant microbial processes occurring in soil. Under aerobic conditions, nitrification—the microbial conversion of ammonium (NH_4^+) into nitrate (NO_3^-)—proceeds efficiently, providing plants with a bioavailable nitrogen source. In contrast, oxygen-limited soils favor denitrification, a process in which nitrate is reduced to nitrous oxide (N_2O) or even nitrogen (N_2) gas, leading to nitrogen losses from the soil. These shifts in nitrogen cycling directly impact soil fertility and the effectiveness of fertilization strategies in agriculture.

Another critical environmental consequence of soil gas exchange is its contribution to greenhouse gas emissions. Soils serve as both sources and sinks for key greenhouse gases, including methane (CH_4) and nitrous oxide (N_2O). The production of these gases is highly dependent on soil conditions, organic matter content, and microbial activity. Methane emissions are particularly pronounced in waterlogged soils such as wetlands, rice paddies, and poorly drained agricultural lands, where methanogenic bacteria thrive under anoxic conditions. Nitrous oxide, a greenhouse gas with a global warming potential almost 300 times greater than CO_2, is released primarily from soils with high nitrogen inputs, particularly in response to fertilization, irrigation, and microbial denitrification. The management of soil aeration, moisture levels, and organic inputs is therefore essential in mitigating emissions and promoting sustainable land use practices. In sections 12.5.1 and 12.5.2, the reader can find more details about the production and emission rate of methane and nitrous oxide in the atmosphere.

Overall, soil gas exchange is a dynamic process that extends beyond plant and microbial interactions, influencing ecosystem functions, agricultural productivity, and global climate systems. Understanding and managing the factors that regulate oxygen diffusion, CO_2 accumulation, and greenhouse gas emissions is critical for maintaining soil health, enhancing crop performance, and reducing the environmental footprint of agricultural practices.

12.5.1 Methane Production and Emission Dynamics

Methane emissions are primarily associated with anaerobic environments, where oxygen depletion forces microbial communities to switch from aerobic respiration to alternative electron acceptors. In this context, methanogenesis becomes a dominant process, carried out by methanogenic archaea that metabolize carbon substrates into methane through either of the following pathways:

$$CO_2 + 4H_2 \rightarrow CH_4 + 2H_2O \tag{12.9}$$

$$CH_3COO^- + H^+ \rightarrow CH_4 + CO_2 \tag{12.10}$$

Equation (12.10) is also known as the acetoclastic pathway of methanogenesis, where acetate is subjected to a dismutation (or disproportionation) reaction, resulting in the simultaneous oxidation of the carboxyl group to carbon dioxide and the reduction of the methyl group to methane[2]. Methane production is strongly influenced by redox potential and soil water content. Typically, E_h values below –200 mV favor methanogenesis, as alternative electron acceptors such as nitrate, iron (III), sulfate, and manganese (IV) are depleted. Waterlogged environments such as peatlands, rice paddies, and compacted agricultural soils majorly contribute to methane emissions.

The emission rate of methane can be modeled using a first-order kinetic equation:

$$E_{CH_4} = k_m \cdot [CH_4] \tag{12.11}$$

where E_{CH_4} is the methane emission rate (in mol m^{-2} s^{-1}), k_m is the methanogenesis rate constant (in s^{-1}), and $[CH_4]$ is the methane concentration (in mol dm^{-3}). Given the high global warming potential of methane, soil management strategies that enhance aeration (e.g., drainage improvements and reduced compaction) are key to mitigating emissions.

12.5.2 Nitrous Oxide Emissions from Nitrification and Denitrification

Nitrous oxide is produced through two main microbial processes:

(a) Nitrification is an aerobic process where ammonium is oxidized to nitrate by autotrophic bacteria.

[2] Kotsyurbenko, O.R., M.V. Glagolev, A.Y. Merkel, A.F. Sabrekov and I.E. Terentieva. 2019. Methanogenesis in soils, wetlands, and peat. *In*: A. Stams, D. Sousa. (eds). Biogenesis of Hydrocarbons. Handbook of Hydrocarbon and Lipid Microbiology. Springer, Cham. https://doi.org/10.1007/978-3-319-78108-2_9

(b) Denitrification is an anaerobic process where nitrate is reduced to nitrous oxide and nitrogen gas under low-oxygen conditions.

The balance between these two processes depends on soil oxygen levels and moisture content. In waterlogged or compacted soils, oxygen availability is reduced, shifting microbial metabolism toward denitrification, leading to greater N_2O emissions. The rate of nitrous oxide release can be modeled as:

$$E_{N_2O} = k_d \cdot \frac{[NO_3^-]}{P_{O_2}} \tag{12.12}$$

Here, E_{N_2O} is the nitrous oxide emission rate (mol m^{-2} s^{-1}), k_d is denitrification rate constant (s^{-1}), $[NO_3^-]$ is the nitrate concentration (in mol dm^{-3}), and P_{O_2} is the oxygen pressure measuring the actual concentration of this gas. Since nitrous oxide is a powerful greenhouse gas, excessive fertilization, and poor drainage conditions can significantly increase emissions. Agricultural strategies such as precision nitrogen management, cover cropping, and controlled irrigation can help mitigate these effects.

12.6 SOIL AERATION AND ITS BIOLOGICAL IMPORTANCE

Soil aeration plays a fundamental role in maintaining plant health, supporting microbial activity, and regulating key biogeochemical processes. The movement of oxygen through the soil matrix influences the respiration of both roots and microorganisms, while the balance between oxygen availability and carbon dioxide accumulation determines whether soils function under aerobic or anaerobic conditions. These dynamics are closely tied to the oxidation-reduction (redox) processes that govern the availability of essential nutrients and the transformation of organic matter, as discussed in Chapter 6, where the fundamental principles of E_h and redox reactions in soils are introduced. To better quantify the efficiency of gas exchange and redox status, indicators such as the oxygen diffusion rate (ODR) and soil redox potential (E_h) provide valuable insights into soil aeration dynamics and oxidative-reductive conditions.

12.6.1 Oxygen Diffusion and Root Health

Oxygen diffusion into the soil is essential for aerobic respiration, enabling plants and microorganisms to metabolize energy efficiently. However, the ability of oxygen to move through the soil is not uniform and depends on factors such as porosity, moisture content, and compaction. The oxygen diffusion rate (ODR) serves as a key measure of how efficiently oxygen reaches plant roots and microbial communities, expressed as:

$$ODR = D \cdot \frac{\Delta C}{L} \tag{12.13}$$

where D is the diffusion coefficient, ΔC is the oxygen concentration gradient, and L is the diffusion path length. A low ODR suggests that the oxygen supply is insufficient, leading to hypoxic conditions that impair root respiration and favor anaerobic microbial activity. This shift in soil conditions can have cascading effects, influencing nutrient transformations and microbial populations.

Closely linked to ODR is the soil redox potential (E_h), which reflects soil's ability to accept or donate electrons, as discussed in Chapter 7, where redox-controlled reactions are analyzed in greater detail. In well-aerated soils, where oxygen is abundant, E_h values typically range between +400 to +700 mV, promoting aerobic microbial activity and essential processes like nitrification. However, when ODR decreases—due to excessive moisture, compaction, or prolonged saturation—oxygen is rapidly consumed, and E_h values drop below +300 mV, creating an environment that favors denitrification, sulfate reduction, and methanogenesis. Under such reducing conditions, alternative electron acceptors, such as nitrate (NO_3^-), sulfate (SO_4^{2-}), and even carbon dioxide, become dominant in microbial respiration, leading to the production of greenhouse gases like nitrous oxide (N_2O) and methane (CH_4). These redox transformations are key components of soil chemistry and have significant environmental implications, as previously detailed in Chapter 7.

Various empirical studies have demonstrated a positive correlation between oxygen diffusion rate and redox potential[3]. This relationship can be approximated by the following empirical equation:

$$E_h = a \cdot \log_{10}(ODR) + b \tag{12.14}$$

where the constants a, and b depend on soil properties and environmental conditions. This relationship highlights that as ODR increases, E_h also rises, reinforcing the idea that enhanced oxygen diffusion promotes an oxidizing environment conducive to plant growth and microbial stability. Conversely, when oxygen movement is restricted, redox potential declines, shifting the soil toward reducing conditions that alter nutrient availability and gas emissions. The role of these redox transformations in nutrient cycling, particularly the fate of nitrogen, sulfur, and iron compounds, has been extensively discussed in Chapter 7, where their influence on soil fertility and environmental quality is explored.

12.6.2 CO_2 Accumulation and Soil Respiration

The exchange of gases in soil is not only defined by oxygen availability but also by carbon dioxide dynamics, which serve as a crucial indicator of biological activity.

---

[3] Callebaut, F., et al. 1982. Redox potential, oxygen diffusion rate, and soil gas composition in relation to water table level in two soils. Soil Science. 134(3): 149–156.
– Melke, J. and St. Uziak. 1989. Dynamics of moisture, redox potential and oxygen diffusion rate of some soils from Calypsostranda, Spitsbergen. Polish Polar Research. 10: 91–104.
– Reiser, R., et al. 2020. System for quasi-continuous simultaneous measurement of oxygen diffusion rate and redox potential in soil. Journal of Plant Nutrition and Soil Science. 183: 316–326.
– Dorau, K., et al. 2021. Soil aeration and redox potential as function of pore connectivity unravelled by X-ray microtomography imaging. European Journal of Soil Science. 73: e13165.

CO_2 is continuously produced through root respiration and microbial decomposition of organic matter, and while its presence is natural, excessive accumulation signals poor aeration and impaired gas exchange. In well-ventilated soils, CO_2 diffuses freely into the atmosphere, maintaining a stable balance. However, in compacted or waterlogged soils, CO_2 can accumulate up to concentrations exceeding 10% (100,000 ppm), suppressing oxygen uptake by roots and leading to physiological stress.

To quantify microbial contributions to CO_2 production, the soil respiration rate (Rs) is used as a measure of biological activity, modeled as:

$$Rs = k \cdot M \cdot T \tag{12.15}$$

where k represents the temperature coefficient, M the microbial biomass, and T the temperature in Kelvin. In particular, M can be quantified either as Microbial Biomass Carbon (MBC)—which is expressed as $mg_{\text{microbial } C} \cdot \text{kg}_{\text{soil}}^{-1}$— or Microbial Biomass Nitrogen (MBN)—which is given as $mg_{\text{microbial } N} \cdot \text{kg}_{\text{soil}}^{-1}$. This means that the amount of microbial biomass present in the soil is commonly assessed through the measurement of its carbon or nitrogen content.

Higher respiration rates generally indicate active microbial decomposition, but under restricted oxygen conditions, CO_2 accumulation can signal an imbalance between aerobic and anaerobic metabolic pathways. When coupled with a declining E_h, excessive CO_2 suggests a transition to anaerobic conditions, promoting processes like fermentation and methanogenesis. These metabolic shifts are intimately connected to the redox-sensitive microbial pathways described in Chapter 7, where the microbial transformations of carbon and nitrogen under varying Eh conditions are explored.

The interplay between ODR, E_h, and CO_2 fluxes provide a comprehensive understanding of soil aeration status. Well-aerated soils maintain high ODR and E_h, ensuring efficient gas exchange, balanced microbial activity, and stable nutrient cycling. In contrast, soils with limited oxygen diffusion experience declining redox potential, increased CO_2 accumulation, and a shift toward anaerobic microbial processes, which can lead to nutrient losses, toxic compound formation, and elevated greenhouse gas emissions. Managing soil structure, moisture, and organic inputs is therefore critical in maintaining optimal aeration and minimizing negative environmental impacts. By integrating knowledge from Chapter 6 on redox principles and Chapter 7 on specific soil redox reactions, a deeper understanding of the dynamic interactions between soil aeration, redox chemistry, and microbial ecology can be achieved.

12.7 SOIL MANAGEMENT FOR OPTIMAL GAS EXCHANGE

The delicate balance of oxygen, carbon dioxide, and other gases in the soil is essential for maintaining plant health, sustaining microbial activity, and preventing the buildup of harmful anaerobic conditions. Effective soil management practices can significantly enhance gas exchange, ensuring an environment that supports root respiration and nutrient cycling while minimizing greenhouse gas emissions.

One of the most critical factors influencing soil aeration is compaction, often caused by heavy machinery and intensive tillage. When soil pores are compressed, oxygen diffusion is severely restricted, accumulating methane and nitrous oxide—potent greenhouse gases produced under oxygen-limited conditions. Strategies such as controlled traffic farming and reduced tillage methods help preserve soil structure, maintaining adequate pore spaces for air movement.

Water management also plays a pivotal role in gas exchange. Waterlogged soils create prolonged anaerobic conditions, favoring denitrification and methanogenesis, processes that contribute to elevated emissions of nitrous oxide and methane. Implementing proper drainage systems and carefully regulating irrigation can help prevent excessive moisture accumulation, maintaining a more oxygen-rich soil environment. Soil structure and porosity can be further improved by adding organic matter, such as compost, mulch, and plant residues. These amendments enhance aggregation, creating macropores facilitating oxygen diffusion and carbon dioxide release. At the same time, optimizing fertilizer application is crucial in preventing excessive nitrogen buildup, which fuels denitrification and associated N_2O emissions. Precision nitrogen management, by tailoring fertilizer application to crop needs, helps minimize nitrogen losses and greenhouse gas emissions without compromising soil fertility. By integrating these soil management strategies, it is possible to create a well-aerated, biologically active soil environment that not only enhances plant growth but also mitigates the negative environmental impacts associated with poor gas exchange.

12.8 IMPLICATIONS FOR CARBON SEQUESTRATION AND CLIMATE CHANGE

Soils are dynamic reservoirs of carbon, acting as both a source and a sink for greenhouse gases. Their role in climate change mitigation is directly linked to their ability to regulate organic matter decomposition, carbon sequestration, and releasing gases such as CO_2, CH_4, and N_2O. Well-aerated soils promote organic matter breakdown under aerobic conditions, forming stable soil organic carbon. In contrast, oxygen-deficient environments slow organic matter decomposition, allowing methane and nitrous oxide to accumulate—further amplifying global warming.

Agricultural practices can be tailored to maximize carbon sequestration while reducing greenhouse gas emissions. One effective strategy is cover cropping, which increases soil carbon inputs, improves aeration, and stimulates root respiration, all of which enhance microbial activity and promote organic matter stabilization. Similarly, reducing tillage preserves soil structure and fosters an environment that supports carbon stabilization within soil aggregates.

The use of biochar amendments (which will be discussed in Chapter 13) presents another promising approach. Biochar, a carbon-rich material from organic matter pyrolysis, improves soil carbon retention and enhances gas exchange, reducing the likelihood of methane and nitrous oxide formation in anaerobic pockets. Additionally, peatland management plays a crucial role in global carbon dynamics. Drained peatlands are major emitters of CO_2 due to the rapid decomposition of accumulated organic matter. However, rewetting these ecosystems can help restore their function as long-term carbon sinks, locking carbon within the soil rather than allowing it to escape into the atmosphere.

By adopting climate-smart soil management practices, agricultural systems can become more resilient to climate change while simultaneously reducing emissions. These strategies not only enhance soil fertility and water retention but also contribute to global efforts to mitigate the impact of anthropogenic greenhouse gases.

EXERCISES

1. A well-aerated soil typically has an oxygen concentration of 18% and a carbon dioxide concentration of 0.5% at a certain depth. However, in compacted or waterlogged soil, oxygen drops to 3%, and carbon dioxide increases to 5%.
   (a) Explain the possible reasons for these changes in gas composition.
   (b) Describe how these altered conditions would affect microbial communities and plant root respiration.
   (a) What soil management strategies could be implemented to improve aeration in this scenario?
2. Using Fick's First Law, the flux of oxygen in soil can be estimated by the equation (1). A researcher measures the oxygen concentration at two different depths in sandy soil:
   - At a depth of 5 cm, the oxygen concentration is 18%.
   - At a depth of 15 cm, the oxygen concentration is 5%.

   Assuming D in the soil is 1.2×10^{-6} m^2s^{-1}, calculate the oxygen flux between these two depths. Interpret the result and discuss how soil texture influences gas diffusion.
3. The solubility of a gas in soil water is governed by Henry's Law [equation (12.7)]. Given that the Henry's constant for carbon dioxide in water at 25°C is 3.3×10^{-2} mol L^{-1} atm^{-1}, determine the concentration of dissolved carbon dioxide in soil water when its partial pressure in soil air is 0.04 atm.
   (a) How does increasing soil moisture affect CO_2 dissolution?
   (b) What happens when the temperature rises, and how might this impact greenhouse gas emissions from soil?
4. Methane is produced in anaerobic soils through methanogenesis according to the reaction (12.9). The methane emission rate can be modeled using a first-order kinetic equation (11). If the methane concentration in an

anaerobic wetland soil is 50 ppm, and the methanogenesis rate constant is 0.02 s^{-1}, calculate the methane emission rate. How would increasing soil aeration influence methane emissions?

5. Redox potential and oxygen diffusion rate (ODR) are correlated through the empirical relationship (14). In well-aerated soil, ODR is 30 mg O_2 cm^{-2} min^{-1}, and the measured redox potential is +600 mV.
   (a) If in a waterlogged soil, ODR drops to 5 mg O_2 cm^{-2} min^{-1}, estimate the new redox potential assuming a = 100 and b = 400.
   (b) How do these changes in ODR and E_h affect microbial processes such as nitrification and denitrification?
   (c) What practical measures can be taken to improve redox conditions in flooded agricultural fields?

FURTHER READING

Jacob, D.J. 1999. Introduction to Atmospheric Chemistry, 1st Ed. Princeton University Press. ISBN: 0691001855

Xue, L., X. Guo and H. Chen. 2020. Fluid Flow in Porous Media, 1st Ed. World Scientific Connect. https://doi.org/10.1142/11805. ISBN: 9789811219528

Weil, R.R. and N.C. Brady. 2017. The Nature and Properties of Soils, 15th Ed. Pearson Prentice-Hall. ISBN: 9781292162232

Chapter **13**

Soil Fertility

13.1 INTRODUCTION

This book has explored the physical and chemical properties that define soils, using a rigorous scientific approach grounded in physical chemistry principles. The discussion began by examining the scientific method applied to soil science, emphasizing the importance of objectivity, reproducibility, and interdisciplinary integration. From there, the fundamentals of soil composition were introduced, presenting soil as an open system where the atmosphere, hydrosphere, lithosphere, and biosphere interact. This provided a foundation for understanding the complex processes that govern soil behavior.

A significant focus has been placed on water and its interactions with soil, covering both its physical and chemical properties, including hydrogen bonding, solvation, and autoionization, as well as the movement, retention, and availability of water within soil matrices. Chemical equilibria, in both ideal and non-ideal conditions, have been analyzed, providing insights into acid-base reactions, solubility equilibria, redox transformations, and buffering systems, all of which play a critical role in soil stability and nutrient cycling.

Considerable attention has also been devoted to soil colloids, their electrochemical properties, adsorption mechanisms, and interactions with nutrients and contaminants. Both the physical behavior of colloids, governed by electrical double-layer models such as Gouy-Chapman and Stern, and their chemical composition, particularly in relation to clay minerals and soil organic matter, have been examined. These insights were further expanded to include cation exchange capacity (CEC), organic matter transformations, and the role of humic substances in soil chemistry.

Another essential aspect addressed in this book is soil aeration and gas dynamics, often referred to as the telluric air. The movement of gases within soil, including oxygen diffusion, carbon dioxide accumulation, microbial respiration, and redox-driven transformations, has been discussed. The role of soil aeration

DOI: 10.1201/9781003780212-13

in greenhouse gas emissions and nutrient cycling has been analyzed, providing a deeper understanding of how physical and chemical interactions regulate biological activity in soils.

With these foundational principles established, the focus now shifts to soil fertility, which represents the practical expression of the physicochemical processes governing soil function. Fertility is not simply a measure of nutrient abundance but rather a dynamic property that reflects the soil's ability to store and release essential elements, sustain biological activity, and maintain an optimal balance between solid, liquid, and gaseous phases. The ability of soil to support plant growth, cycle nutrients, and regulate environmental interactions is ultimately governed by the chemical equilibria, colloidal properties, and water retention dynamics discussed in the preceding chapters.

This chapter aims to provide a scientifically rigorous definition of soil fertility, rooted in the physical chemistry of soils, while also offering insights into how fertility can be maintained and enhanced through an understanding of fundamental soil processes. Rather than approaching fertility solely from an agronomic perspective, it will be examined through the lens of physical and chemical principles, demonstrating that soil fertility is an emergent property arising from the intricate interplay of soil composition, structure, and reaction dynamics.

13.2 WHAT DOES SOIL FERTILITY MEAN?

Soil fertility represents soil inherent capacity to sustain plant growth and facilitate productive agricultural systems. It is a dynamic property governed by the interplay of physical, chemical, and biological factors, each of which has been explored throughout the preceding chapters.

At its core, fertility is not just about the presence of nutrients but also about their availability, retention, and cycling within the soil matrix. The physical properties of soil, including texture, structure, and porosity, dictate how well it can retain water and air—both essential for root function and microbial activity. The chemical composition of soils, from *pH* levels to cation exchange capacity, determines how plants store, release, and absorb nutrients. Meanwhile, the biological component, including microbial life and soil fauna, plays a crucial role in organic matter decomposition, nitrogen fixation, and overall soil regeneration.

Fertile soils not only provide nutrients but also buffer environmental stresses, regulating water supply, resisting erosion, and maintaining a stable chemical equilibrium. However, fertility is not a static attribute—it is continuously shaped by natural processes and human interventions. Agricultural practices, climate variability, and soil management strategies all influence soil's ability to remain productive over time. Understanding these interactions is essential for maintaining and enhancing soil fertility through informed management practices.

13.3 KEY NUTRIENTS AND THEIR ROLE IN PLANT GROWTH

Soil fertility is intrinsically linked to the availability and balance of essential nutrients, which plants require in varying amounts to complete their life cycle. These nutrients, derived from soil minerals, organic matter decomposition, and atmospheric deposition, are classified into macronutrients and micronutrients based on the quantities needed by plants. While carbon, nitrogen, phosphorus (P), and sulfur cycles in soil have been discussed in Chapter 11, this section focuses on the role of primary and secondary nutrients in sustaining plant growth.

13.3.1 Macronutrients: The Structural and Functional Pillars of Plant Growth

Macronutrients are required in large quantities (typical content intervals present in plant tissues are reported in Table 13.1) and play fundamental roles in plant metabolism, energy transfer, and structural integrity. The roles of the different macronutrients are reported below.

Table 13.1 Typical amounts of macronutrients in plant tissues. Carbon, hydrogen, and oxygen have been kept away from this table

Nutrient	Typical Range in Plant Tissues (% of Dry Weight)
Nitrogen	1.5–5.0
Phosphorus	0.1–0.5
Potassium	1.0–5.0
Calcium	0.2–3.0
Magnesium	0.1–1.0
Sulfur	0.1–0.5

- **Nitrogen:** A critical component of amino acids, proteins, and nucleic acids, nitrogen is indispensable for plant growth. It is the central element in chlorophyll molecules, facilitating photosynthesis and biomass accumulation. Nitrogen deficiency leads to stunted growth, chlorosis (yellowing of leaves), and reduced crop yields, while excessive nitrogen application can result in nutrient imbalances, leaching, and environmental pollution.
- **Phosphorus:** It is essential for energy transfer, and phosphorus is a key element in ATP (adenosine triphosphate) and nucleotides (Figure 13.2), supporting cell division, root development, and flowering. Since phosphorus availability is strongly influenced by soil *pH* and mineral interactions, its deficiency often manifests as poor root growth and delayed plant maturity.
- **Potassium:** Unlike nitrogen and phosphorus, potassium is not incorporated into organic molecules but functions as a regulator of osmotic balance, improving plant drought resistance, enzyme activation, and disease

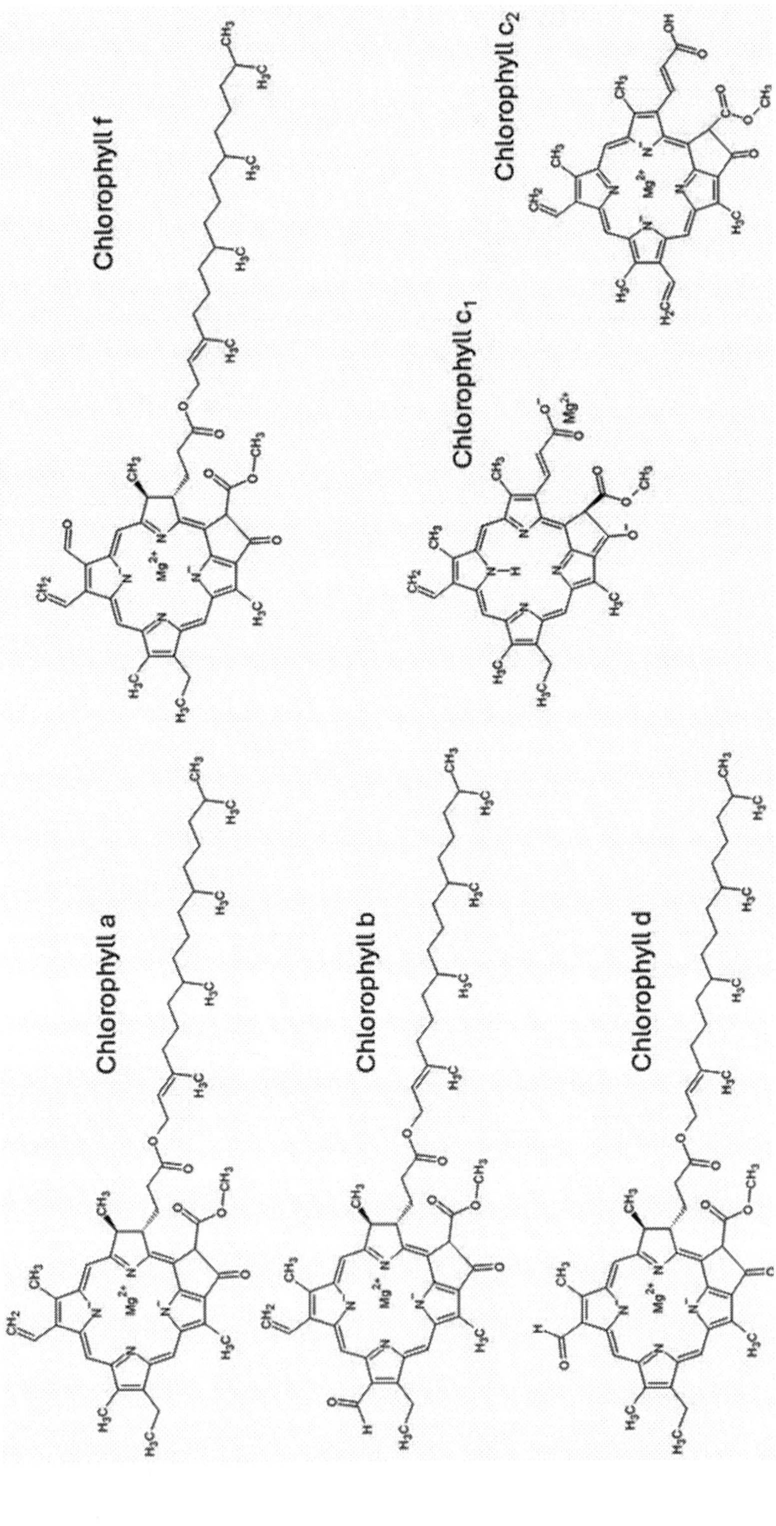

Figure 13.1 Structures of the different types of chlorophylls.

tolerance. Soils rich in clay minerals typically have a higher potassium retention capacity, while sandy soils often require supplementation.

- **Calcium:** As a key component of cell walls, calcium provides structural stability to plant tissues and is vital for cell division and membrane permeability. Calcium deficiencies, common in acidic and highly leached soils, result in deformed new growth, weak root systems, and blossom-end rot in fruits.
- **Magnesium:** Magnesium is the central atom in chlorophyll (Figure 13.1), and it plays an indispensable role in photosynthesis. It is also involved in enzyme activation, protein synthesis, and phosphorus metabolism. Its deficiency, particularly in highly weathered or acidic soils, leads to interveinal chlorosis and reduced energy efficiency in plants.
- **Sulfur:** Required for protein synthesis and enzyme function, sulfur contributes to chlorophyll formation and nitrogen assimilation. It is particularly significant in leguminous plants, which rely on sulfur for nitrogen fixation. Deficiencies result in pale green young leaves and poor overall growth.

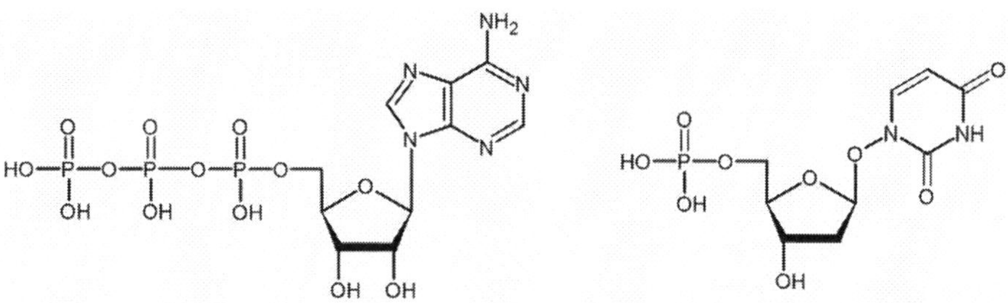

Adenosine triphoshpate (ATP) Deoxycytidylic acid (dCMP)

Figure 13.2 Examples of phosphorus-containing biomolecules. ATP (left) is a nucleotide that serves as the primary energy carrier in cells, driving metabolic reactions and cellular processes through the hydrolysis of its phosphate bonds. dCMP (right) is a deoxyribonucleotide that serves as one of the building blocks of DNA. It contains the nitrogenous base cytosine, a deoxyribose sugar, and a phosphate group. While dCMP is specifically a cytosine-containing nucleotide, different types of nucleotides, such as dAMP (deoxyadenylate), dGMP (deoxyguanylate), and dTMP (deoxythymidylate), also contribute to the genetic material of plants and other organisms.

13.3.2 Micronutrients: The Catalysts of Physiological Processes

Although needed in smaller amounts (typical micronutrient amount intervals present in plant tissues are given in Table 13.2), micronutrients are critical for enzymatic reactions, photosynthesis, and hormonal regulation. Even in trace amounts, deficiencies can severely affect plant health and yield. A list and role of the main micronutrients is given below.

Table 13.2 Typical amounts of micronutrients in plant tissues

Nutrient	Typical Range in Plant Tissues (mg kg^{-1})
Iron	50–250
Zinc	20–100
Copper	5–20
Manganese	20–300
Boron	10–100
Molybdenum	0.1–5

- **Iron:** A crucial element in chlorophyll synthesis and electron transport during photosynthesis, iron plays a central role in redox reactions and metabolic pathways. Iron availability is often limited in alkaline soils with high carbonate content, leading to iron chlorosis, characterized by yellowing between leaf veins.
- **Zinc:** Essential for enzyme activation, protein synthesis, and hormone regulation, zinc contributes to auxin production, which is responsible for plant growth and elongation. Zinc deficiency often causes stunted growth and leaf distortion, particularly in calcareous soils.
- **Copper:** Integral to photosynthetic enzymes and lignin formation, copper enhances plant respiration and reproductive processes. Copper-deficient soils result in twisted young leaves, delayed flowering, and weak stems.
- **Manganese:** This element is a cofactor for multiple enzymatic reactions involved in photosynthesis, nitrogen metabolism, and lignin synthesis. Manganese deficiencies lead to chlorotic spots on leaves and poor disease resistance.
- **Boron:** A critical element for cell wall formation, sugar transport, and pollen germination, boron ensures fruit and seed development. Deficiencies are evident in deformed leaves, hollow stems, and reduced flower viability.
- **Molybdenum:** An essential component of enzymes responsible for nitrogen fixation and nitrate reduction, molybdenum supports protein synthesis and root growth. It is particularly vital in legume crops, where its deficiency can impede symbiotic nitrogen fixation, leading to poor nodulation and yellowed leaves.

13.4 STRATEGIES FOR MAINTAINING AND IMPROVING SOIL FERTILITY: FERTILIZERS AND AMENDMENTS

Soil fertility management is a complex and dynamic process that relies on the appropriate use of fertilizers and amendments, both of which play essential roles in sustaining plant growth, maintaining soil health, and ensuring long-term agricultural productivity. One of the fundamental challenges in agriculture is that crop production inherently depletes soil nutrients. When plants grow, they extract essential elements from the soil to build their tissues, and when crops are harvested, those nutrients are removed from the system. Unlike in natural ecosystems, where organic matter eventually decomposes and nutrients return to the soil, in agricultural settings, harvested products are transported elsewhere—consumed, processed, or used for other purposes—without replenishing the original soil. Over time, this continuous removal of nutrients leads to soil exhaustion, making it necessary to restore fertility through external inputs.

Farmers rely on fertilizers and amendments to counteract this loss and sustain soil productivity, which, though sometimes used interchangeably, have distinct roles in soil science.

Fertilizers are primarily designed to supply essential nutrients directly to plants, enhancing their growth and improving crop yields. They can be broadly categorized into synthetic (chemical) fertilizers and organic fertilizers. Chemical fertilizers are formulated to provide nutrients in highly bioavailable forms, allowing for rapid plant uptake and precise nutrient management. They typically contain macronutrients such as nitrogen (N), phosphorus (P), and potassium (K) in well-defined ratios, ensuring that plants receive the necessary elements for their metabolic processes. However, while these fertilizers quickly replenish soil nutrients, they do not contribute to soil organic matter or microbial life, which are essential for long-term soil health. In contrast, organic fertilizers, such as manure and compost, release nutrients more gradually as they undergo microbial decomposition. Beyond their nutritional contribution, they also improve soil organic matter content, enhance microbial diversity, and boost soil structure and water retention, making them valuable for both immediate nutrient supply and long-term soil improvement.

While fertilizers primarily function as nutrient sources, amendments are applied with the specific goal of modifying and enhancing the soil's physical, chemical, or biological properties. Rather than directly providing plant nutrients, amendments improve the overall soil environment, making it more conducive to plant growth. Some common amendments include biochar, lime, gypsum, and organic matter-based materials. Lime is widely used to neutralize soil acidity, thereby improving nutrient availability for plants, while gypsum is often applied to enhance soil structure and permeability, promoting root penetration and reducing the risks of compaction and waterlogging.

However, the distinction between fertilizers and amendments is not always clear-cut. Some materials can function as both, depending on their composition and how they are applied. Biochar, for instance, is traditionally considered an amendment due to its ability to improve soil porosity, water retention, and cation exchange capacity. However, when it is chemically functionalized, such as by enriching it with nitrogen or phosphorus, it can also act as a slow-release fertilizer, gradually supplying essential nutrients while simultaneously improving soil structure and microbial habitat.

Understanding the interactions between fertilizers, amendments, and soil processes is fundamental to sustainable soil management. The way these materials are used can significantly influence soil fertility dynamics, microbial activity, and nutrient cycling, making it essential to adopt an integrated approach that balances nutrient input with soil enhancement. In the following sections, we will explore different types of fertilizers and amendments in greater detail, examining their benefits, limitations, and contributions to the long-term health and productivity of soils.

13.4.1 Chemical Fertilizers: Synthesis and Function

Chemical fertilizers are synthetically produced materials specifically designed to provide plants with nutrients in highly bioavailable forms, ensuring rapid absorption and efficient use. Unlike organic sources, which rely on gradual microbial decomposition to release nutrients, chemical fertilizers deliver precise amounts of essential macronutrients directly to the soil, making them a key component in modern agriculture. They are generally classified based on their primary nutrient content, with the three major categories being nitrogen-based fertilizers, phosphorus-based fertilizers, and potassium-based fertilizers. Each of these plays a crucial role in plant metabolism, supporting growth, energy transfer, and physiological functions.

Nitrogen fertilizers, among the most widely used, are produced through the Haber-Bosch process, a landmark industrial method that revolutionized agricultural productivity. This process involves the reaction of atmospheric nitrogen (N_2) with hydrogen (H_2) under high pressure and temperature in the presence of an iron-based catalyst, yielding ammonia (NH_3):

$$N_2 + 3H_2 \rightarrow 2NH_3$$

Ammonia serves as a foundational compound in nitrogen fertilizer production and can be further processed into urea ($CO(NH_2)_2$) or ammonium nitrate (NH_4NO_3), both of which are highly soluble and readily available to plants. These compounds provide the necessary nitrogen for protein synthesis, chlorophyll formation, and enzymatic functions, making them indispensable in plant nutrition.

Phosphorus fertilizers originate from phosphate rock, primarily composed of fluorapatite ($Ca_5(PO_4)_3F$), which must be chemically treated to render its phosphorus content bioavailable. The most common industrial method involves the application of sulfuric acid, resulting in superphosphate:

$$Ca_5(PO_4)_3\,F + H_2SO_4 \rightarrow Ca(H_2PO_4)_2 + CaSO_4 + HF$$[1]

[1] One of the key byproducts of the reaction between fluorapatite and sulfuric acid in the production of phosphate fertilizers is hydrofluoric acid. This reaction, essential for making superphosphate fertilizers, releases *HF* as a gaseous or aqueous byproduct, which, if not properly managed, poses significant health and environmental risks due to its extreme corrosiveness and toxicity. However, rather than being wasted, *HF* is an important industrial compound with numerous applications across different sectors. In the chemical industry, *HF* serves as a crucial precursor for synthesizing a wide range of fluorine-containing compounds, including fluoropolymers such as polytetrafluoroethylene (PTFE, commonly known as Teflon), which is widely used

This reaction breaks down the stable mineral structure, releasing soluble phosphorus compounds that plants can absorb efficiently. Phosphorus is vital for energy transfer within plant cells, as it is a key component of ATP (adenosine triphosphate), nucleic acids, and membrane lipids, playing an essential role in root development, flowering, and seed production.

Unlike nitrogen and phosphorus sources, potato fertilizers are not synthetically produced but instead extracted from natural potash deposits. The most common forms are potassium chloride (KCl) and potassium sulfate (K_2SO_4), both of which undergo purification and refinement to remove impurities before being used as fertilizers. Potassium is crucial in regulating water balance within plant cells, activating enzymes, and enhancing disease resistance, contributing significantly to overall plant health and stress tolerance.

While chemical fertilizers provide a highly efficient means of replenishing soil nutrients, their excessive or unbalanced application can lead to negative environmental and soil health consequences. Overuse of nitrogen fertilizers, for example, can result in soil acidification, nitrate leaching into groundwater, and disruptions to microbial communities. Similarly, phosphorus runoff contributes to eutrophication, leading to algal blooms in water bodies, while excessive potassium can affect cation exchange dynamics, potentially altering soil structure. Therefore, the integration of responsible fertilization strategies, such as precision application,

---

for non-stick coatings, high-performance plastics, and chemical-resistant materials. Additionally, *HF* is fundamental in producing fluorinated refrigerants, such as hydrofluorocarbons (HFCs) and hydrofluoroolefins (HFOs), essential in air conditioning and refrigeration systems. Another major application of *HF* is in glass etching and semiconductor manufacturing, where its ability to dissolve silicates and silicon dioxide (SiO_2) makes it indispensable for producing precision glass, microchips, and electronic components. The same property is also exploited in the petroleum refining industry, where HF acts as a catalyst in alkylation processes, producing high-octane gasoline components. In the metal industry, HF is used for stainless steel pickling, a process that removes oxide layers from the metal surface, enhancing its resistance to corrosion. Beyond these applications, HF also plays a significant role in nuclear technology, where it is used in the production of uranium hexafluoride (UF_6), a key compound in the enrichment of nuclear fuel. Furthermore, its incorporation into the pharmaceutical and agrochemical industries has led to the development of fluorinated drugs and pesticides, as fluorine can significantly improve the bioavailability, metabolic stability, and efficacy of active compounds. Given its hazardous nature, handling HF requires strict safety protocols, and many fertilizer production plants capture and convert HF into fluorosilicic acid (H_2SiF_6). This derivative is widely used for the fluoridation of drinking water, the production of aluminum fluoride (AlF_3)—a key additive in the aluminum smelting process—and the synthesis of cryolite (Na_3AlF_6), essential in the electrolytic refining of aluminum. Although hydrofluoric acid is often perceived as a dangerous and challenging byproduct, its widespread use in industrial chemistry, materials science, energy, and pharmaceutical applications highlights its importance as a strategically recovered and repurposed resource rather than a mere waste product. The careful management of *HF* in phosphate fertilizer production not only mitigates environmental risks but also contributes to the supply chain of high-value fluorine-based compounds, reinforcing its role in modern industry.

slow-release formulations, and complementary use of organic amendments, is crucial to maximizing benefits while minimizing detrimental impacts. Sustainable soil fertility management requires an integrated approach that considers both nutrient efficiency and long-term soil health, ensuring that chemical fertilizers are used in a way that supports, rather than undermines, agricultural sustainability.

13.4.2 Organic Fertilizers: Manure and Compost

Unlike synthetic fertilizers, which supply nutrients in readily soluble forms for immediate plant uptake, organic fertilizers follow a different path, releasing nutrients gradually as they decompose in the soil. This slow but sustained release not only nourishes crops over time but also improves soil structure, enhances microbial diversity, and contributes to long-term fertility. Among the most widely used organic fertilizers, manure and compost play a pivotal role in maintaining soil health, yet each undergoes distinct biochemical and physicochemical transformations that influence how nutrients become available to plants.

13.4.2.1 Manure: Physical Chemistry and Soil Interactions

Manure, composed of animal excreta often mixed with bedding materials such as straw or sawdust, is a complex and heterogeneous material. Its nutrient composition varies depending on the animal species, diet, and storage conditions, but it typically contains a mix of organic matter, macronutrients (N, P, K, Ca, Mg, and S), micronutrients (Fe, Zn, Cu, Mn, B, and Mo), and a diverse microbial community that drives its decomposition in the soil.

Once applied, manure undergoes microbial decomposition, triggering a series of biochemical reactions that influence nutrient availability, soil *pH*, and overall soil structure. One of the most crucial transformations is nitrogen mineralization, where the organic nitrogen contained in proteins, amino acids, and nucleic acids is broken down into ammonium (NH_4^+). This ammonium can then be oxidized by nitrifying bacteria into nitrate (NO_3^-), the form most readily absorbed by plants:

$$RNH_2 \rightarrow NH_4^+ \rightarrow NO_3^-$$

While nitrate is essential for plant nutrition, it is also highly soluble and prone to leaching, particularly in sandy soils or regions with heavy rainfall. Proper manure management is therefore essential to prevent nitrogen losses and reduce environmental risks such as groundwater contamination.

Another fundamental process occurring in manure decomposition is carbon breakdown, particularly involving cellulose, hemicellulose, and lignin. As microbes metabolize these organic compounds, carbon dioxide (CO_2) is released, and humic substances form, enriching the soil and increasing its cation exchange capacity, which enhances nutrient retention.

Phosphorus, another key element in manure, is often locked within organic molecules and must be solubilized by microbial activity before plants can absorb it. Phosphatase enzymes released by soil microorganisms break down these organic phosphorus compounds, releasing bioavailable phosphate ions ($H_2PO_4^-$, HPO_4^{2-}) that contribute to plant growth.

Despite its clear benefits, the use of fresh manure comes with potential risks, including nutrient runoff, methane emissions, and the presence of pathogens. Excessive application can also lead to soil salinity and microbial imbalances. For these reasons, manure is often composted before use, a process that stabilizes nutrients and enhances its effectiveness as a soil amendment.

13.4.2.2 Compost: Physical Chemistry and Composting Reactions

Composting is an aerobic decomposition process that transforms raw organic matter into a nutrient-rich, humus-like material. Unlike manure, which decomposes in an uncontrolled manner when applied directly to the soil, composting allows for a controlled breakdown of organic substances, reducing nutrient loss and eliminating pathogens while improving the material's overall stability.

During composting, microbial oxidation of carbon compounds plays a critical role. The decomposition of sugars, starches, and proteins by thermophilic microorganisms generates carbon dioxide (CO_2), water (H_2O), and heat, significantly accelerating the breakdown process:

$$C_6H_{12}O_6 + O_2 \rightarrow CO_2 + H_2O + \text{heat}$$

This reaction illustrates the degradation of simple saccharides such as glucose or fructose, representing a fundamental step in the conversion of organic waste into compost. The heat generated in this process drives compost through three distinct decomposition phases, each characterized by different microbial communities and biochemical activities.

The first phase, the mesophilic phase, is initiated by mesophilic bacteria, actinomycetes, fungi, and protozoa, which thrive in moderate temperatures ranging between 10–45°C. These microorganisms rapidly degrade easily decomposable organic compounds, such as sugars and amino acids, triggering oxidative exothermic reactions that lead to an increase in temperature and *pH*. As the microbial metabolism intensifies, the compost pile shifts toward the second and most active phase of decomposition.

The thermophilic phase begins when temperatures exceed 45–50°C, favoring the proliferation of thermophilic bacteria and fungi that continue to degrade complex organic materials. This high-temperature phase, which can last several weeks, is crucial as it eliminates pathogens, weed seeds, and parasites while accelerating the breakdown of more resistant organic compounds, such as lignin. However, if the temperature surpasses 60°C, microbial diversity decreases significantly, slowing the decomposition of recalcitrant materials and requiring proper aeration management to prevent overheating.

The final stage of composting, known as the cooling and maturation phase, begins once most of the easily degradable organic matter has been consumed. As microbial activity slows, the compost gradually cools down, allowing mesophilic microorganisms to recolonize the material. At this stage, complex organic structures such as lignin undergo further degradation, and pseudo-humic substances begin to form. This phase is essential, as it ensures that the compost stabilizes and matures, preventing the excessive release of nutrients that could otherwise lead to soil imbalances when applied.

Throughout the composting process, nitrogen undergoes multiple transformations, starting with the microbial breakdown of proteins, urea, and uric acid, which release ammonium. Under high temperatures and alkaline conditions ($pH > 7$), ammonium may volatilize as ammonia (NH_3), potentially leading to nitrogen losses. As composting progresses, nitrification occurs, in which ammonium is oxidized into nitrate (NO_3^-), a more plant-accessible form of nitrogen. However, if the compost pile experiences anaerobic conditions, denitrification may take place, resulting in nitrogen loss as nitrous oxide (N_2O) or molecular nitrogen (N_2).

Carbon also undergoes significant transformations during composting. In an aerobic environment, carbon is primarily converted into CO_2 and pseudo-humic substances, enhancing soil stability and nutrient retention. However, in oxygen-limited conditions, certain microbial communities, particularly methanogenic bacteria, facilitate the production of methane (CH_4), a potent greenhouse gas. For this reason, proper aeration management is crucial to prevent the formation of anaerobic zones and minimize methane emissions.

In addition to carbon dioxide and methane, several volatile organic compounds (VOCs) are emitted during composting, including aldehydes, ketones, alcohols, amines, nitriles, mercaptans, and sulfur compounds. These compounds contribute to the distinct odors associated with composting, and while they are generally harmless in open environments, excessive VOC emissions may pose environmental and health concerns. To mitigate these effects, composting facilities often implement controlled aeration and biofiltration systems to manage gaseous emissions effectively.

13.4.2.3 Environmental Considerations and Soil Benefits

Compost application to soils significantly enhances soil structure, microbial activity, and nutrient cycling, making it a powerful tool for sustainable agriculture. However, environmental risks must be considered. Compost-amended soils release nitrogen oxides (N_xO_y) and ammonia (NH_3), contributing to air pollution, while leaching of nitrate (NO_3^-) and ammonium (NH_4^+) into water systems may pose contamination risks if not properly managed. Studies indicate that nitrogen loss through ammonia volatilization can reach 92% of the nitrogen content in compost, highlighting the importance of precision compost application to prevent nutrient depletion.

Despite these challenges, compost remains a key player in sustainable agriculture, not only for its role in nutrient recycling and soil regeneration but also for its potential in bioremediation. Composting techniques have been successfully applied to degrade organic contaminants, such as polycyclic aromatic hydrocarbons (PAHs) and industrial pollutants, contributing to soil detoxification and rehabilitation.

13.4.3 Biochar: From Amendment to Fertilizer[2]

Biochar, a carbonaceous material produced through the pyrolysis of biomass in oxygen-limited conditions, has been widely recognized as a versatile soil amendment due to its ability to improve soil structure, enhance water retention, and support microbial activity. However, its role in soil fertility and plant nutrition extends far beyond these functions. Depending on its physicochemical properties, biochar can act not only as an amendment but also as a fertilizer, particularly when it is functionalized, enriched with nutrients, or co-composted with organic materials.

The dual role of biochar as both an amendment and a potential fertilizer arises from its high porosity, surface reactivity, and ability to retain and gradually release nutrients. These properties are not fixed but are strongly influenced by the feedstock used, the pyrolysis temperature, and the quenching process applied during biochar production. Studies have shown that different quenching methods

[2] All the information about biochar's physical chemistry has been obtained from:

– Kammann, C.I., et al. 2015. Plant growth improvement mediated by nitrate capture in co-composted biochar. Scientific Reports. 5: 11080. https://doi.org/10.1038/srep11080.

– Hagemann, N., et al. 2017. Organic coating on biochar explains its nutrient retention and stimulation of soil fertility. Nature Communications. 8: 1089. https://doi.org/10.1038/s41467-017-01123-0.

– Joseph, S., et al. 2018. Microstructural and associated chemical changes during the composting of a high temperature biochar: Mechanisms for nitrate, phosphate and other nutrient retention and release. Science of the Total Environment. 618: 1210–1223. https://doi.org/10.1016/j.scitotenv.2017.09.200.

– Dieguez-Alonso, A., et al. 2019. Designing biochar properties through the blending of biomass feedstock with metals: Impact on oxyanions adsorption behavior. Chemosphere 214: 743e753. https://doi.org/10.1016/j.chemosphere.2018.09.091.

– Conte, P., et al. 2021. Recent Developments in Understanding Biochar's Physical–Chemistry. Agronomy. 11: 615. https://doi.org/10.3390/agronomy11040615.

– Fellet, G., et al. 2022. Changes in physicochemical properties of biochar after addition to soil. Agriculture. 12: 320. https://doi.org/10.3390/agriculture12030320.

– Pipíška, M., et al. 2022. Biochar from Wood Chips and Corn Cobs for Adsorption of Thioflavin, T. and B. Erythrosine. Materials. 15: 1492. https://doi.org/10.3390/ma15041492.

– Fellet, G., et al. 2023. Biochar effects on Ce leaching and plant uptake in lepidium sativum L. grown on a ceria nanoparticle spiked soil. Applied Sciences. 13: 6846. https://doi.org/10.3390/app13116846

– Di Vincenzo, A., et al. 2025. A Study on the oxidative functionalization of a poplar biochar. Molecules. 30: 1048. https://doi.org/10.3390/molecules30051048.

– Conte, P., et al. 2025. Wood-biochar influence on rill erosion processes and hydrological connectivity in amended soils. Hydrological Processes. 39: e70093. https://doi.org/10.1002/hyp.70093.

– Bagarello, V., et al. 2025. How rilling and biochar addition affect hydraulic properties of a clay-loam soil. European Journal of Soil Science. 76: e70034. https://doi.org/10.1111/ejss.70034.

(dry vs. wet quenching[3]) significantly impact biochar's chemical stability, surface oxidation, water retention capacity, and overall effectiveness in soil.

Beyond its agronomic functions, biochar has gained considerable attention as a tool for carbon sequestration and environmental remediation. It enhances nutrient cycling and soil microbial dynamics, modulates soil *pH*, improves cation exchange capacity, and serves as a sink for organic and inorganic contaminants . Despite its well-documented benefits, biochar's interaction with soil systems remains highly dynamic, undergoing physicochemical transformations upon soil incorporation. These transformations, including surface oxidation, porosity changes, and modifications in hydrophilicity, directly affect its capacity to influence soil fertility and nutrient availability over time.

Given its structural variability, biochar can be optimized for use as a fertilizer substitute or supplement through functionalization processes that enhance its nutrient retention and slow-release capabilities. When enriched with essential macronutrients such as nitrogen, phosphorus, and potassium, biochar can reduce reliance on synthetic fertilizers while promoting long-term soil fertility and sustainable nutrient management.

This section explores the physical chemistry of biochar, its role as an amendment, and the mechanisms through which it can be transformed into a fertilizer. Special attention is given to the processes governing nutrient retention and release, including cation exchange, anion adsorption, microbial interactions, and oxidation-reduction reactions. Furthermore, the impact of biochar functionalization and co-composting on its agronomic effectiveness is examined, shedding light on its potential as a sustainable solution for soil fertility management.

[3] Quenching is the cooling process to terminate pyrolysis and stabilize biochar after production. The two primary methods—wet and dry quenching—lead to distinct differences in biochar's physicochemical properties, affecting its porosity, surface chemistry, and water retention capacity. In dry quenching, biochar can cool gradually at ambient temperature in an oxygen-limited environment. This process promotes continued carbonization, increasing aromaticity and greater thermal stability. Dry quenching also increases surface area and porosity, as volatile compounds have more time to escape. Additionally, it produces more hydrophobic biochar since fewer oxygen-containing functional groups form on the surface. Another key characteristic is that dry-quenched biochar typically has a higher pH due to the more complete breakdown of organic acids during cooling. In wet quenching, biochar is rapidly cooled by adding water or another liquid immediately after pyrolysis. This method abruptly interrupts thermal degradation, leading to a higher retention of labile organic compounds. However, wet quenching results in biochar with lower surface area and porosity, as pores may become clogged with water or residual organic matter. Unlike dry quenching, wet quenching enhances hydrophilicity, as the surface undergoes partial oxidation, forming oxygen-containing functional groups such as carboxyl (–COOH) and hydroxyl (–OH) groups. This process typically produces biochar with a lower pH, depending on the extent of interaction with water and dissolved compounds. As a general summary, dry quenching biochar is generally more stable, more porous, and hydrophobic, making it better for long-term carbon sequestration and soil aeration, whereas wet quenching biochar is more hydrophilic, with greater nutrient retention potential, making it useful for water management and improving soil fertility.

13.4.3.1 The Physical and Chemical Properties of Biochar

The physical and chemical properties of biochar are dictated by a combination of feedstock characteristics, pyrolysis conditions, and post-production treatments. These factors influence biochar's porosity, surface chemistry, elemental composition, and stability, all of which play a crucial role in determining its effectiveness as both a soil amendment and a nutrient carrier.

13.4.3.1.1 Porosity, Surface Area, and Structural Features

Biochar exhibits a highly porous architecture, with a structure inherited mainly from its precursor biomass. The extent and distribution of porosity depend on pyrolysis temperature and quenching methods, which influence both micro- and mesopore formation. Higher temperatures typically result in greater porosity and surface area, enhancing water retention and adsorption of nutrients and organic compounds. Conversely, biochar produced at lower temperatures retains more of its original biomass structure, leading to higher volatile matter content and lower thermal stability.

The surface area of biochar is commonly quantified using BET (Brunauer–Emmett–Teller) analysis, which measures gas adsorption on biochar particles (see Chapter 9, specifically Section 9.5). Studies have shown that dry-quenched biochar tends to exhibit higher BET-specific surface areas than wet-quenched biochar, as the latter retains a greater fraction of organic material, reducing overall porosity. Figure 13.3 compares biochar porosity and surface area based on different quenching treatments, highlighting the structural differences between dry- and wet-quenched biochar.

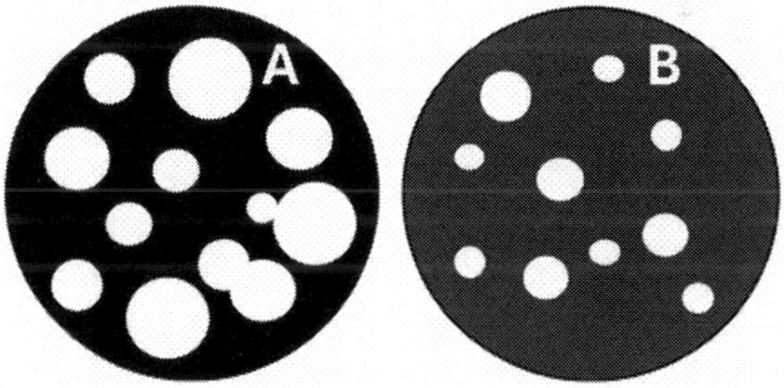

Figure 13.3 Schematic representation of dry-quenched (**A**) and wet-quenched (**B**) biochars, highlighting differences in porosity and hydrophilicity. Dry quenching results in a more porous, highly stable biochar with enhanced surface area, while wet quenching leads to a biochar with reduced porosity but greater hydrophilic surface properties. The grey shading in (**B**) signifies the increased presence of oxygen-containing functional groups that improve water retention, a key factor in soil applications.

13.4.3.1.2 Cation Exchange Capacity and Surface Functional Groups

Biochar plays a critical role in nutrient retention and exchange due to its cation exchange capacity, which governs its ability to adsorb and gradually release essential soil nutrients. Unlike mineral soils, which derive their CEC primarily from clay minerals and organic matter, biochar's CEC is largely dependent on its surface functional groups and porosity, both of which evolve over time as biochar interacts with the soil environment.

CEC in biochar is inherently low at the time of production, particularly in freshly pyrolyzed materials, as the thermal degradation of biomass results in the loss of labile organic matter and the formation of predominantly aromatic carbon structures. However, biochar's CEC increases over time, especially when exposed to oxygen, soil microbes, and moisture, leading to the gradual oxidation of its surface. This oxidation process promotes the formation of oxygen-containing functional groups, including carboxyl (–COOH), hydroxyl (–OH), and carbonyl (–C=O) groups, which enhance material ability to bind and exchange cations. The development of these functional groups significantly improves nutrient retention, particularly for essential plant macronutrients such as calcium (Ca^{2+}), magnesium (Mg^{2+}), potassium (K^+), and ammonium (NH_4^+).

Pyrolysis conditions and post-production modifications strongly influence the extent of CEC development in biochar. Biochars produced at low-to-moderate pyrolysis temperatures (300–500°C) tend to have higher initial CEC due to the partial preservation of acidic functional groups, whereas biochars pyrolyzed at higher temperatures (>600°C) exhibit lower CEC initially but gain greater exchange capacity as they weather in soil. Additionally, the feedstock composition plays a crucial role—biochars derived from lignocellulosic materials, such as wood and crop residues, generally develop higher CEC than those derived from manure or sludge due to differences in their organic molecular structures.

Beyond natural oxidation, post-production chemical treatments can further modify biochar's surface chemistry, tailoring its cation and anion retention properties to specific soil and agricultural needs. Acid treatments, such as nitric acid (HNO_3) or sulfuric acid (H_2SO_4) oxidation, enhance the formation of negatively charged functional groups, improving the adsorption of metal cations like Ca^{2+}, Mg^{2+}, and Fe^{2+}, which are critical for plant growth. Conversely, alkaline modifications, often using potassium hydroxide (KOH) or calcium hydroxide ($Ca(OH)_2$), neutralize surface acidity, increasing biochar's affinity for phosphate (PO_4^{3-}) and nitrate (NO_3^-) anions, making it particularly effective in retaining plant-available phosphorus and preventing nutrient leaching.

Furthermore, functionalization with transition metals or nanomaterials has been explored to enhance biochar's sorption properties for specific nutrients. For example, iron-modified biochar (Fe-biochar) has been shown to increase phosphate adsorption, while magnesium-enriched biochar (Mg-biochar) can improve sulfur and potassium retention. Such modifications extend the role of biochar beyond a conventional amendment, allowing it to function as an engineered fertilizer with controlled nutrient release properties.

A schematic representation of biochar surface functionalization (Figure 13.4) visually demonstrates how different treatments modify functional group availability and enhance nutrient interactions, offering a clearer understanding of how biochar's chemical properties influence soil fertility and nutrient dynamics.

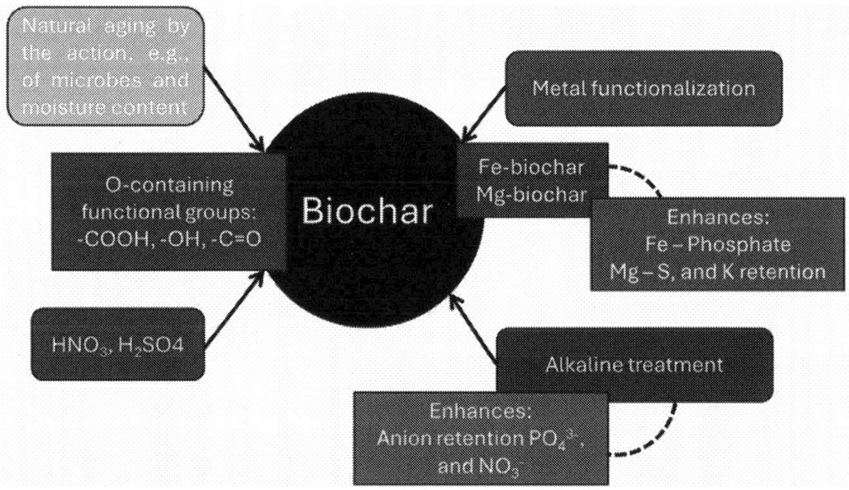

Figure 13.4 Biochar surface functionalization through oxidation, metal doping, and alkaline treatments. Oxygen-containing functional groups (–COOH, –OH, –C=O) increase cation exchange capacity (CEC), improving nutrient retention. Metal-enriched biochars enhance phosphate and potassium adsorption, while alkaline-treated biochar facilitates anion retention. These modifications directly impact biochar's role as a slow-release fertilizer and soil amendment, optimizing its agronomic benefits.

13.4.3.1.3 Elemental Composition and Thermal Stability

The elemental composition of biochar is primarily dictated by its feedstock source and pyrolysis temperature, both of which influence its carbon stability, nutrient content, and overall soil interactions. Biochars produced from woody biomass—such as hardwoods, softwoods, and forestry residues—typically have higher fixed carbon content, resulting in greater thermal stability and longer persistence in soil. These biochars are predominantly composed of aromatic carbon structures, making them highly resistant to microbial degradation and leaching. As a result, woody biochars contribute more effectively to long-term carbon sequestration and play a significant role in improving soil structure, water retention, and microbial habitat formation. Conversely, biochars derived from herbaceous materials and agricultural residues, including straw, husks, and manure, tend to be richer in nitrogen, phosphorus, and potassium. These biochars are more readily degradable, releasing nutrients into the soil more quickly than their woody counterparts. While they enhance soil fertility in the short term, they also decompose faster, reducing their long-term contribution to carbon sequestration. The higher mineral ash content of these biochars makes them particularly effective in modifying soil *pH*, buffering acidity, and increasing nutrient availability, but they lack the structural integrity and longevity of high-carbon biochars.

The stability of biochar in soil is closely linked to its degree of aromatic condensation, a factor that increases with higher pyrolysis temperatures. At low pyrolysis temperatures (300–400°C), biochar retains a higher proportion of labile organic compounds, including cellulose, hemicellulose, and volatile matter, making it more biodegradable. In contrast, biochars produced at high temperatures (>600°C) undergo extensive graphitization, forming highly condensed polyaromatic structures more resistant to microbial and chemical degradation. This high-temperature biochar remains in the soil matrix for decades to centuries, functioning as a long-term carbon sink while influencing water retention, microbial colonization, and nutrient dynamics.

While high-temperature biochars provide greater stability, they also lose a significant portion of their nitrogen and oxygen-containing functional groups during pyrolysis, reducing their ability to retain cations and interact with soil nutrients. This trade-off between biochar longevity and nutrient availability is a critical consideration when determining its optimal application in agricultural and environmental settings.

13.4.3.1.4 Hydrophobic vs. Hydrophilic Surface Behavior

Biochar's interaction with water is strongly governed by its surface chemistry, porosity, and aging processes, all of which influence its hydrophobic or hydrophilic characteristics. Freshly produced biochar, particularly obtained from high-temperature pyrolysis (>600°C), often exhibits hydrophobic behavior due to the predominance of aromatic carbon structures and the loss of oxygen-containing functional groups during thermal degradation. This hydrophobicity limits water adsorption and can initially reduce biochar's ability to enhance soil moisture retention, especially in sandy or degraded soils. However, as biochar undergoes oxidation, microbial colonization, and exposure to moisture over time, it progressively becomes more hydrophilic, improving its ability to absorb and retain water within the soil matrix.

Beyond natural oxidation and microbial interactions, post-production treatments and quenching methods significantly influence biochar's surface wettability. While wet-quenching biochar leads to partial oxidation, incorporating oxygen-containing functional groups (–COOH, –OH, –C=O) that enhance water retention and cation exchange capacity, dry-quenching biochar preserves a more hydrophobic nature due to the absence of water-driven oxidation during cooling. This distinction is particularly relevant in agricultural applications where immediate water retention benefits are required—wet-quenched biochar is generally preferred in such cases due to its enhanced affinity for soil moisture.

The feedstock composition and pyrolysis conditions also play a crucial role in determining biochar's hydrophobic or hydrophilic properties. Biochars produced from woody biomass and lignin-rich materials tend to be more hydrophobic due to their high aromatic carbon content, while those derived from herbaceous biomass and manure contain higher ash and polar organic matter, leading to greater initial hydrophilicity. Additionally, biochars pyrolyzed at lower temperatures (<500°C) retain more polar functional groups, making them more naturally hydrophilic, whereas those produced at higher temperatures become progressively more hydrophobic until they undergo surface oxidation.

These hydrophobic-hydrophilic transitions in biochar have significant implications for soil applications. In water-limited environments, hydrophilic biochar can help improve soil water-holding capacity and reduce drought stress by enhancing moisture retention and redistribution. Meanwhile, hydrophobic biochar may act as a barrier to rapid water infiltration, potentially influencing soil aeration and gas diffusion processes.

Ultimately, the ability to engineer biochar's surface wettability through controlled oxidation, functionalization, or blending with hydrophilic organic amendments presents a valuable opportunity to tailor biochar for specific agronomic and environmental applications. Understanding the complex relationship between biochar's production parameters and its interaction with water is therefore essential for optimizing its performance in sustainable soil management.

13.4.3.1.5 Biochar as a Soil Amendment

Biochar role as a soil amendment extends beyond its ability to supply nutrients, acting primarily to modify the physical structure, water dynamics, and microbial ecology of the soil. By altering bulk density, improving porosity, and increasing water-holding capacity, biochar enhances soil aeration, moisture availability, and resistance to degradation, making it particularly effective in improving marginal or degraded soils.

One of biochar's most notable properties is its high surface area and porosity, which allow it to adsorb and retain both water and nutrients. This characteristic reduces nutrient leaching losses, particularly for mobile elements like nitrogen and potassium, ensuring that these nutrients remain available for plant uptake over extended periods. The extent of biochar's water retention capacity depends on feedstock type, pyrolysis temperature, and quenching method, with wet-quenched and low-temperature biochars generally displaying greater hydrophilicity and moisture-holding capacity. This property makes biochar an effective tool for mitigating drought stress in water-limited environments.

Beyond its physical and chemical effects, biochar significantly influences soil microbiology, fostering beneficial microbial communities that contribute to nutrient cycling, organic matter decomposition, and disease suppression. Studies have demonstrated that biochar additions enhance microbial populations involved in nitrogen fixation (e.g., *Rhizobia* and free-living *diazotrophs*) and organic matter decomposition (e.g., *actinomycetes* and *fungi*). These microbial processes play a crucial role in sustaining long-term soil fertility, particularly in low-fertility or degraded soils. Moreover, biochar provides a habitable microenvironment for microbial colonization, as its porous structure offers protective niches against desiccation and predation.

The interactions between biochar and soil microbial life also extend to its influence on greenhouse gas emissions. By modifying soil redox conditions and microbial respiration, biochar can reduce nitrous oxide (N_2O) emissions by promoting complete denitrification pathways. Furthermore, its stabilization of organic carbon in soil can mitigate CO_2 emissions by slowing the decomposition of labile carbon pools, effectively contributing to long-term carbon sequestration (Figure 13.5).

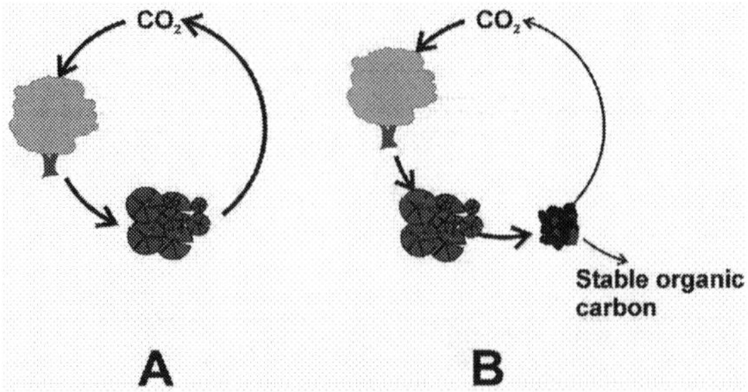

Figure 13.5 The role of biochar in carbon sequestration compared to conventional organic matter decomposition. (**A**) In unmanaged soil systems, plant residues decompose, rapidly releasing CO_2 back into the atmosphere. (**B**) Through pyrolysis, biomass is converted into biochar, stabilizing carbon in a highly recalcitrant form that remains in the soil for centuries. This process mitigates greenhouse gas emissions and enhances soil fertility, linking biochar to both climate change adaptation and sustainable soil management[4].

While the benefits of biochar as a soil amendment are well-documented, its effects vary depending on soil type, biochar characteristics, and application rate. In sandy soils, biochar significantly improves water retention and nutrient availability, while in clay-rich soils, it enhances aggregate stability and aeration, reducing compaction issues. However, excessive biochar application in highly fertile soils may sometimes alter nutrient balance or interfere with microbial dynamics, emphasizing the importance of site-specific application strategies.

In summary, biochar serves as a multifunctional soil amendment, improving physical structure, water dynamics, and microbial activity, while simultaneously reducing nutrient leaching, stabilizing organic matter, and mitigating greenhouse gas emissions. Its effectiveness, however, depends on its physicochemical properties and environmental conditions, reinforcing the need for tailored biochar applications based on specific soil and crop requirements.

Figure 13.6 illustrates how the natural oxidation of biochar—previously described as a key factor in enhancing nutrient retention and improving cation exchange capacity (CEC)—also plays a crucial role in soil structure formation. This structural transformation directly supports both the living and non-living mechanisms discussed in this section, contributing to improved soil aeration, water dynamics, and microbial habitat formation.

13.4.3.1.6 Biochar as a Fertilizer: Functionalization and Nutrient Retention

While biochar in its raw form primarily acts as a soil amendment, its chemical modification and functionalization can transform it into a direct nutrient source,

4 This figure has been reproduced from Conte, P., et al. 2021. Recent developments in understanding biochar's physical–chemistry. Agronomy. 11(4): 615. https://doi.org/10.3390/agronomy11040615

extending its role from a structural soil enhancer to an active component in plant nutrition. By tailoring its surface chemistry, biochar can be optimized to retain and gradually release essential nutrients, reducing leaching losses and improving long-term soil fertility. This transformation is particularly valuable in nutrient-poor soils, where conventional fertilizers often fail to provide sustained nutrient availability.

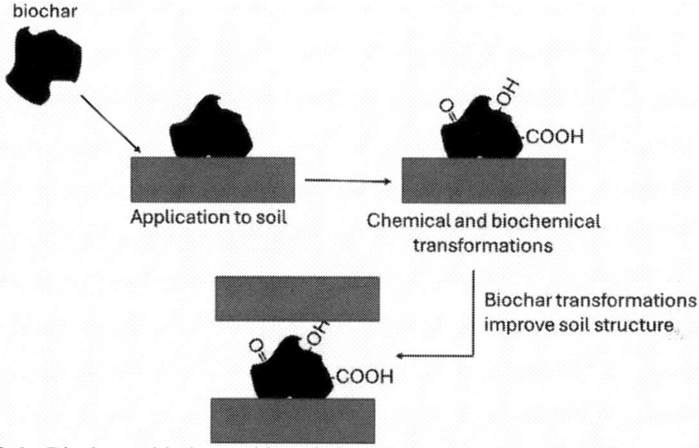

Figure 13.6 Biochar oxidation and its role in soil structure formation. Upon application to soil, biochar undergoes chemical and biochemical transformations, leading to the formation of oxygen-containing functional groups. These changes enhance cation exchange capacity (CEC) and promote aggregation of soil particles, improving aeration, water retention, and microbial habitat. The resulting structural stability supports key biogeochemical interactions, making biochar a long-term soil fertility enhancer[5].

One of the most effective ways to enhance biochar's fertilizing capacity is through oxidative functionalization, which involves treatments with strong acids such as nitric acid (HNO_3), sulfuric acid (H_2SO_4), or sulfonitric mixtures. These oxidative processes introduce oxygen-containing functional groups (–COOH, –OH, –C=O) onto the biochar surface, increasing its negative surface charge and improving its ability to adsorb and retain positively charged nutrients such as calcium (Ca^{2+}), magnesium (Mg^{2+}), potassium (K^+), and ammonium (NH_4^+). Additionally, oxidation enhances the retention of phosphate (PO_4^{3-}), a crucial macronutrient often lost due to runoff, thereby improving phosphorus availability for plants. However, excessive oxidation can lead to extreme acidity, potentially limiting biochar's buffering capacity in soil. Careful control of the oxidation process ensures that biochar retains both its adsorption capabilities and its structural integrity, making it an efficient slow-release fertilizer.

Beyond oxidation, metal doping further enhances biochar's role in nutrient retention. The incorporation of metal ions such as iron, magnesium, calcium, and aluminum during biochar synthesis or post-treatment improves its ability

[5]This figure is based on the results contained in Fellet, G., et al. 2022. Changes in physical-chemical biochar properties following addition to soil. Agriculture. 12: 320.

to adsorb anionic nutrients such as phosphate and nitrate, reducing their loss through leaching. For instance, iron-functionalized biochar has been shown to significantly increase phosphate retention, preventing eutrophication in nearby water bodies while enhancing phosphorus availability in agricultural soils. Similarly, magnesium-enriched biochar has demonstrated an increased capacity to retain sulfur and potassium, making it a useful amendment for nutrient-depleted or alkaline soils. These modifications extend biochar's utility beyond a passive soil conditioner, transforming it into a tailored material capable of addressing specific soil fertility challenges.

Another highly effective approach to enhancing biochar's fertilizing potential is co-composting, where biochar is blended with organic waste materials and subjected to microbial decomposition. This process not only improves biochar's capacity to retain nitrogen but also promotes microbial activity, leading to increased bioavailability of phosphorus and other nutrients. By adsorbing ammonium and nitrate during composting, biochar helps to minimize nitrogen volatilization and leaching losses, ensuring that these essential nutrients remain in the soil over extended periods. Additionally, biochar's porous structure provides microhabitats for beneficial soil microbes involved in nutrient mineralization and organic matter decomposition, contributing to overall soil fertility. The resulting biochar-enriched compost offers improved water retention, reduced greenhouse gas emissions, and enhanced humic substance formation, making it a more stable and nutrient-dense organic amendment compared to conventional composts.

Through oxidation, metal doping, and co-composting, biochar can be engineered to function as a slow-release fertilizer, offering a more sustainable alternative to conventional nutrient inputs. These modifications not only improve soil fertility and nutrient retention but also help mitigate environmental impacts such as phosphorus runoff, nitrogen leaching, and greenhouse gas emissions. By strategically modifying biochar's surface properties, it becomes possible to develop biochar-based fertilizers tailored for specific crops and soil conditions, providing long-term agricultural benefits while reducing reliance on synthetic fertilizers.

13.4.3.1.7 Reactions and Mechanisms of Nutrient Retention

Biochar interacts with soil nutrients through a combination of physical, chemical, and biological mechanisms, each contributing to nutrient retention, availability, and cycling within the soil matrix. Its high surface area, porosity, and surface functionalization enable it to act as both a reservoir and a regulator of essential plant nutrients, influencing their mobility, uptake efficiency, and long-term stability in soil.

One of the most critical properties governing biochar's nutrient interactions is its cation exchange capacity, which arises from the presence of negatively charged oxygen-containing functional groups on the biochar surface. As already stated above, these functional groups facilitate the adsorption and exchange of essential cations, including ammonium, potassium, calcium, and magnesium. This process helps to regulate the availability of macronutrients while preventing their loss through leaching, particularly in sandy or highly weathered soils. The development of CEC is influenced by feedstock composition, pyrolysis conditions, and post-production modifications, with oxidative treatments further enhancing biochar's capacity to bind and release plant-available cations.

Beyond cation exchange, biochar also plays a crucial role in anion adsorption, particularly for phosphate and nitrate. Unlike cations, which are retained through electrostatic interactions, anions are primarily adsorbed onto biochar through ligand exchange mechanisms, surface precipitation, and interactions with metal-enriched biochars. Functionalization with iron, magnesium, or aluminum significantly enhances biochar's ability to bind phosphate, preventing its rapid leaching from agricultural soils and ensuring a more sustained release for plant uptake. Similarly, biochar can reduce nitrate leaching, though its retention capacity for NO_3^- is generally lower than for phosphate due to the weaker interactions between negatively charged surfaces and anionic species. The ability to control biochar's surface chemistry through alkaline or acid modifications allows for the targeted retention of specific nutrients, optimizing soil fertility based on agronomic needs.

In addition to its direct chemical interactions with nutrients, biochar also has a profound influence on soil microbial communities, which further enhance nutrient retention and cycling. Its high porosity and surface reactivity provide protective microenvironments for beneficial microorganisms, including nitrogen-fixing bacteria, mycorrhizal fungi, and phosphorus-solubilizing microbes. These microbial interactions enhance biological nitrogen fixation, facilitating the conversion of atmospheric nitrogen into plant-available forms, while mycorrhizal fungi improve phosphorus uptake through symbiotic root associations. Additionally, biochar's ability to stabilize organic matter contributes to slow nutrient mineralization, ensuring a continuous supply of macronutrients and micronutrients over time.

These combined physicochemical and biological mechanisms make biochar a versatile tool for sustainable soil management, offering both immediate and long-term benefits in nutrient retention, plant growth promotion, and environmental protection. By modifying biochar's structural and chemical properties, its interactions with essential soil nutrients can be optimized to reduce fertilizer losses, improve nutrient-use efficiency, and enhance soil health, making it a valuable component in regenerative and climate-resilient agriculture.

13.4.3.1.8 Biochar Functionalization and Soil Applications

The potential of biochar-based fertilizers extends beyond simple nutrient retention, offering a sustainable alternative to synthetic fertilizers by providing slow-release nutrient delivery while simultaneously enhancing soil health. Functionalized biochar, enriched with nitrogen (N) and phosphorus (P) through oxidative, microbial, or mineral-doping treatments, has demonstrated greater efficiency in plant growth promotion, reducing the need for frequent fertilizer applications. This enhanced efficiency is particularly valuable in intensive agricultural systems, where rapid nutrient depletion and leaching often lead to low fertilizer-use efficiency and environmental degradation.

One of the key advantages of biochar functionalization is its ability to regulate nutrient availability over extended periods, preventing the sudden influx of nutrients that often accompanies synthetic fertilizer applications. The gradual release of nitrogen and phosphorus from functionalized biochar aligns with plant uptake rates, ensuring that essential nutrients remain available when needed rather than being lost through leaching or volatilization. For example, oxidatively treated biochar with –COOH and –OH functional groups improves ammonium and phosphate retention, reducing nitrogen losses to the atmosphere and minimizing nitrate contamination of groundwater. Similarly, biochar co-composted with organic matter undergoes microbial transformations that increase its bioavailability, enhancing soil microbial activity and improving nutrient cycling.

Beyond its direct effects on fertility, biochar offers long-term benefits in soil carbon sequestration, playing a significant role in climate change mitigation. Unlike traditional fertilizers, which contribute to greenhouse gas emissions, biochar stabilizes organic carbon in soils, effectively removing atmospheric CO_2 and enhancing soil organic matter content. This dual role as a nutrient carrier and carbon sink makes biochar a strategic solution for sustainable agriculture, helping to balance productivity with environmental conservation.

The practical application of functionalized biochar varies based on soil type, crop requirements, and regional climate conditions. In sandy soils, biochar primarily enhances water retention and nutrient availability, while in clay-heavy soils, it improves aeration and prevents compaction. The application rate and method of biochar-based fertilizers must be carefully optimized, as excessive use can alter soil pH and microbial dynamics. Additionally, field studies have shown that the co-application of biochar with organic amendments—such as manure, compost, or microbial inoculants—further enhances its benefits, creating synergistic effects on soil fertility and plant health.

By integrating functionalized biochar into soil management strategies, it is possible to reduce dependency on chemical fertilizers, enhance nutrient efficiency, and contribute to long-term soil sustainability. As research progresses, the development of tailored biochar formulations will enable more precise nutrient management, optimizing crop productivity while mitigating environmental impacts. The ability to engineer biochar-based fertilizers for specific soil and crop needs represents a significant advancement in regenerative and climate-resilient agriculture.

13.4.3.1.9 Biochar: A Multifunctional Tool for Sustainable Soil Management

Biochar represents a unique intersection between soil amendment and fertilizer, offering a versatile solution for improving soil fertility through both physical and chemical pathways. Its ability to enhance soil structure, regulate moisture retention, support microbial life, and retain essential nutrients makes it a promising tool for sustainable agriculture. Unlike traditional fertilizers, which often contribute to nutrient leaching, soil acidification, and greenhouse gas emissions, biochar provides a long-term approach to soil improvement, stabilizing organic carbon while minimizing environmental degradation.

The functionalization and co-composting of biochar further expand its agronomic potential, transforming it from a passive soil conditioner into an active nutrient provider. By tailoring its surface chemistry, cation exchange capacity, and anion retention properties, biochar can be optimized for specific soil conditions, addressing nutrient deficiencies, improving fertilizer-use efficiency, and reducing dependency on synthetic inputs. These modifications not only support plant growth and yield but also enhance soil resilience to climate variability, reinforcing biochar's role as a key component in regenerative agriculture.

Despite the growing body of research highlighting biochar's benefits, its widespread agricultural adoption requires further investigation into its long-term field performance, interactions with diverse soil microbiomes, and economic viability in large-scale farming systems. Future studies should focus on optimizing biochar formulations based on soil type, crop requirements, and environmental conditions, ensuring that its application is both efficient and sustainable. By integrating biochar-based strategies into soil fertility management, agriculture can move toward a more circular, low-input system that balances productivity with ecological responsibility, unlocking biochar's full potential as a multifunctional soil enhancer for the future of sustainable farming.

13.4.4 Other Essential Soil Amendments

In the paragraphs above the amendment function of biochar has been described. However, this is only one of many soil amendments that address specific challenges related to soil chemistry, structure, and nutrient availability. Among these, lime and alkaline amendments play a fundamental role in managing soil acidity, which can severely impact nutrient availability and microbial activity. Materials such as calcium carbonate ($CaCO_3$), dolomite ($CaMg(CO_3)_2$), and calcium hydroxide ($Ca(OH)_2$) are widely used to raise soil *pH* by neutralizing hydrogen (H^+) and aluminum (Al^{3+}) ions, ensuring optimal conditions for plant nutrient uptake. When calcium carbonate reacts with soil acidity, it forms calcium ions and bicarbonate, stabilizing *pH* through the following reaction:

$$CaCO_3 + H_2O + CO_2 \rightarrow Ca^{2+} + 2HCO_3^{2-}$$

Similarly, dolomite is particularly beneficial for magnesium-deficient soils, supplying both calcium and magnesium to support plant growth. However, in alkaline soils where high *pH* limits micronutrient availability, amendments such as elemental sulfur, iron sulfate ($FeSO_4$), and aluminum sulfate ($Al_2(SO_4)_3$) are used to acidify the soil. Through microbial oxidation, sulfur is converted into sulfuric acid (H_2SO_4), which reacts with alkaline soil components, thereby lowering *pH* and enhancing the solubility of essential micronutrients like iron, manganese, and zinc:

$$S + O_2 + H_2O \rightarrow H_2SO_4$$

This process is particularly useful in calcareous soils, where iron deficiency often leads to chlorosis in sensitive crops. Additionally, in sodic soils, where high sodium (Na^+) levels lead to poor structure and low permeability, gypsum ($CaSO_4{\bullet}2H_2O$) serves as a critical amendment by supplying calcium ions, which replace sodium on soil exchange sites, facilitating its removal through leaching:

$$CaSO_4 + 2Na^+ \rightarrow Ca^{2+} + Na_2SO_4$$

The displaced sodium is then washed away with irrigation, restoring soil structure and improving aeration, which is essential for root growth and microbial

Table 13.3 Comparative Overview of Soil Amendments. This table summarizes the primary functions, optimal application conditions, and key considerations for various soil amendments

Amendment	Primary Function	Best Applied in	Application Considerations
Biochar	Improves soil structure, water retention, and nutrient retention	Sandy, degraded, or low-organic matter soils	May require functionalization for maximum nutrient retention
Lime ($CaCO_3$)	Neutralizes soil acidity, improves calcium availability	Acidic soils with low calcium availability	Over-application may lead to excessive alkalinity
Dolomite ($CaMg(CO_3)_2$)	Corrects acidity while also supplying magnesium	Acidic and magnesium-deficient soils	Should be carefully dosed to balance Ca:Mg ratio
Gypsum ($CaSO_4 \cdot 2H_2O$)	Reclaims sodic soils, improves soil structure	Sodic soils with high sodium content	Requires adequate irrigation for sodium leaching
Elemental Sulfur (S)	Lowers soil pH, improves micronutrient availability	Alkaline or calcareous soils	Overuse may cause excessive acidification
Iron(II) Sulfate ($FeSO_4$)	Lowers soil pH, supplies iron to plants	Iron-deficient, alkaline soils	Best used in iron-deficient soils
Compost	Enhances organic matter, microbial activity, and nutrient cycling	All soil types, particularly degraded soils	Needs proper decomposition before application
Manure	Provides macronutrients (N, P, K) and improves soil fertility	Nutrient-depleted soils needing organic matter	Risk of nutrient leaching if improperly managed
Silicate Minerals	Supplies silicon for plant resilience and stress resistance	Soils where plant silicon uptake is beneficial	Enhances plant drought and disease resistance

activity. Apart from mineral amendments, organic matter-based amendments such as compost and manure are essential in improving soil fertility by increasing cation exchange capacity, moisture retention, and microbial diversity. Compost undergoes controlled aerobic decomposition, leading to the formation of humic substances that stabilize nutrients and enhance soil structure, while manure provides an immediate source of nitrogen, phosphorus, and potassium, supporting plant growth. However, improper manure management can result in nutrient leaching, methane emissions, and pathogen contamination, necessitating its careful application to maintain soil health. Recent research has also highlighted the potential of silicon-based amendments, such as silicate minerals and bio-silicates, in improving soil resilience. Silicon (Si) not only strengthens plant cell walls but also enhances resistance to drought stress and heavy metal toxicity, particularly benefiting crops such as rice and wheat, which have high silicon requirements. These combined soil amendments—mineral-based, organic, or bio-silicate—maintain soil fertility by regulating *pH*, improving nutrient retention, and supporting biological activity. The careful selection and application of these materials based on specific soil conditions and crop requirements ensure long-term sustainability, improved productivity, and enhanced environmental resilience, reinforcing the need for a holistic approach to soil fertility management beyond biochar. Table 13.3 provides a comparative overview of various soil amendments, summarizing their primary functions, optimal application conditions, and key considerations for effective soil management.

13.5 SOIL FERTILITY: A PILLAR OF SUSTAINABILITY AND GLOBAL RESILIENCE

The management of soil fertility is far more than a technical aspect of agricultural science. It is a cornerstone of environmental sustainability, food security, and climate resilience. As the global population continues to rise, ensuring stable and sustainable food production becomes an urgent challenge, one that is inextricably linked to the health and functionality of our soils. Fertile soils are not just the foundation of productive agriculture—they are dynamic ecosystems that regulate nutrient cycling, store carbon, and influence water availability. In an era of increasing land degradation, biodiversity loss, and resource depletion, preserving and enhancing soil fertility is no longer an option but a necessity.

The impacts of soil mismanagement extend far beyond individual fields. Unsustainable practices, such as excessive fertilizer use, intensive tillage, and deforestation, contribute to nutrient imbalances, erosion, and declining soil organic matter, exacerbating environmental instability. Meanwhile, climate change intensifies these threats, with shifting precipitation patterns and rising temperatures placing additional stress on already vulnerable soils. To address these challenges, a paradigm shift in soil management is essential—one that integrates chemical, physical, and biological approaches to create resilient agricultural systems.

By leveraging scientific innovation, responsible land stewardship, and interdisciplinary collaboration, we can transition toward regenerative soil practices that enhance fertility without compromising ecological integrity. Sustainable nutrient management, the strategic use of soil amendments, and the adoption of biochar and organic matter-based solutions can all contribute to mitigating climate impacts while fostering long-term productivity. Moreover, soil restoration and conservation efforts must be reinforced at policy levels, ensuring that agricultural and environmental frameworks prioritize soil health as a critical global resource.

A profound understanding of soil chemistry and physics is indispensable in shaping the future of agriculture. Scientific research and field applications must work in tandem to develop precision soil management strategies tailored to specific regions, climates, and cropping systems. The future of food security, climate resilience, and environmental preservation depends on how effectively we manage the hidden yet powerful world beneath our feet. By investing in soil knowledge, responsible innovation, and sustainable practices, we lay the groundwork for a more resilient and productive agricultural system, safeguarding the well-being of both current and future generations.

EXERCISES

1. Explain the differences between soil amendments and fertilizers, providing two examples. How do soil amendments indirectly improve plant growth?
2. A farmer applies 500 kg of calcium carbonate ($CaCO_3$) per hectare to an acidic soil with a pH of 5.0. Assuming complete neutralization of H^+ ions, how much calcium carbonate would be needed to raise the soil pH to 6.5 in an area of 10 hectares? Show your calculations.
3. Compare the effect of gypsum ($CaSO_4 \cdot 2H_2O$) and lime ($CaCO_3$) in improving soil conditions. Write the key chemical reactions involved in their respective applications.
4. Discuss the role of biochar in soil fertility management. What are its advantages and potential limitations in different soil types? How does biochar compare to compost in terms of long-term soil health benefits?
5. Find a recent scientific article on soil amendments (published in the last five years). Summarize the key findings and discuss how the study's results align with or challenge the information presented in this chapter.

FURTHER READING

Guo, M., Z. He and S.M. Uchimiya (eds). 2016. Agricultural and Environmental Applications of Biochar. Advances and Barriers, 1st Ed. SSSA. Special Publication. 63. ISBN: 9780891189640

Havlin, J., S. Tisdale, W. Nelson and J. Beaton. 2014. Soil Fertility and Fertilizers: An Introduction to Nutrient Management, 8th Ed. Pearson. ISBN: 9780135033739

Lal, R. 2014. Soil Quality and Soil Erosion. 1st Indian preprint. CRC Press. ISBN: 1574441000

R. Lal. (ed.). 2020. Soil and Fertilizers. Managing the Environmental Footprint, 1st Ed. CRC Press. ISBN: 9781138600072

Weil, R.R. and N.C. Brady. 2017. The Nature and Properties of Soils, 15th Ed. Pearson Prentice Hall. ISBN: 9781292162232

White, R.E. 2006. Principles and Practice of Soil Science. The Soil as a Natural Resource, 4th Ed. Blackwell Publishing. ISBN: 9780632064557

Index

A

B

C

Made in the USA
Las Vegas, NV
19 September 2026

4bac8b5d-0840-47cc-99c2-c4e190216ad8R01